NANOTECHNOLOGY SCIENCE AND TECHNOLOGY

CAPILLARY ELECTROPHORESIS (CE)

PRINCIPLES, CHALLENGES AND APPLICATIONS

NANOTECHNOLOGY SCIENCE AND TECHNOLOGY

Additional books in this series can be found on Nova's website under the Series tab.

Additional e-books in this series can be found on Nova's website under the e-book tab.

NANOTECHNOLOGY SCIENCE AND TECHNOLOGY

CAPILLARY ELECTROPHORESIS (CE)

PRINCIPLES, CHALLENGES AND APPLICATIONS

CHRISTIAN REED
EDITOR

nova publishers
New York

Library of Congress Cataloging-in-Publication Data

Capillary electrophoresis (CE) (Nova Science Publishers)
Capillary electrophoresis (CE) : principles, challenges and applications / [editor] Christian Reed.
pages cm. -- (Nanotechnology science and technology)
Includes index.
ISBN 978-1-63483-122-2 (hardcover)
1. Capillary electrophoresis. I. Reed, Christian, editor. II. Title.
TP248.25.C37C367 2015
541'.372--dc23
2015020602

Published by Nova Science Publishers, Inc. † New York

CONTENTS

PREFACE

This book examines challenges and applications, as well as principles of capillary electrophoresis. Some of the topics discusses include the preparation and application of photosensitive capillary electrophoresis coatings; the application of capillary zone electrophoresis to trace analyses of inorganic anions in seawater; theoretical principles and applications of high performance capillary electrophoresis; and the application of capillary zone electrophoresis methods for polyphenols and organic acids to separate different extracts.

Chapter 1 - Novel methods for the preparation of covalently linked capillary coatings of anti-protein-fouling polymers were demonstrated using photosensitive diazoresin (DR) as coupling agents. Layer by layer (LBL) self-assembly films of DR and anti-protein-fouling polymers based on hydrogen or ionic bonding were fabricated on the inner wall of capillary, then the hydrogen or ionic bonding was converted into covalent bonding after treatment with UV light through the unique photochemistry reaction of DR. The covalently bonded coatings suppressed basic protein adsorption on the inner surface of capillary, and thus a baseline separation of proteins was achieved using capillary electrophoresis (CE). Compared with bare capillary or non-covalently bonded coatings, the covalently linked capillary coatings not only improved the CE separation performance for proteins, but also exhibited good stability and repeatability. Due to the replacement of highly toxic and moisture sensitive silane coupling agent by DR in the covalent coating preparation, these methods may provide a green and easy way to make the covalently coated capillaries for CE.

Chapter 2 - Capillary zone electrophoresis (CZE) is an environmentally friendly analytical method that provides high separation capability with minimum consumption of samples and reagents. The authors have been developing CZE methods for the determination of cationic and anionic substances in environmental waters (e.g., seawater, river water, sewage) and in biospecimens (e.g., serum, cerebrospinal fluid, urine, and vegetables such as spinach (*Spinacia oleracea*) and ice-plant (*Mesembryanthemum crystallinum* L.)). Seawater contains salts of high concentrations, with many elements, species, and concentration ranges. It is a difficult task to apply a high-resolution method to seawater analyses. Nevertheless, the sensitivity of CZE with a UV detector, which is usually used as the detector, is insufficient for low concentrations of analytes because of the short light-path length of the capillary inner diameter. Therefore, in addition to protecting the influence of high concentrations of salts, some on-line concentration procedure is necessary to determine the concentrations of trace substances in seawater. The authors used artificial seawater as the background electrolyte for this study to decrease the influence of salts, and used transient isotachophoresis as the on-line concentration procedure. This chapter presents a summary of the analytical procedures and the results for the determination of trace inorganic anions such as nitrite and nitrate, phosphate, iodide and iodate, and bromate in seawater (and salt).

Chapter 3 - In the last decades, miniaturized separation techniques have rapidly gained popularity in different areas of analysis such as pharmaceutical, biopharmaceutical, clinical, biological, environmental, and forensics. The great advantages presented by the analytical miniaturized techniques, including high separation efficiency and resolution, rapid analysis and minimal consumption of reagents and samples, make them an attractive alternative to the conventional chromatographic methods.

In this sense, capillary electrophoresis (CE) is a family of related techniques that employs narrow-bore capillaries to perform highly efficient separations from large to small molecules. Different modes are applied in CE. Capillary zone electrophoresis (CZE) using a simple buffer as electrolyte, is widely used for the analysis of inorganic and organic ions. Another CE mode is electrokinetic chromatography (EKC), in which the separation principle is based on the differential partition between the analytes and a pseudostationary phase as well as the migration behavior of the analytes. Several nanostructures are used as pseudostationary phases like micelles, microemulsion droplets, and polymers, increasing the selectivity and versatility of the analytical system.

CE advantages with respect to other analytical techniques comprise very high resolution in short time of analysis, versatility, the possibility to analyze molecules without chromophore groups, simultaneous analysis of compounds with different hydrophobic characteristics, small sample volume, and low cost. Moreover, it is possible to adapt this technique to the analysis of numerous types of compounds like biological macromolecules, chiral compounds, inorganic ions, organic acids, DNA fragments and even whole cells and virus particles. An increasing number of CE applications are in progress in many clinical laboratories. As this technique employs small sample volumes, is ideal for the analysis of biological fluids in which the limited amount of sample represents a challenge. In pharmaceutical quality control, it is possible to determine active ingredients in the presence of related substances with different physicochemical characteristics, especially chiral impurities in the final products using the same analytical system with a relatively simple instrumental. Moreover, numerous applications are reported in the analysis of inorganic ions. Also, the determination of macromolecules such as polysaccharides, therapeutic proteins, and flavonoids present in plant extracts, biological and biopharmaceutical products may also be analyzed with this technique.

In summary, CE has become an important analytical tool in the field of research, clinics and pharmaceutical industry offering a large number of applications, in biological, natural and pharmaceutical samples, as an alternative or complementary option to traditional analytical techniques to implement in the routine laboratory.

Chapter 4 - Capillary electrophoresis (CE) comprises a family of related separation techniques in which an electric field is used to achieve the separation of components in a mixture. Electrophoresis in a capillary is differentiated from other forms of electrophoresis in that separation is carried out within the confines of narrow-bore capillaries, from 20 to 200 μm inner diameter (i.d.), which are usually filled only with a solution containing electrolytes (typically, although not always necessary, a buffer solution). One of the key features of CE is the simplicity of the instrumentation required, and today this technique allows working in various modes of operation. Among them, capillary zone electrophoresis (CZE) is the most widely used due to its simplicity of operation and its versatility. The use of high electric fields results in short analysis times and high efficiency and resolution. In addition, the minimal sample volume requirement (in general few nanoliters), the on-capillary detection, the potential for both qualitative and quantitative analysis, the automation, and the possibility of hyphenation with other techniques such as mass spectrometry (MS) is allowing CZE to become one of the premier separation techniques in multiple fields, such as bio-analysis, food safety and environmental applications.

However, one of CZE handicaps is sensitivity due to the short path length (capillary inner diameter) when on-capillary detection is carried out, and the low amount of samples injected. For these reasons, many CZE applications will require of off-line and/or on-line preconcentration methods in order to improve limits of detection (LOD). Many different techniques have been developed to improve LODs in CZE. Among them, on-line electrophoretic-based preconcentration techniques are becoming very popular because no special requirement but a CE instrument is necessary for their application. These on-line preconcentration methods are designed to compress analyte bands within the capillary, thereby increasing the volume of sample that can be injected without losing separation efficiency. So, these methods are based on the principle of stacking analytes in a narrow band between two separate zones in the capillary where the compounds have different electrophoretic mobilities (for instance at the boundary of two buffers with different resistivities).

This chapter will address the principles of on-line electrophoretic-based preconcentration methods in capillary zone electrophoresis. Coverage of all kind of on-line electrophoretic-based preconcentration methods is beyond the scope of the present contribution, so the authors will focus on the most frequently used in CZE such as sample stacking, large-volume sample stacking (LVSS), field-amplified sample injection (FASI), pH-mediated sample stacking, and electrokinetic supercharging (EKS). Relevant applications of these preconcentration methods in several fields (bio-analysis, food safety, environmental analysis) will also be presented.

Chapter 5 - Micellar electrokinetic capillary chromatography (MECC or MEKC) is maybe the most intriguing mode of capillary electrophoresis (CE) techniques for the determination of small molecules, and it is considered a hybrid of electrophoresis and chromatography. The use of micelle-forming surfactant solutions can give rise to separations that resemble reversed-phase liquid chromatography (LC) with the benefits of CE techniques. Introduced by Professor Shigeru Terabe in 1984, MECC is today, together with capillary zone electrophoresis (CZE), one of the most widely used CE modes, and its main strength is that it is the only electrophoretic technique that can be used for the separation of neutral analytes as well as charged ones.

In MECC, a suitable charged or neutral surfactant, such as sodium dodecyl sulfate (SDS), is added to the separation buffer in a concentration sufficiently high to allow the formation of micelles. Surfactants are long chain molecules (10-50 carbon units) and are characterized as possessing a long hydrophobic tail and a hydrophilic head group. When surfactant concentration in the buffer solution reach a certain level (known as critical micelle concentration), they aggregate into micelles which are, in the case of normal micelles, arrangements that will have a hydrophobic inner core and a hydrophilic outer surface. Micelles are dynamic and constantly form and break apart, constituting a pseudo-stationary phase in solution within the capillary. It is the interaction between the micelles and the solutes (neutral or charged ones) that causes their separation.

However, as in the case of other CE techniques, one of MECC handicaps is sensitivity due to the short path length (capillary inner diameter) when on-capillary detection is performed, and the low volume of samples frequently used. In order to improve MECC sensitivity, off-line and/or on-line preconcentration methods can be employed. Among them, on-line electrophoretic-based preconcentration techniques are also becoming very popular in MECC because no special requirement but a CE instrument is necessary. These on-line preconcentration methods are designed to compress analyte bands within the capillary, thereby increasing the volume of sample that can be injected without an important loss in electrophoretic efficiency. In MECC, these on-line preconcentration methods are based on either the manipulation of differences in the electrophoretic mobility of analytes at the boundary of two buffers with differing resistivities and the partitioning of analytes into a micellar pseudostationary phase.

This chapter will address the principles of on-line electrophoretic-based preconcentration methods in micellar electrokinetic capillary chromatography. Coverage of all kind of on-line electrophoretic-based preconcentration methods is beyond the scope of the present contribution, so only the most frequently used in MECC such as sweeping, field-amplified sample injection (FASI), ion-exhaustive sample injection-sweeping (IESI-sweeping) and dynamic pH junction-sweeping will be discussed. Relevant applications of these preconcentration methods in several fields (bio-analysis, food safety, environmental analysis) will also be presented.

Chapter 6 - The capillary electrophoretic (CE) analysis of inorganic ions in parenteral nutrition solution in association with capacitively coupled contactless conductivity detection (C^4D) is a simple, flexible, economic and eco-friendly method. The aim of this study was to improve the repeatability and linearity properties as well as the application of this validated method to estimate the quantity of each inorganic cation in commercial samples. The method is carried out on an uncoated fused silica capillary with 50 μm i.d. and 365 μm o.d. and 60 cm length of which 50 cm is the effective length. Before the actual analysis, the capillary is rinsed sequentially with 0.05 M H_3PO_4 for 10 minutes followed by water for 20 minutes. To ensure a stable baseline, an additional rinsing of the capillary by 0.1 M NaOH, water and background electrolyte (BGE) consisting of 8 mM of L-arginine and 5 mM of DL-malic acid has been performed. Both constant current (CC) and constant voltage (CV) CE separation show acceptable linearity ($R^2 > 0.995$) for all cations in concentration ranges up to 100 μg/mL. The CC separation mode gives lower migration time (MT), better resolution and peak integration than the CV mode. Repeatability of peak area for individual cations is increased further by employing the rinsing sequence in-between sample injections as well as by using lithium chloride (LiCl) as internal standard. Although the CC mode is found to improve repeatability of peak area, it exhibits more day-to-day variability. The %RSD of the MT and the relative peak area (RPA) of sodium and potassium are however always within the specified limit. The CV mode shows good repeatability for calcium and magnesium. The sample quantification by calibration curve shows out-of-the-limit values for all analytes due to marked matrix interference. The standard addition method, in the same way proved ineffective to approximate the actual quantity of analytes in parenteral nutrition (PN) solutions. Finally, a single point calibration technique proved fruitful in the assay of cations by the use of simulated standard solution.

Chapter 7 - This book chapter is aimed at addressing the theoretical principles and applications of capillary electrophoresis (CE) for the separation of high intensity artificial sweeteners. Electrophoresis is a technique in which solutes are separated by their movement with different rates of migration in the presence of an electric field. Capillary electrophoresis emerged as a combination of the separation mechanism of electrophoresis and instrumental automation concepts in chromatography. Its separation mainly depends on the difference in the solutes migration in an electric field caused by the application of relatively high voltages, thus generating an electro-osmotic flow (EOF) within the narrow-bore capillaries filled with the background electrolyte. Currently capillary electrophoresis is a very powerful analytical technique with a major and outstanding importance in separations of compounds such as amino acids, chiral drugs, vitamins, pesticides etc., because of simpler method development, minimal sample volume requirements and lack of organic waste.

The main advantage of capillary electrophoresis over conventional techniques is the availability of the number of modes with different operating and separation characteristics include free zone electrophoresis and molecular weight based separations (capillary zone electrophoresis), micellar based separations (micellar electrokinetic chromatography), chiral separations (electrokinetic chromatography), isotachophoresis and isoelectrofocusing makes it a more versatile technique being able to analyse a wide range of analytes.

The ultimate goal of the analytical separations is to achieve low detection limits and CE is compatible with different external and internal detectors such as UV or photodiode array detector (DAD) similar to HPLC. CE also provides an indirect UV detection for analytes that do not absorb in the UV region. Besides the UV detection, CE provides five types of detection modes with special instrumental fittings such as Fluorescence, Laser-induced Fluorescence, Amperometry, Conductivity and Mass spectrometry. Infact, the lowest detection limits attained in the whole field of separations are for CE with laser induced fluorescence detection.

Regarding the applications of CE, the separation and determination of high intensity sweeteners were discussed in this chapter. The materials which show sweetness are divided into two types (i) nutritive sweeteners and (ii) non-nutritive sweeteners. The main nutritive sweeteners include glucose, crystalline fructose, dextrose, corn sweeteners, honey, lactose, maltose, invert sugars, concentrated fruit juice, refined sugars, high fructose corn syrup and various syrups. Non-nutritive sweeteners are sub-divided into two groups of artificial sweeteners and reduced polyols.

On the other hand, based on their generation; artificial sweeteners can further be divided into three types as (a) first generation artificial sweeteners which includes saccharin, cyclamate and glycyrrhizin (b) second generation artificial sweeteners are aspartame, acesulfame K, thaumatin and neohesperidinedihydrochalcone (c) neotame, sucralose, alitame and steviol glycosides falls under third generation artificial sweeteners. Artificial sweeteners are also classified into three types based on their synthesis and extraction: (i) synthetic (saccharin, cyclamate, aspartame, acesulfame K, neotame, sucralose, alitame) (ii) semi-synthetic (neohesperidinedihydrochalcone) and (iii) natural sweeteners (steviol glycosides, mogrosides and brazzein protein). Polyols are other groups of reduced-calorie sweeteners which provide bulk of the sweetness, but with fewer calories than sugars.

The commonly used polyols are: erythritol, mannitol, isomalt, lactitol, maltitol, xylitol, sorbitol and hydrogenated starch hydrolysates (HSH).

The studies revealed that capillary electrophoresis was successfully used for the separation of high intensity artificial sweeteners such as neotame, sucralose and steviol glycosides.

Additionally, the available methods for the other artificial sweeteners using capillary electrophoresis were reviewed besides the above indicated sweeteners.

Chapter 8 - Monoamines are chemicals containing an amine group and they possess enormous biological importance. They include most of the amino acids, the catecholamines, the indoleamines among the most important molecules. Polyamines are aliphatic chains containing multiple amine groups that generally originate from the amino acid arginine. They include citrulline, agmatine, ornithine, putrescine, spermine, spermidine and cadaverine. In general, they are concentrated in the micromolar to picomolar range. They participate in proliferation, differentiation, development, and cell signaling. Due to the lack of highly sensitive analytical techniques, most of the studies on mono and polyamines have been confined to tissue homogenates and very few studies have been carried out in extracellular fluids such as plasma, cerebral spinal fluid (CSF), or microdialysates of several tissues. The development of analytical techniques based on Capillary Zone Electrophoresis and Laser Induced Fluorescence Detection (CZE-LIFD) has been crucial to opening fields of studies in the aforementioned extracellular fluids and the physiological, as well as pathological role of polyamines. In the last two decades the author have successfully applied CZE-LIFD to the study of meningitis, preeclampsia, the mechanism of memory circuits of the brain, schizophrenia, Parkinson's disease (PD) and neuro-development. For such goals the authors have developed analytical techniques based on CZE-LIFD capable of detecting down to 2 nanomolar concentrations of glutamine, glutamate, arginine, agmatine, citrulline and putrescine in extracellular fluids. In CSF of meningitis-stricken children the authors found low glutamine levels, particularly when the etiological agent was *Haemophylus influenzae*. These levels increased to normal during the convalescence of the patient. This finding suggests that *H. influenzae* uses large amounts of glutamine probably because it lacks the first two enzymes of the Krebs cycle. In patients suffering preeclampsia low levels of arginine and high levels of agmatine in CSF and plasma were found. These results suggest that arginine might be an essential amino acid in preeclampsia patients and that it might be of therapeutic value. By means of brain microdialysis, 90 nanomolar concentration of agmatine were found in the *stratum radiatum* of the hippocampus in rats. The agmatine in the extracellular fluid of the hippocampus was nerve impulse and calcium dependent, suggesting an exocytotic origin and possible involvement in memory processes. Injecting agmatine by reverse microdialysis in the striatum it was found that extracellular dopamine increased, suggesting a role for agmatine in the control of automatic movements and a role in schizophrenia. Lately, the authors developed a method to measure putrescine and found that PD patients have higher levels of putrescine both in red cells and plasma from blood, providing a biological marker for PD and suggesting a role of putrescine and other polyamines in the degeneration of *substantia nigra* dopaminergic neurons, which is the hallmark of PD. Recently the authors found low levels of arginine and citrulline and a lack of correlation between arginine and citrulline in the plasma of preterm babies, as compared with fully developed neonates. These findings suggest that arginine and citrulline might be essential amino acids in premature babies; that they should be supplemented in their diets and that premature babies might have a disarray of the nitric oxide metabolic pathway. These findings show that CZE-LIFD is becoming a useful tool that could lead to a better understanding of the physiological and pathological roles of bioamine and to the development of therapeutic resources for several conditions.

Chapter 9 - Capillary electrophoresis (CE) was used to study residues of steroid hormones in influent and effluent waters of drinking water treatment plants. Steroids were of special interest, because they are slightly water-soluble. In general, their concentrations are at ng/ L level in environmental waters, but cannot be totally purified from drinking waters.

In this research, a partial-filling micellar electrokinetic chromatographic (PF-MEKC) method was developed and optimized for separation and determination of neutral steroids and their metabolites. The micelle solution contained 1.5 mM sodium taurocholate and 29.5 mM SDS in 20 mM ammonium acetate (pH 9.68). The CE separations were detected with an UV detector at the steroid specific wavelength 247 nm. The optimization was made with six steroid standards.

The samples from water treatment plants were concentrated to 6:1000 (v/v) with solid-phase extraction (SPE) in nonpolar sorbents. The PF-MEKC method was very repeatable (r^2 0.99), which was detected from the migration times of the studied compounds. The relative standard deviations of electroosmosis and the steroids were 0.01-0.04% and 0.01-0.07%, respectively. Concentration ranges for the steroids were linear at 0.5-10 ng/μL range. The influent waters contained 3.22-68.3 ng/L of 4-androsten-17β-ol-3-one glucosiduronate, androstenedione, and progesterone. On the contrary, the effluent waters after the treatment contained those analytes at 2.72-27.9 ng/ L level.

Chapter 10 - In CE, the synchronization of three major elements - injection, separation, and detection – is responsible for successful analyte determination. All these parts are indispensable and failure of either of them spoils the whole analysis.

The current goal of determination of extremely low concentrations in extremely low sample amounts leads to developments especially in detection part of the setup. It is most commonly realized by the UV/Vis photometric detection; however, its drawback is in a relatively low sensitivity. Nevertheless, by utilization of fluorescence detection even picomolar levels can be reached. Currently, a variety of both covalent and non-covalent labeling probes from the area of either small organic molecules or nanomaterial-based labels with high quantum yields is available.

Besides the development of fluorescent labels, also instrumental advances in the field of detector design enhance the sensitivity and applicability of this detection mode. In hard competition with other techniques, especially mass spectrometry, the fluorescence detection remains important player with significant advantages.

In this chapter is summarized not only the state of the art of the instrumental developments but also labeling strategies utilizing well-established and modern fluorescent tags. Finally, selected applications of capillary electrophoresis with laser-induced fluorescence detection are highlighted.

Chapter 11 - Capillary electrophoresis has proved to be a good alternative technique to high performance liquid chromatography for the investigation of various compounds due to its good resolution, versatility, simplicity, short analysis time and low consumption of chemicals and samples.

This chapter presents a synthesis of the author's work regarding applications of capillary electrophoretic methods (capillary zone electrophoresis with diode array detection): the separation of small-chain organic acids from plants extracts, wines, lactic bacteria fermentation products, and the separation of polyphenolic compounds from propolis extracts, plant extracts and wines.

Quantitative evaluation of organic acids in plants and foodstuff is important for flavour and nutritional studies, and also could be used as marker of bacterial activity. Organic acids occurring in foods are additives or end-products of carbohydrate metabolism of lactic acid bacteria. A good selection of lactic acid bacteria, in terms of content in organic acids, allows the control of mould growth and improves the shelf life of many fermented products and, therefore, reduces health risks due to exposure to mycotoxins.

On the other side, the largely studied group of phytochemicals is polyphenols, an assembly of secondary metabolites with various chemical structures and functions and biological activities, which are produced during the physiological plant growth process as a response to different forms of environmental conditions.

The methods for separation and quantification of organic acids and polyphenolic compounds were validated in terms of linearity of response, limit of detection, limit of quantification, precisions (i.e., intra-day, inter-day reproducibility) and recovery. The methods are simply, rapid, reliable and cost effective.

In: Capillary Electrophoresis (CE)
Editor: Christian Reed

ISBN: 978-1-63483-122-2

Chapter 1

Preparation and Application of Photosensitive Capillary Electrophoresis Coatings

Hailin Cong[1,2,*], Bing Yu[1,2], Xin Chen[1], Ming Chi[1], Peng Liu[1] and Mingming Jiao[1]

[1]College of Chemical Engineering, Qingdao University, Qingdao, China

[2]Laboratory for New Fiber Materials and Modern Textile, Growing Base for State Key Laboratory, Qingdao University, China

Abstract

Novel methods for the preparation of covalently linked capillary coatings of anti-protein-fouling polymers were demonstrated using photosensitive diazoresin (DR) as coupling agents. Layer by layer (LBL) self-assembly films of DR and anti-protein-fouling polymers based on hydrogen or ionic bonding were fabricated on the inner wall of capillary, then the hydrogen or ionic bonding was converted into covalent bonding after treatment with UV light through the unique photochemistry reaction of DR. The covalently bonded coatings suppressed basic protein adsorption on the inner surface of capillary, and thus a baseline separation of proteins was achieved using capillary electrophoresis (CE). Compared with bare capillary or non-covalently bonded coatings, the covalently linked capillary coatings not only improved the CE separation performance for proteins, but also exhibited good stability and repeatability. Due to the replacement of highly toxic and moisture sensitive silane coupling agent by DR in the covalent coating preparation, these methods may provide a green and easy way to make the covalently coated capillaries for CE.

Keywords: Capillary electrophoresis, coated capillary column, capillary coatings, diazoresin, proteins

---

[*] Corresponding author: hailincong@yahoo.com.

1. INTRODUCTION

Having the advantages of high efficiency, high sensitivity, speediness and economy, capillary electrophoresis (CE) is a powerful separation tool for biomacromolecule analysis [1–5]. However, when analyzing proteinaceous samples, protein adsorption onto fused-silica capillary walls is one of the major challenges of CE [6–9]. This leads to sample loss, peak broadening, poor resolution, unstable electroosmotic flow (EOF), and long migration times [10–13]. In order to minimize protein adsorption onto the capillary surface, surface modification with capillary coatings has become a research hotspot [14–20].

Capillary coatings can be divided into covalent coatings and non-covalent coatings [21–25]. The non-covalent coating can be produced simply by flushing the capillary with coating solutions. The coating molecules absorb on capillary surface by weak interactions such as electrostatic, van der Waals, and hydrogen bonding [26–28]. Furthermore, the layer-by-layer (LBL) self-assembly technique can also be used to prepare the non-covalently bonded capillary coatings, which provides the coating with new structures and functions [29–33]. For example, Haselberg et al. [34] prepared polybrene-dextran sulfate-polybrene (PB-DS-PB) triple layer coatings by the LBL self-assembly technique, and the coatings were fully compatible with mass spectrometry (MS) detection, causing no background signals and ionization suppression. The coatings were used for the analysis of α-chymotrypsinogen, ribonuclease A, Cyt-c and Lys by CE-MS, and the detection limits for them were 16, 11, 14 and 19 nM, respectively. Tang et al. prepared a non-covalent capillary coating by self-assembly of hexadimethrine bromide with enzyme for separation of enzyme inhibitors [35]. Compared with the non-covalently bonded coatings, the covalently bonded coatings are very stable and robust. For instance, Xu et al. [36] prepared chemically bonded polyvinyl alcohol (PVA) coatings which were used for high efficiency separation of cationic proteins (Cyt-c and Lys) and anionic proteins (myoglobin and trypsin inhibitor). Timperman et al. [37] prepared chemically bonded polyethylene glycol (PEG) coatings which were used for high efficiency separation of four basic proteins (BSA, alcohol dehydrogenase, carbonic anhydrase and trypsin inhibitor). Tuma et al. prepared a covalently polyacrylamide (PAA) capillary coating for the separation of biomolecules [38]. The covalently linked coatings of PVA, PEG or PAA not only showed very good anti-protein fouling properties, but also demonstrated excellent stabilities for repeatable separations. However, the preparation process of covalent coatings is usually complicated which includes multi-steps such as capillary pretreatment, introducing coupling agents, and inserting target coating reagents [39, 40]. Moreover, highly toxic and moisture sensitive silane coupling agents are traditionally used in the covalent coatings, which often cause environmental and quality problems during the manufacture and application [41–44].

In the fabrication process of capillary coatings with high quality and performance, how to combine the advantages of the non-covalently and covalently bonded coatings together, and avoid their disadvantages, is becoming one of the main development directions. In this chapter, we reported novel methods for the preparation of covalently linked PVA, PEG and poly(*N*-vinyl aminobutyric acid) (PVAA) capillary coatings using the LBL self-assembly technique combined with photochemistry reactions. The fabrication, structure and property of the coatings were studied and discussed preliminarily.

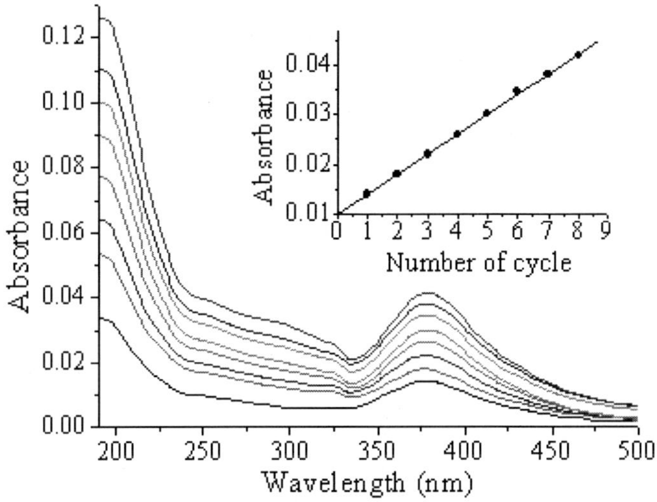

Figure 1. The UV–vis spectra of the assembly from the DR and PVA. Number of assembly cycles (bottom to top): 1, 2, 3, 4, 5, 6, 7 and 8. The inset plot shows that the absorbance of the films at 380 nm changes linearly with the number of assembly cycles.

2. Covalently Bonded PVA Capillary Coating

2.1. Coating Preparation

UV-vis spectroscopy is used to monitor the assembly process. The absorbance of the DR/PVA film at 380 nm, which derives from the characteristic π–π^* transition absorption of the diazo group of DR, increases linearly with the number of assembly cycles (Figure 1). This indicates that the LBL assembly is carried out successfully and uniformly. The driving force of the assembly comes from the hydrogen bond between the diazo group of DR and hydroxyl group of PVA.

DR is a non-toxic photoactive component often used as cell culture supports [45, 46], and the diazo groups involved in the DR/PVA multilayer films will be decomposed under UV irradiation, which results in a gradual decrease in the absorbance of the film at 380 nm (Figure 2). The photoreaction that takes place in the multilayer films, which originates from the diazo decomposition, is a first-order reaction: $\ln[(A_0–A_e)/(A_t–A_e)]$ changes linearly with irradiation time (Figure 2, inset), where A_0, A_t and A_e represent the absorbance of the film before irradiation, after irradiating for time t, and at the end of irradiation (30 s), respectively.

As illustrated in Figure 3, following the decomposition of the diazo group in the film under UV irradiation, the hydrogen bonds convert into covalent bonds [47]. The unique photo-crosslinking reaction of DR has been applied to the fabrication of covalently attached self-assembly films [48], hollow microcapsules [49], and biochips [50]. For example, Shi and co-workers reported the fabrication of stable, multilayer ultrathin films by self-assembly of DR with single-walled carbon nanotubols (SWNTols) followed by cross-linking under UV irradiation [51]. Yang et al. fabricated stable DR/chiral polyaniline composites (CPAC) shell on polystyrene (PS) colloids by self-assembly and UV cross-linking. After the PS core was removed by chemical etching, stable DR/CPAC hollow spheres were obtained [52]. Yu et al. prepared stable ultrathin DR/deoxyribonucleic acid (DNA) micropatterns by self-assembly and photolithography, which could find important application in biochips intended for gene therapy and drug identification [53].

As can be seen in Figure 4, the spectrum of the irradiated coating does not change after immersion in DMF for 30 min (Figure 4a), due to its covalently crosslinked structure. However, the spectrum of the nonirradiated film (Figure 4b) changes dramatically because of the etching by the DMF.

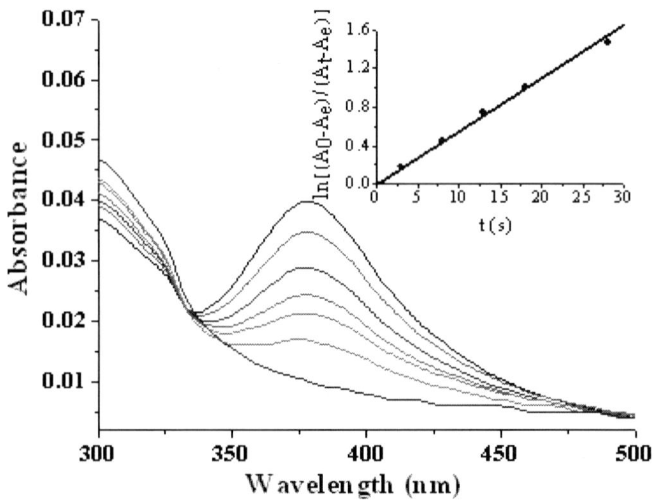

Figure 2. UV-vis spectra of DR/PVA multilayer coatings at different irradiation times. Irradiation time (s) (top to bottom): 0, 3, 8, 13, 18, 28 and 30; Irradiation intensity (at 365 nm): 350 μW/cm^2. Inset: relationship between ln[(A_0–A_e)/(A_t–A_e)] and irradiation time.

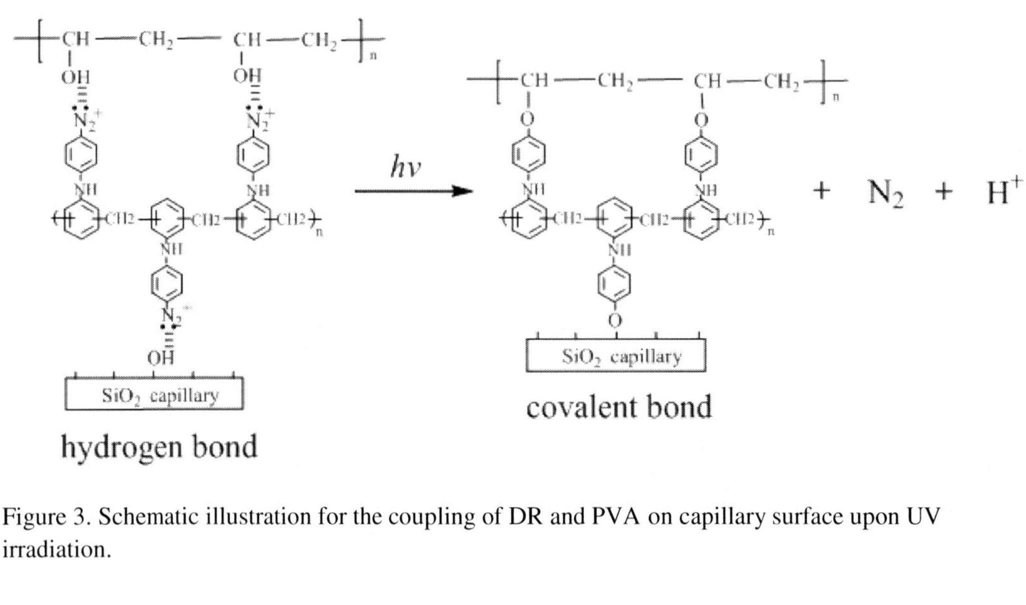

Figure 3. Schematic illustration for the coupling of DR and PVA on capillary surface upon UV irradiation.

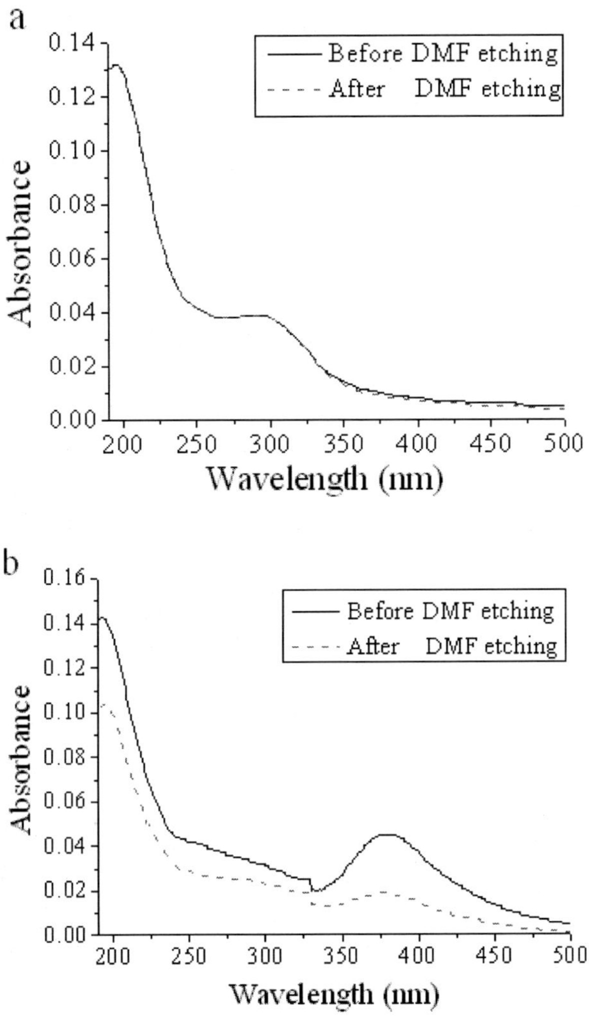

Figure 4. UV-vis spectra of irradiated (a) and nonirradiated (b) DR/PVA multilayer coatings before (solid lines) and after (dash lines) etching with DMF at 25 °C for 30 min.

2.2. Coating Performance

Figure 5 shows CE separation results of three proteins by using bare capillary, DR/PVA non-covalent, and DR/PVA covalent capillary coatings in the optimized conditions, respectively. The bare capillary performs a strong adsorption to the proteins, and thus a bad separation result with only two characteristic peaks is obtained. Although the separation performance of DR/PVA non-covalent capillary coating is better than that of bare capillary, baseline separation of the proteins cannot be achieved, and the stability of the coating is very poor due to lack of strong bondings to the capillary. Compared with them, the PVA covalent capillary coating has the best separation performance, and a stable and baseline separation of the Cyt-c, Lys, and BSA is achieved within 7 minutes.

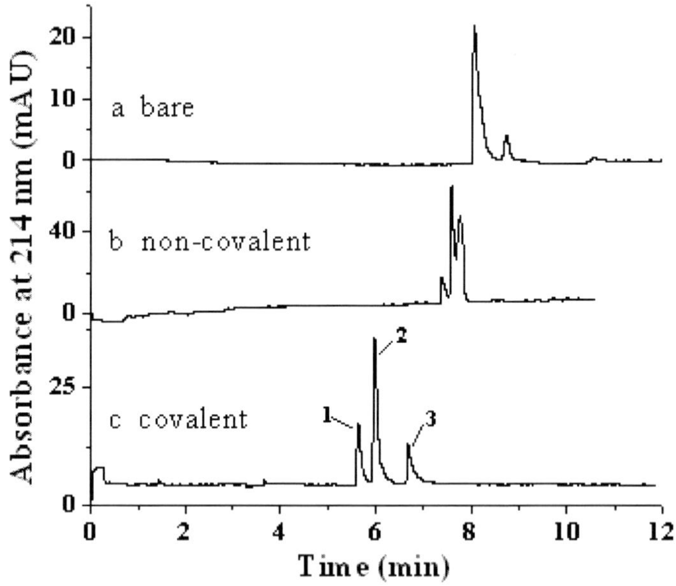

Figure 5. Separation of three proteins using the bare capillary (a), 4-layer PVA non-covalent coated capillary (b) and 4-layer PVA covalently coated capillary (c). Separation conditions: buffer, 40 mM phosphate (pH = 4.0); injection, 20 s with a height difference of 20 cm; applied voltage, +15 kV; UV detection, 214 nm; sample, 0.5 mg/mL for each protein; capillary, 75 μm ID × 50 cm (41 cm effective); capillary temperature, 25 °C. Peak identification: 1, Cyt-c; 2, Lys; 3, BSA.

The 4-layer DR/PVA covalent coatings prepared by this method have very good stability and repeatability. Table 1 shows that the run-to-run (n = 5) RSD of migration time for the proteins is less than 2%, day-to-day (n = 3) RSD is less than 3%, and capillary-to-capillary (n = 3) RSD is less than 4%. The limitation of detection (LOD) for BSA, Lys and Cyt-c is 0.213, 0.252 and 0.662 μM, respectively.

3. Covalently Bonded PEG Capillary Coating

3.1. Coating Preparation

As shown in Figure 6a, UV-vis spectroscopy is used to monitor the LBL self-assembly process. The absorbance of the DR/PEG film at 380 nm, which derives from the characteristic π–π^* transition absorption of the diazo group of DR, increases linearly with the number of assembly cycles (Figure 6a, inset). This indicates that the LBL assembly is carried out successfully and uniformly. The driving force of the assembly comes from the hydrogen bond between the diazo group of DR and hydroxyl group of PEG.

The diazo groups involved in the DR/PEG multilayer films will be decomposed under UV irradiation, which results in a gradual decrease in the absorbance of the film at 380 nm (Figure 6b). The photoreaction that takes place in the multilayer films, which originates from the diazo decomposition, is a first-order reaction: $\ln[(A_0-A_e)/(A_t-A_e)]$ changes linearly with irradiation time (Figure 6b, inset), where A_0, A_t and A_e represent the absorbance of the film before irradiation, after irradiating for time t, and at the end of irradiation (35 s), respectively. As illustrated in Figure 7, following the decomposition of the diazo group in the film under UV irradiation, the hydrogen bonds convert into covalent bonds.

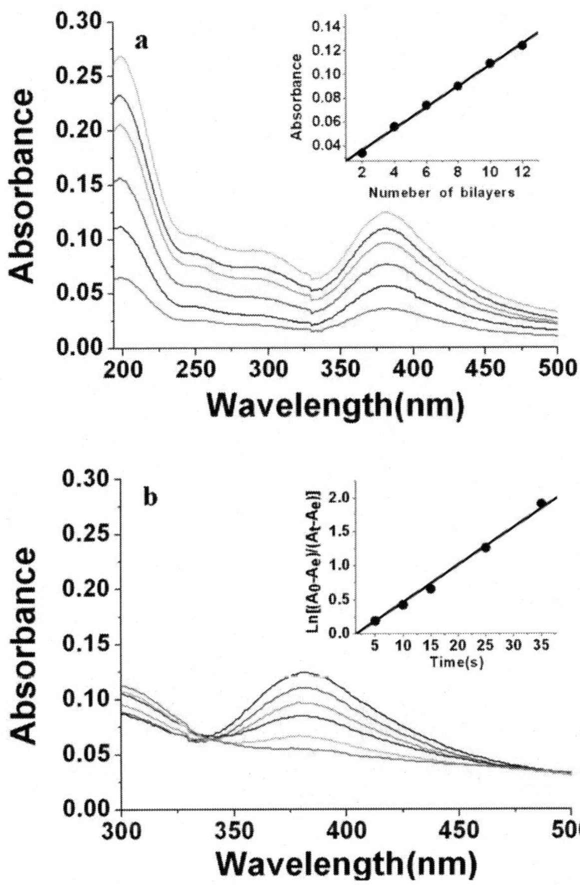

Figure 6. (a) UV–vis spectra of the assembly from DR and PEG. Number of assembly cycles (bottom to top): 1, 2, 3, 4, 5 and 6. The inset plot shows that the absorbance of the films at 380 nm changes linearly with the number of assembly cycles; (b) UV-vis spectra of DR/PEG multilayer coatings at different UV irradiation times. Irradiation time (s) (top to bottom): 0, 5, 10, 15, 25 and 35. Irradiation intensity (at 365 nm): 350 μW/cm^2. Inset: relationship between ln[(A_0–A_e)/(A_t–A_e)] and irradiation time.

Table 1. Separation performance of the 4-layer DR/PVA covalent capillary coatings

Protein	Detection limit (μM)	Migration time RSD (%)			
		run to run (n = 5)	day to day (n = 3)	capillary to capillary (n = 3)	continuous 60 times running
Cyt-c	0.662	0.82	2.01	3.14	2.61
Lys	0.252	0.63	2.68	3.51	2.28
BSA	0.213	1.87	2.17	3.88	2.58

Separation conditions: the same as Figure 5.

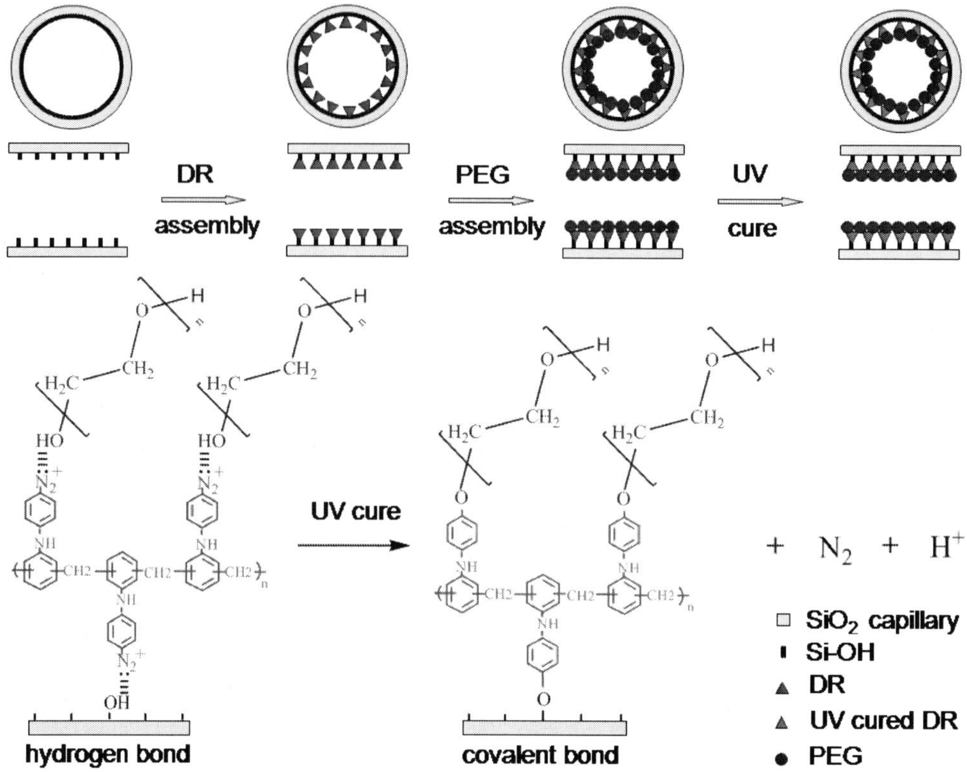

Figure 7. Schematic illustration of the preparation of DR/PEG covalent coating on capillary surface.

3.2. Coating Performance

Figure 8a–8c show CE separation results of four proteins by using bare capillary, DR/PEG non-covalent, and DR/PEG covalent capillary coatings in the optimized conditions, respectively. The bare capillary performs a strong adsorption to the proteins, and thus a bad separation result with only two characteristic peaks is obtained. Although the separation performance of DR/PEG non-covalent capillary coating is better than that of bare capillary, effective separation of the proteins cannot be achieved, and the stability of the coating is very poor due to lack of strong bondings to the capillary. Compared with them, the PEG covalent capillary coating has the best separation performance, and a stable and baseline separation of the Cyt-c, Lys, BSA and RNase A is achieved within 10 minutes.

Table 2 shows that the run-to-run (n = 5) RSD of migration time for the proteins is less than 1 %, day-to-day (n = 7) RSD is less than 2.5 %, and capillary-to-capillary (n = 5) RSD is less than 3.5 %. After a continuous 200 times running in a coating column, the RSD of migration time for the proteins are all less than 2.5 % (Table 2), and the separation performance of the DR/PEG covalent coatings is not deteriorated. Therefore, the DR/PEG covalently coated capillaries are robust and may be used in heavy duty analysis.

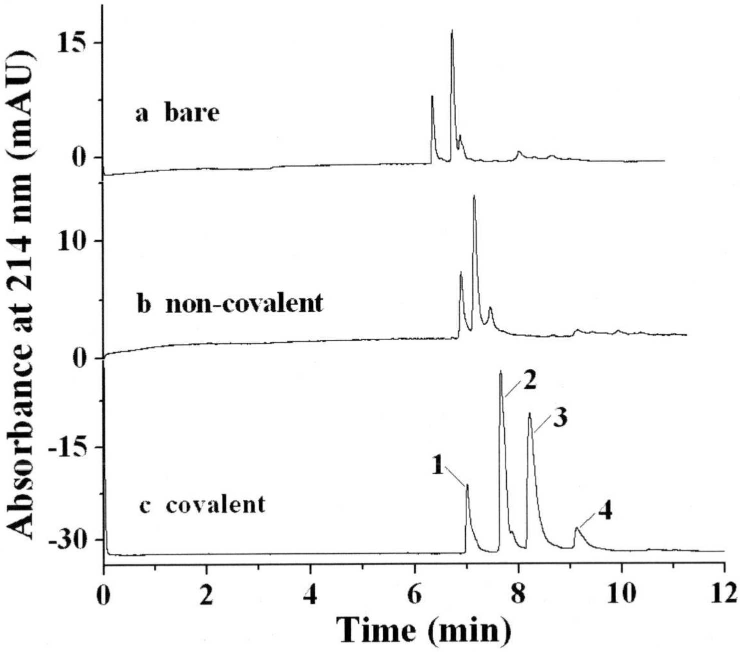

Figure 8. Separation of proteins using the bare capillary (a), PEG non-covalently coated capillary (b) and 2-layer PEG covalently coated capillary (c). Separation conditions: buffer, 40 mM phosphate (pH = 3.0); injection, 20 s with a height difference of 20 cm; applied voltage, +18 kV; UV detection, 214 nm; sample, 0.5 mg/mL for each protein; capillary, 75 μm ID × 50 cm (41 cm effective); capillary temperature, 25 °C. Peak identification: 1, Cyt-c; 2, Lys; 3, BSA; 4, RNase A.

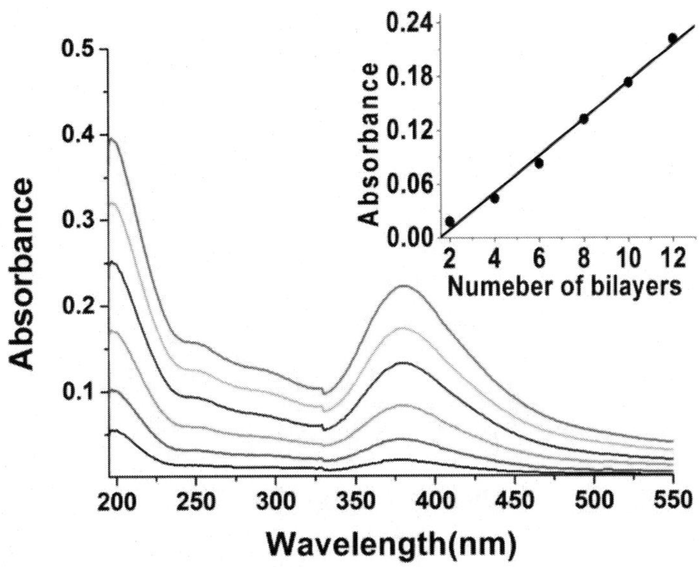

Figure 9. The UV–vis spectra of the assembly from the DR and PVAA. Number of assembly cycles (bottom to top): 1, 2, 3, 4, 5 and 6. The inset plot shows that the absorbance of the films at 380 nm changes linearly with the number of assembly cycles.

Table 2. Separation performance of the 2-layer DR/PEG covalent coatings

Protein	Migration time RSD (%)			
	run to run(n = 5)	day to day(n = 7)	capillary to capillary(n = 5)	Continuous 200 times running
Cyt-c	0.98	1.51	2.32	1.89
Lys	0.46	1.56	2.37	1.65
BSA	0.56	2.23	3.41	2.40
RNase A	0.79	2.18	3.39	2.38

Separation conditions: the same as Figure 8.

4. Covalently Bonded PVAA Capillary Coating

4.1. Coating Preparation

UV-vis spectroscopy is used to monitor the assembly process. The absorbance of the DR/PVAA film at 380 nm, which derives from the characteristic π–π* transition absorption of the diazonium group of DR, increases linearly with the number of assembly cycles (Figure 9). This indicates that the LBL assembly is carried out successfully and uniformly. The driving force of the assembly comes from the electrostatic interaction between the positive diazonium group ($-N_2^+$) of DR and the negative carboxyl group of PVAA.

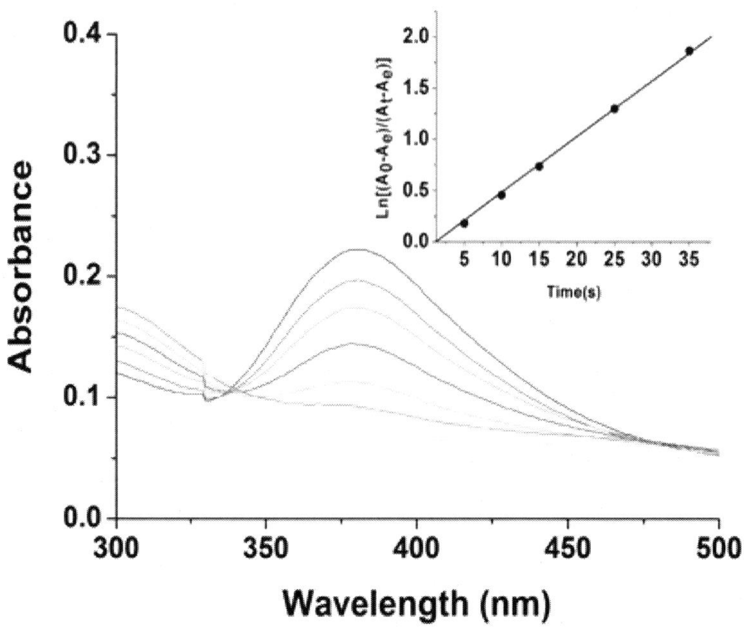

Figure 10. UV-vis spectra of DR/PVAA multilayer coatings at different irradiation times. Irradiation time (s) (top to bottom): 0, 5, 10, 15, 25 and 35; Irradiation intensity (at 365 nm): 350 μW/cm^2. Inset: relationship between ln[(A_0–A_e)/(A_t–A_e)] and irradiation time.

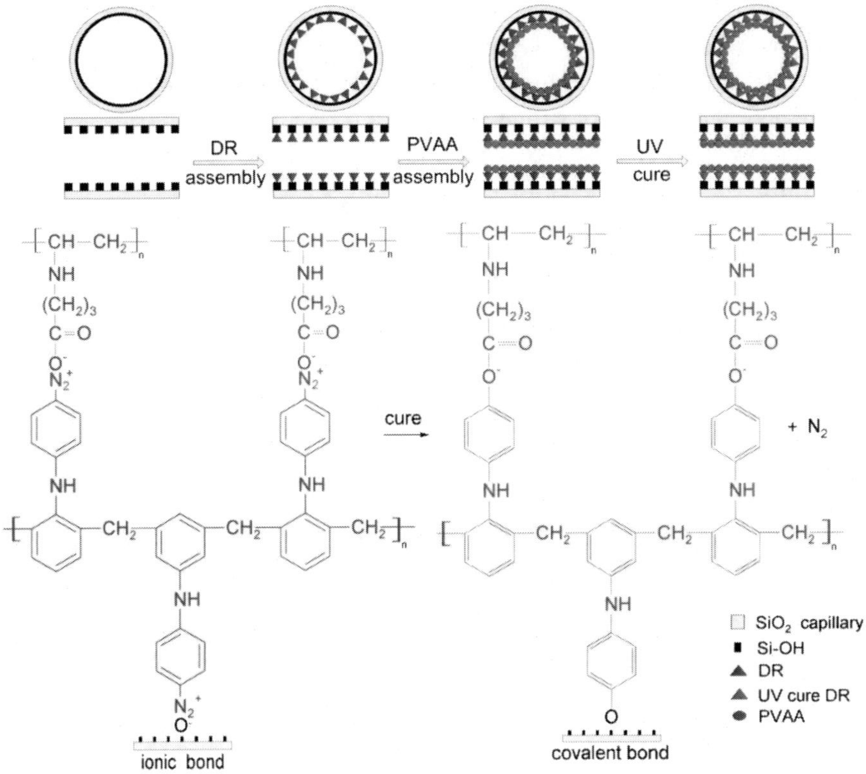

Figure 11. Schematic illustration of the preparation of DR/PVAA covalent coating on capillary surface.

The diazonium groups involved in the DR/PVAA multilayer films will be decomposed under UV irradiation, which results in a gradual decrease in the absorbance of the film at 380 nm (Figure 10). The photoreaction that takes place in the multilayer films, which originates from the $-N_2^+$ decomposition, is a first-order reaction: $\ln[(A_0-A_e)/(A_t-A_e)]$ changes linearly with irradiation time (Supporting Information Figure S1, inset), where A_0, A_t and A_e represent the absorbance of the film before irradiation, after irradiating for time t, and at the end of irradiation (35 s), respectively. Following the decomposition of the $-N_2^+$ group in the coating, the ionic bonds convert into covalent bonds. Referring to previous studies, the nature of the bond conversion can be represented schematically as show in Figure 11.

As can be seen in Figure 12, the spectrum of the irradiated coating does not change after immersion in DMF for 30 min (Figure 12a), due to having covalently crosslinked structure. However, the spectrum of the non-irradiated film (Figure 12b) changes dramatically because of the etching by the DMF.

4.2. Coating Performance

Figure 13a–13c show CE separation results of four proteins using bare capillary, DR/PVAA non-covalent, and DR/PVAA covalent capillary coatings in the optimized conditions, respectively. The bare capillary performs a strong adsorption to the proteins, and thus a bad separation result with only two characteristic peaks is obtained. Although the separation performance of DR/PVAA non-covalent capillary coating is better than that of bare capillary, effective separation of the proteins cannot be achieved, and the stability of the coating is very poor due to lack of strong bonding to the capillary. Compared with them, the PVAA covalent capillary coating has the best separation performance, and a stable and baseline separation of the Cyt-c, Lys, BSA and RNase A is achieved within 10 minutes. The detection limit of BSA with and without the covalent coating is 8 and 50 μg/mL, respectively.

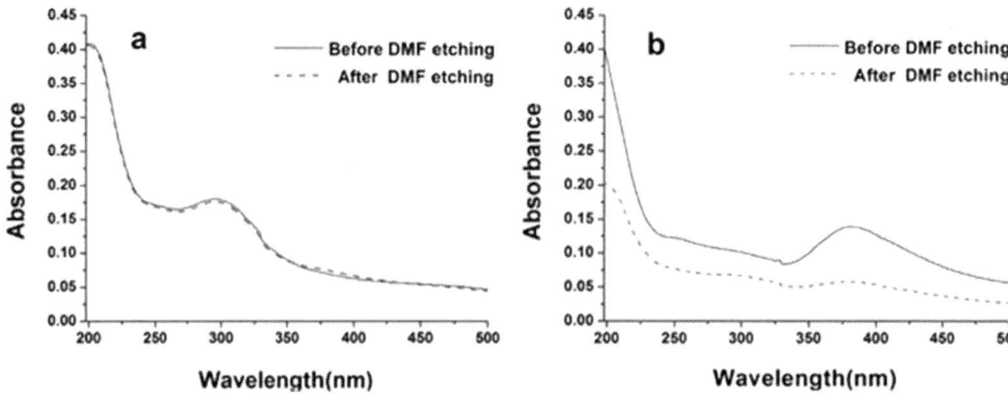

Figure 12. UV-vis spectra of irradiated (a) and non-irradiated (b) DR/PVAA multilayer coatings before (solid lines) and after (dash lines) etching with DMF at 25 °C for 30 min.

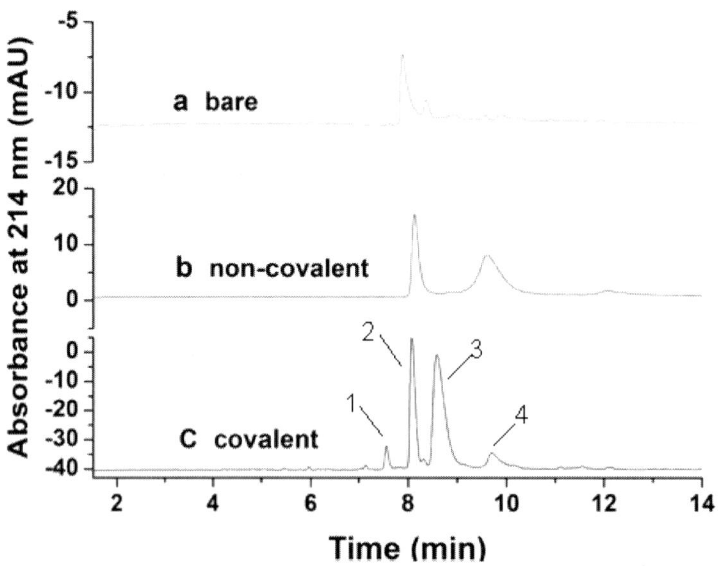

Figure 13. Separation of four proteins using the bare capillary (a), 2-layer PVAA non-covalently coated capillary (b) and 2-layer PVAA covalenty coated capillary (c). Separation conditions: buffer, 40 mM phosphate (pH = 3.0); injection, 20 s with a height difference of 20 cm; applied voltage, +15 kV; UV detection, 214 nm; sample, 0.5 mg/mL for each protein; capillary, 75 μm ID × 50 cm (41 cm effective); capillary temperature, 25 °C. Peak identification: 1, Cyt-c; 2, Lys; 3, BSA; 4, RNase A.

Table 3. Separation performance of the 2-layer DR/PVAA covalent capillary coatings

Protein	Migration time RSD (%)			
	run to run (n = 5)	day to day (n = 3)	capillary to capillary (n = 3)	continuous 60 times running
Cyt-c	0.56	1.87	2.09	1.82
Lys	0.84	1.33	2.38	1.87
BSA	0.56	2.01	3.09	1.96
RNase A	0.48	1.29	3.58	2.10

Separation conditions: the same as Figure 13.

Table 3 shows that the run-to-run (n = 5) RSD of migration time for the proteins is less than 1%, day-to-day (n = 3) RSD is less than 2.5%, and capillary-to-capillary (n = 3) RSD is less than 3.6%. After a continuous 60 times running in a coating column, the RSD of migration time for the proteins are all less than 2.5%, and the separation performance of the DR/PVAA covalent coatings is not deteriorated. Therefore, the DR/PVAA covalently coated capillaries are robust and may be used in heavy duty analysis.

CONCLUSION

In this work, new types of covalently linked capillary coatings are prepared successfully using photosensitive DR as coupling agents combined with the LBL self-assembly technique. The hydrogen or ionic bonding between the DR and anti-protein-fouling polymers is converted into covalent bonding after treatment with UV light through the unique photochemistry reaction of DR. The covalently bonded coatings suppress protein adsorption on the inner surface of capillary, and thus baseline separation of Lys, Cyt-c, BSA, and RNase A is achieved within 10 minutes at optimized separation conditions. Compared with bare capillary or non-covalently bonded coatings, the covalently linked capillary coatings not only improve the CE separation performance for proteins, but also exhibit good stability and repeatability. Moreover, for the replacement of highly toxic and moisture sensitive silane coupling agent by DR in the covalent coating preparation, this method may provide a green and easy way to make the covalently coated capillaries for all kinds of CE applications.

ACKNOWLEDGMENTS

This work is financially supported by the National Key Basic Research Development Program of China (973 special preliminary study plan, 2012CB722705), the Natural Science Foundation of China (21375069, 21404065), the Fok Ying Tong Education Foundation (131045), the Natural Science Foundation for Distinguished Young Scientists of Shandong Province (JQ201403), the Graduate Education Innovation Project of Shandong Province (SDYY14028), the Scientific Research Foundation for the Returned Overseas Chinese Scholars of State Education Ministry (20111568), the Science and Technology Program of

Qingdao (1314159jch), the China Postdoctoral Science Foundation (2014M561886) and the Doctoral Scientific Research Foundation of Qingdao.

REFERENCES

[1] Haselberg, R., Jong, G. J. & Sonmsen, G. M. (2007). *J. Chromatogr. A*, *1159*, 81.
[2] Chen, W. J., Mou, K., Xu, B., Ling, X. M., Cui, J. R. & Xu, P. (2009). *Analytical Biochemistry*, *394*, 62.
[3] Mei, S., Xu, G. & Wu, C. (2000). *Chinese J. Anal. Chem.*, *28*, 507.
[4] Prokhorova, A. F., Kuznetsov, M. A., Shapovalova, E. N., Staroverov, S. M. & Shpigun, O. A. (2010). *Procedia Chem.*, *2*, 9.
[5] Doherty, E. A. S., Meagher, R. J., Albarghouthi M. N. & Barron, A. E. (2003). *Electrophoresis*, *24*, 34.
[6] Kraly, J., Abul Fazal, M., Schoenherr, R. M., Bonn R., Harwood, M. M., Turner, E., Jones, M. & Dovichi, N. J. (2006). *Anal. Chem. 78*, 4097.
[7] Liu, C. & Kang, J. (2012). *J. Chromatogr. A*, *1238*, 146.
[8] Ahmadzadeh, H., Dovichi, N. J. & Krylov S. (2004). *Methods Mol. Biol.*, *276*, 15.
[9] DeJong, S. & Krylov, S. N. (2012). *Anal. Chem.*, *84*, 453.
[10] Khaledi, M. G. (1998). *High-Performance Capillary Electrophoresis: Theory, Techniques, and Applications*, Wiley, New York.
[11] Rush, R. S., Cohen, A. S. & Karger, B. L. (1991). *Anal. Chem.*, *63*, 1346.
[12] Hult, E. L., Emmer, Å., Roeraade, J. (1997). *J. Chromatogr. A*, *757*, 255.
[13] Araya, F., Huchet, G., McGroarty, I., Skellern, G. G. & Waigh, R. D. (2007). *Methods*, *42*, 141.
[14] Zhang, Z. X., Yan, B., Liu, K. L., Liao, Y. P. & Liu, H. W. (2009). *Electrophoresis*, *30*, 379.
[15] Znaleziona, J., Petr J., Knob, R., Maier, V. & Sevcik, J., (2008). *Chromatographia*, *67*, S5.
[16] Fu, X. F., Huang, L., Gao, F., Li, W., Pang, N. N.; Zhai, M. L., Liu, H. W. & Wu, M. T. (2007). *Electrophoresis*, *28*, 1958.
[17] Yone, A., Carballo, R. R., Grela, D. A., Rezzano, I. N. & Vizioli, N. M. (2011). *Electrophoresis*, *32*, 2840.
[18] Bachmann, S., Vallant, R., Bakry, R., Huck, C. W., Corradini D. & Bonn, G. K. (2010). *Electrophoresis*, *31*, 618.
[19] Zhu, X. X. & Wang, Y. M. (2011). *J. Funct. Polym.*, *24*, 1.
[20] Xu, W. C., Chu, Y. J., Lin, D., Yang, W., Lin, B. & Shou, C. Q. (2010). *Chinese J. Anal. Chem.*, *38*, 1167.
[21] Cao, F. H., Luo, Z. F., Zhou, D., Zeng, R. J. & Wang, Y. M. (2011). *Electrophoresis*, *32*, 1148.
[22] Gulcev, M. D. & Lucy, C. A. (2008). *Anal. Chem.*, *80*, 1806.
[23] Liu, Y., Fu, X. F., Bai, Y., Zhai, M. l., Liao, Y. P., Liao, J. & Liu, H. W. (2011). *Anal. Bioanal. Chem.*, *399*, 2821.
[24] Yu, B., Liu, P., Cong, H., Tang, J., Zhang, L. (2012). *Electrophoresis*, *33*, 3066.
[25] Yu, B., Cui, W., Cong, H., Jiao, M., Liu, P. & Yang, S. (2013). *RSC Adv.*, *3*, 20010.

[26] Yu, B., Jiao, M., Cong, H., Su, X. & Yang, S. (2014). *J. Sep. Sci.*, *37*, 725.

[27] Zhou, D., Tan, L., Xiang, L., Zeng, R., Cao, F., Zhu, X. & Wang, Y. (2011). *J. Sep. Sci.*, *34*, 1738.

[28] Li, J., Han, H., Wang, Q., Liu, X. & Jiang, S. (2011). *J. Sep. Sci.*, *34*, 1555.

[29] Zheng, C., Liu, Y. P., Zhou, Q. H. & Di, X. (2010). *J. Chromatogr. B*, *878*, 2933.

[30] Yu, B., Cong, H. L., Liu, H. W., Li, Y. Z. & Liu, F. (2005). *J. Sep. Sci.*, *28*, 2390.

[31] Nilsson, C., Becker, K., Harwigsson, I., Bulow, L., Birnbaum, S. & Nilsson, S. (2009). *Anal. Chem.*, *81*, 315.

[32] Catai, J. R., Torano, J. S., Jongen, P. M. J. M. & Jong, G. J. D., Somsen, G. W. (2007). *J. Chromatogr. B*, *852*, 160.

[33] Sultan, Y., Walsh, R., Monreal, C. & DeRosa, M. C. (2009). *Biomacromolecules*, *10*, 1149.

[34] Haselberg, R., Jong, G. J. D. & Somsen, Govert, W. (2010). *Analytica Chimica Acta*, *678*, 128.

[35] Tang, Z. & Kang, J. (2006). *Anal. Chem.*, *78*, 2514.

[36] Xu, L., Dong, X. Y. & Sun, Y. (2010). *Biochemical Engineering Journal*, *53*, 137.

[37] Razunguzwa, T. T., Warrier, M. & Timperman, A. T. (2006). *Anal. Chem.*, *78*, 4326.

[38] Tuma, P., Samcova, E. & Stulik, K. (2011). *Anal. Chim. Acta.*, *685*, 84.

[39] Mei, J., Tian, Y. P., He, W., Xiao, Y. X., Wei, J. & Feng, Y. Q. (2010). *J. Chromatogr. A*, *1217*, 6979.

[40] Zhou, S. Y., Tan, J. J., Chen, Q. H., Lin, X. C., Lu, H. X. & Xie, Z. H. (2010). *J. Chromatogr. A*, *1217*, 8346.

[41] Omae, K., Sakai, T., Sakurai, H., Yamazaki, K., Shibata, T., Mori, K., Kudo, M., Kanoh, H. & Tati, M. (1992). *Arch. Toxicol.*, *66*, 750.

[42] Clayton, G. D. & Clayton, F. E. (Eds.), *Patty's industrial hygiene and toxicology*, Wiley, New York 1981.

[43] Lokajova, J., Tiala, H. D., Viitala, T., Riekkola, M. L. & Wiedmer, S. K. (2011). *Soft Matter*, *7*, 6041.

[44] Gao, J., Latep, N., Ge, Y., Tian, J., Wu, J. Q. & Qin, W. D. (2013). *J. Sep. Sci.*, *36*, 1575.

[45] Plewa, A., Niemiec, W., Filipowska, J., Osyczka, A. M., Lach, R., Szczubialka, K. & Nowakowska, M. (2011). *European Polymer Journal, 47*, 1503.

[46] Fu, Y., Yao, G. J., Dong, Y. P., Cao, Y. J. & Duan, E, K. (2003). *Chemical Journal of Chinese Universities, 24*, 182.

[47] Chen, J. & Cao, W. (1999). *Chem. Commun.*, 1711.

[48] Yang, Z., Chen, J. & Cao, W. (2004). *Polym. Int.*, *53*, 815.

[49] Zhu, H. & Mcshane, M. J. (2005). *Langmuir, 21*, 424.

[50] Cao, T., Wei, F., Jiao, X., Chen, J., Liao, W., Zhao X. & Cao, W. (2003). *Langmuir*, *19*, 8127.

[51] Shi, J., Qin, Y., Luo, H., Guo, Z., Woo H. & Park, D. (2007). *Nanotechnology*, *18*, 365704.

[52] Yang, L., Yang Z. & Cao, W. (2005). *J. Colloid Interface Sci.*, *292*, 503.

[53] Yu, B., Cong, H. L., Liu. H. W., Lu, C. H., Wei, F. & Cao, W. X. (2006). *Anal. Bioanal. Chem.*, *384*, 385.

In: Capillary Electrophoresis (CE)
Editor: Christian Reed

ISBN: 978-1-63483-122-2

Chapter 2

APPLICATION OF CAPILLARY ZONE ELECTROPHORESIS TO TRACE ANALYSES OF INORGANIC ANIONS IN SEAWATER

Keiichi Fukushi*

Kobe University Graduate School of Maritime Sciences, Japan

ABSTRACT

Capillary zone electrophoresis (CZE) is an environmentally friendly analytical method that provides high separation capability with minimum consumption of samples and reagents. We have been developing CZE methods for the determination of cationic and anionic substances in environmental waters (e.g., seawater, river water, sewage) and in biospecimens (e.g., serum, cerebrospinal fluid, urine, and vegetables such as spinach (*Spinacia oleracea*) and ice-plant (*Mesembryanthemum crystallinum* L.)). Seawater contains salts of high concentrations, with many elements, species, and concentration ranges. It is a difficult task to apply a high-resolution method to seawater analyses. Nevertheless, the sensitivity of CZE with a UV detector, which is usually used as the detector, is insufficient for low concentrations of analytes because of the short light-path length of the capillary inner diameter. Therefore, in addition to protecting the influence of high concentrations of salts, some on-line concentration procedure is necessary to determine the concentrations of trace substances in seawater. We used artificial seawater as the background electrolyte for this study to decrease the influence of salts, and used transient isotachophoresis as the on-line concentration procedure. This chapter presents a summary of the analytical procedures and the results for the determination of trace inorganic anions such as nitrite and nitrate, phosphate, iodide and iodate, and bromate in seawater (and salt).

* E-mail: fukushi@maritime.kobe-u.ac.jp.

1. Introduction

Seawater sample analysis is important to elucidate relations among numerous geochemical, biological, and anthropogenic activities and to assess the ionic composition of marine matrices [1]. Capillary zone electrophoresis (CZE) is a suitable analytical method for application to the environmental water samples such as seawater because most substances exist as ionic species in seawater. Moreover, CZE is an environmentally friendly analytical method because of its minimal consumption of samples and reagents. Nevertheless, the sensitivity of CZE with a UV detector, which is usually used as the detector, is insufficient for low concentrations of analytes because of the short light-path of the capillary inner diameter. The analytes must be enriched when CZE is applied to trace component analyses in seawater samples. Transient isotachophoresis (tITP) is typically used as the on-line concentration procedure. Avoiding the influence of salts is also necessary, just as it is with other instrumental analytical methods. Artificial seawater [2] was used as the background electrolyte (BGE) to resolve the issue. This report introduces the analytical procedures and results for the determination of nitrite and nitrate, phosphate, iodide and iodate, and bromate in seawater using tITP-CZE.

2. Transient Isotachophoresis

Isotachophoresis (ITP) has been used as an analytical method for the determination of ionic species in water samples. The separation mechanism of ITP is explainable as follows. An analyte ion is sandwiched between an ion which has greater mobility than the analyte mobility (leading ion) and an ion which has less mobility than the analyte mobility (terminating ion) in a capillary. Voltage is applied between the electrode in an electrolyte (leading electrolyte) containing the leading ion and another electrode in an electrolyte (terminating electrolyte) containing the terminating ion. When the analyte concentrations are lower than the concentrations of leading and terminating ions, the analyte ions are enriched between both electrolytes. After a steady state is established, the leading, analyte, and terminating ions migrate with the same speed to the electrode which has opposite polarity of these ions.

The ITP state can be generated temporarily to concentrate analyte ions in CZE. The analyte ions migrate in the state of zone electrophoresis after the concentration is finished. The analytes are separated during migration and detected. This on-line concentration technique is designated as transient isotachophoresis (tITP). The tITP mechanism is presented in Figure 1a)–1d) [3]. Sample ions (S_1 and S_2) are injected into the capillary after the BGE is fulfilled. Then the terminating ion (T) is introduced (Figure 1a). The leading ion (L) is involved in the BGE in this case. When voltage is applied, the analytes are enriched as arranged according to the magnitude of the effective mobility (μ) (Figure 1b). The μ for the analyte S_1 (μ_{S1}) is larger than the μ for the analyte S_2 (μ_{S2}). When L migrates into the terminating electrolyte to form a mixed zone, the tITP state disappears and the concentration is completed (Figure 1c). Finally, all ions migrate in the state of zone electrophoresis (Figure 1d). The tITP is the only on-line concentration procedure which is useful for the enrichment of trace analytes in highly saline samples such as seawater.

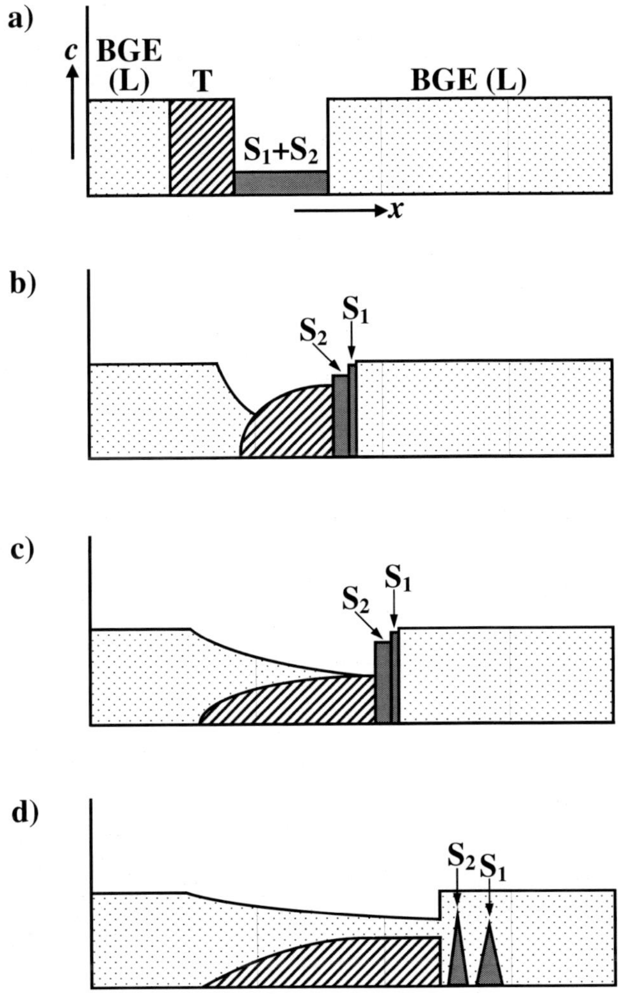

Figure 1. Schematic of tITP: *c*, concentration; *x*, capillary length; BGE (L), background electrolyte (BGE) containing leading ion (L); S_1, S_2, sample ions ($\mu_{S1} > \mu_{S2}$); T, terminating ion. Reproduced with permission from Wiley-VCH Verlag GmbH & Co.

3. Simultaneous Determination of Nitrite and Nitrate [4]

3.1. Outline

tITP-CZE using artificial seawater as the BGE was improved to lower the limit of detection (LOD) further for the determination of nitrite and nitrate in seawater. By lowering the pH of BGE, the difference between the effective mobility of nitrite and that of nitrate increased, thereby permitting increased sample volumes to be tolerated and decreasing their LOD values. Artificial seawater containing no bromide, adjusted to pH 3.0 with phosphate buffer, was adopted for use as the BGE. To reverse the electroosmotic flow (EOF), a capillary was flushed with 0.1 mM dilauryldimethylammonium bromide (DDAB) for 3 min before the capillary was filled with the BGE. The respective LODs for nitrite and nitrate were 2.7 and 3.0 μg/l (as nitrogen). The LODs were obtained at a signal-to-noise ratio (S/N) of 3. The respective values of the relative standard deviation (RSD) of the peak area for these ions were 2.0 and 0.75% when the nitrite concentration was 0.05 mg/l and the nitrate concentration was 0.5 mg/l. The RSDs of peak height were 4.4 and 2.3%. The RSD values of migration time for these ions were 0.19 and 0.17%. The proposed method was applied to the determination of nitrite and nitrate in a proposed certified reference material for nutrients in seawater, MOOS-1, distributed by the National Research Council of Canada (NRC). Results agreed with the assigned tolerance interval. This method was also applied to the determination of these ions in seawater collected from Osaka Bay in Japan. Results closely approximated those obtained using a conventional spectrophotometric method.

3.2. Procedure

Nitrite and nitrate in MOOS-1 and real seawater samples were determined using the following procedure. Seawater samples were filtered through a 0.45-μm membrane before analysis. No pretreatment procedure or sample cleanup was necessary, except for filtration. The detection wavelength was set at 210 nm for CZE determination of nitrite and nitrate. The capillary was thermostated at 30°C. A new capillary was washed with 1 M sodium hydroxide for 40 min and then with water for 10 min. The capillary was rinsed with 0.1 mM DDAB [5] for 3 min to reverse the EOF. Subsequently, the capillary was filled with BGE (artificial seawater containing no bromide, adjusted to pH 3.0 with phosphate buffer) by vacuum for 3 min. After a sample was vacuum injected into the CE apparatus for 4 s (84 nl) the terminating ion solution, 600 mM acetate was injected for 17 s (357 nl). The injection period of 1 s corresponds to the sample volume of 21 nl when the total capillary length ($L_{tot.}$) = 72 cm. Voltage of 8 kV was applied with the sample inlet side as the cathode. Each step was run automatically. Calibration graphs were prepared using synthetic standards.

3.3. Calibration Graphs

Standard solutions for nitrite and nitrate were prepared using artificial seawater containing 68 mg/l bromide. Calibration graphs for nitrite and nitrate were linear using both the peak area and peak height. The regression equations relating the area response to the concentrations for nitrite (x, 0–0.1 mg/l) and nitrate (x, 0–0.5 mg/l) were $y = 2.51\times10^4x + 566$ ($r = 0.9963$) and $y = 6.31\times10^4x + 352$ ($r = 0.9999$), respectively. The regression equations relating peak height were $y = 3.10\times10^4x + 900$ ($r = 0.9901$) and $y = 2.92\times10^4x + 197$ ($r = 0.9999$). Nitrite has lower correlation because of the lower nitrite concentrations in the sample solutions. The values of the RSD and the LOD for nitrite and nitrate are presented in Table 1, where, $L_{det.}$, the effective length, denotes the capillary length from the end of a capillary at the sample inlet side to the detector.

Table 1. Precision and detection limits of determination of nitrite and nitrate

	RSD (%)			LOD (NO_2^-–N, NO_3^-–N)
	Area	Height	Time	(μg/l, S/N = 3)
NO_2^-	2.0	4.4	0.19	2.7
NO_3^-	0.75	2.3	0.17	3.0

Electrophoretic conditions: capillary, total length ($L_{tot.}$) = 72 cm, effective length ($L_{det.}$) = 50 cm, 75 μm i.d.×375 μm o.d., pre-rinsed with 0.1 mM DDAB; BGE, artificial seawater without Br^- adjusted to pH 3.0 with 40 mM phosphate buffer; voltage, –8 kV; wavelength for detection, 210 nm. Sample, artificial seawater containing 68 mg/l Br^-, 0.05 mg/l NO_2^-–N, and 0.5 mg/l NO_3^-–N, eight determinations for RSD; vacuum (16.9 kPa) injection period, 4 s (84 nl). Terminating ion solution, 600 mM acetate; vacuum injection period, 17 s (357 nl). Reproduced with permission from Elsevier.

Table 2. Analytical results[a] for nitrite and nitrate in MOOS-1[b]

Bottle No.	NO_2^-–N (mg/l)	NO_3^-–N (mg/l)	NO_2^-–N+ NO_3^-–N (mg/l)
1	0.0469	0.277	0.324
2	0.0423	0.262	0.305
3	0.0432	0.268	0.311
Assigned tolerance interval	0.0386 ± 0.0081	–	0.325 ± 0.034

[a]Electrophoretic conditions are identical to those in Table 1; results for duplicate analyses using the peak area. [b]MOOS-1, a proposed certified reference material for nutrients in seawater distributed by the National Research Council of Canada (NRC). Reproduced with permission from Elsevier.

3.4. Analytical Results

The proposed method was applied to the determination of nitrite and nitrate in MOOS-1. The results are presented in Table 2. Duplicate analyses were performed on each of three bottles. Then the average values were calculated. The method was found to be accurate: The results for nitrite and the sums of nitrite and nitrate closely approximated with the assigned tolerance intervals, which were determined by NRC. Figure 2A depicts an electropherogram of MOOS-1. The sharp peaks for nitrite and nitrate with baseline separation were detected within 7 min. The method was also applied to the determination of these anions in seawater samples taken from the surface and the seabed around the coastal area of Osaka Bay on 11 July and 1 August 2002. The seawater samples were also analyzed using naphthylethylenediamine spectrophotometry (NS) [6], which is conventionally used for nitrite and nitrate analyses. The results are presented in Table 3; the values are the averages of duplicate analyses. The CZE results for nitrite and nitrate closely approximated those obtained using the NS method (r = 0.9647 for nitrite, r = 0.9790 for nitrate). It was noteworthy that the concentrations of nitrite and nitrate in the seawater sample taken from the seabed in the Rokko Island on July 11 were approximately equal to those concentrations in the seawater sample taken from the same sampling site on August 1. Most concentrations of nitrite and nitrate in the seawater samples taken from other sampling sites on July 11 were higher than those on August 1. Figure 2B presents an electropherogram of surface seawater from the pond at our university (Kobe University).

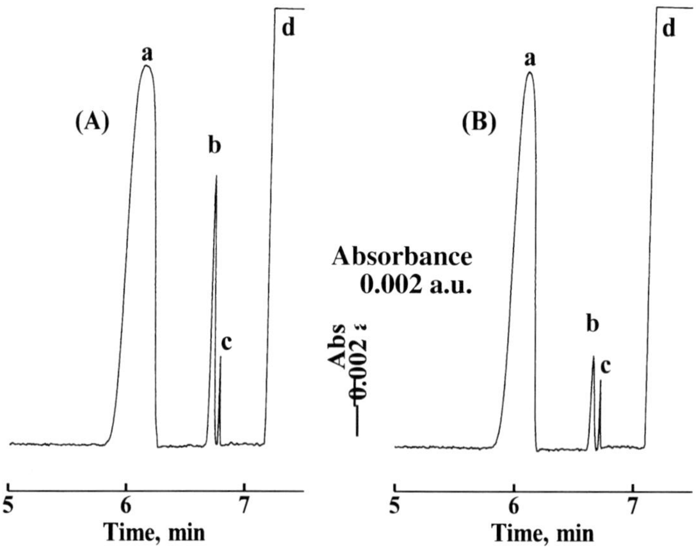

Figure 2. Electropherograms of MOOS-1 and surface seawater from the pond at Kobe University (KU): (A) Sample, MOOS-1; (B) Sample, surface seawater from the pond at KU. Electrophoretic conditions are identical to those of Table 1. Peaks: a = Br^-, b = NO_3^-, c = NO_2^-, and d = CH_3COO^-. Reproduced with permission from Elsevier.

Table 3. Analytical results for seawater nitrite and nitrate

Sampling site	Depth (m)	NO_2^-–N (mg/l)		NO_3^-–N (mg/l)	
		CZE[a]	NS[b]	CZE[a]	NS[b]
Port of Kobe[c]	0	0.016	0.011	0.023	0.038
Port of Kobe[d]	0	–	0.006	0.005	0.003
Port of Kobe[c]	9.5	0.015	0.010	0.012	0.019
Port of Kobe[d]	8.0	0.019	0.019	0.023	0.017
Rokko Island[c]	0	0.016	0.014	0.103	0.111
Rokko Island[d]	0	0.008	0.004	0.026	0.025
Rokko Island[c]	11	0.019	0.017	0.014	0.029
Rokko Island[d]	11	0.019	0.018	0.014	0.012
Pond at KU[c,e]	0	0.031	0.033	0.093	0.119
Pond at KU[d,e]	0	0.010	0.006	0.012	0.008
Pond at KU[c,e]	5.0	0.002	0.003	0.002	0.005
Pond at KU[d,e]	4.5	–	0.002	–	0.002

[a]Electrophoretic conditions are identical to those of Table 1; results for duplicate analyses using peak area. [b]NS, naphthylethylenediamine spectrophotometry. [c]Sampling date: 11 July 2002. [d]Sampling date: 1 August 2002. [e]KU, Kobe University. Reproduced with permission from Elsevier.

4. Determination of Phosphate [7]

4.1. Outline

We developed CZE with indirect UV detection for the determination of phosphate in seawater using tITP as an on-line concentration procedure. The following optimum conditions were established: BGE, 5 mM 2,6-pyridinedicarboxylic acid (PDC) containing 0.01%(w/v) hydroxypropyl methylcellulose (HPMC) adjusted to pH 3.5; detection wavelength, 200 nm; vacuum injection period of sample, 3 s (45 nl); terminating ion solution, 500 mM 2-(*N*-morpholino)ethanesulfonic acid (MES) adjusted to pH 4.0; vacuum injection period of the terminating ion solution, 30 s (450 nl); and applied voltage of 30 kV with the sample inlet side as the cathode. The LOD for phosphate was 16 µg/l (PO_4^{3-}–P) at S/N of three. The respective values of the RSD of the peak area, peak height, and migration time for phosphate were 2.6, 2.3, and 0.34%. The proposed method was applied to the determination of phosphate in a seawater certified reference material for nutrients, MOOS-1. The results closely resembled certified values. The method was also applied to the determination of phosphate in coastal seawaters. The results agreed with those obtained using a molybdenum blue spectrophotometry (MBS) [8].

4.2. Procedure

Phosphate contents in MOOS-1 and actual seawater samples were ascertained using the following procedure. No pretreatment procedure was necessary except for filtration. The detection wavelength was set at 200 nm for CZE determination of phosphate. The capillary was thermostated at 30°C. The capillary was filled with BGE (a mixture of 5 mM PDC and 0.01%(w/v) HPMC adjusted to pH 3.5 with 1 M sodium hydroxide) by vacuum for 4 min. After a sample was vacuum injected into the CZE apparatus for 3 s (45 nl), the terminating ion solution (500 mM MES adjusted to pH 4.0 with 1 M sodium hydroxide) was injected for 30 s (450 nl). The injection period of 1 s corresponds to the sample volume of 15 nl when the $L_{tot.}$ = 97 cm. Voltage of 30 kV was applied with the sample inlet side as the cathode.

4.3. Calibration Graphs

Calibration graphs for phosphate were linear using both the peak area and peak height. Regression equations relating area and height responses to concentration for phosphate (x, 0–0.2 mg/l) were $y = 1.02\times10^4x + 3.19\times10^2$ ($r = 0.9993$) and $y = 8.83\times10^3x+3.19 \times 10^2$ ($r = 0.9993$), respectively, in the case of lower phosphate concentrations. However, those for higher phosphate concentrations (x, 0–1.0 mg/l) were $y = 1.29\times10^4x + 5.72\times10^2$ ($r = 0.9999$) and $y = 1.11\times10^4x + 6.24 \times 10^2$ ($r = 0.9998$), respectively. Table 4 presents the RSDs and LOD for phosphate using the proposed method.

Table 4. Precision and detection limits of determination of phosphate

	RSD (intraday, %, n = 8)			RSD (interday, %, n = 6)			LOD (PO_4^{3-}–P)
	Area	Height	Time	Area	Height	Time	(µg/l, S/N = 3)
PO_4^{3-}	2.6	2.3	0.34	8.7	6.0	0.48	16

Electrophoretic conditions: capillary, $L_{tot.}$ = 97 cm, $L_{det.}$ = 75 cm, 75 µm i.d.×375 µm o.d.; BGE, 5 mM PDC containing 0.01%(w/v) HPMC adjusted to pH 3.5 with 1 M sodium hydroxide; voltage, –30 kV; wavelength for detection, 200 nm. Sample, artificial seawater containing 0.4 mg/l PO_4^{3-}–P; vacuum (16.9 kPa) injection period, 3 s (45 nl). Terminating ion solution, 500 mM MES adjusted to pH 4.0 with 1 M sodium hydroxide; vacuum injection period, 30 s (450 nl). Reproduced with permission from Wiley-VCH Verlag GmbH & Co.

4.4. Analytical Results

The proposed method was applied to the determination of phosphate in MOOS-1. Table 5 presents those results: triplicate analyses were performed on each of 12 bottles. Results for phosphate closely approximated the certified values as determined by NRC. Figure 3A depicts an electropherogram of MOOS-1. A small but sharp phosphate peak was detected within 16 min with baseline separation from high concentrations of chloride and sulfate.

Table 5. Analytical results for phosphate in MOOS-1

Bottle no.	PO_4^{3-}–P (mg/l)
1	0.043 ± 0.007
2	0.046 ± 0.007
3	0.048 ± 0.007
4	0.046 ± 0.005
5	0.048 ± 0.004
6	0.051 ± 0.006
7	0.045 ± 0.002
8	0.046 ± 0.002
9	0.049 ± 0.005
10	0.043 ± 0.002
11	0.046 ± 0.002
12	0.046 ± 0.007
Certified values	0.048 ± 0.002

Electrophoretic conditions are identical to those of Table 4; results for triplicate analyses using peak area. Reproduced with permission from Wiley-VCH Verlag GmbH & Co.

The method was also applied to the determination of phosphate in seawater samples (salinity, 27.4–31.2) taken from the surface and the seabed around the coastal area of Osaka Bay, between Nishinomiya Harbor and the Port of Kobe, on 11 September 2006. Seawater samples were also analyzed using MBS, which is used conventionally for phosphate analysis. Table 6 presents these results. The CZE results for phosphate agreed with those obtained using the MBS method (r = 0.9936 for phosphate). It was interesting that the concentrations of phosphate in the seabed seawater samples were approximately two-fold to three-fold higher than those in the surface seawater samples, except for the samples taken from the Port of Kobe. However, concentrations of dissolved oxygen (DO) in the seabed seawaters were

0.01–0.04 mg/l, except for the Port of Kobe. Anoxic conditions prevailed on the seabed. Progressive anoxic conditions at the surface sediment reportedly engender the reduction of Fe (III) to Fe (II), thereby causing the transformation of insoluble $FePO_4$ into more soluble $Fe_3(PO_4)_2$, in turn releasing PO_4^{3-} ions into the overlying seawater [9]. It was inferred from these results that phosphate was released from the surface sediments at the sampling sites listed above, except for the Port of Kobe. In addition, the extensive surface in Nishinomiya Harbor was cloudy, with a strong smell of hydrogen sulfide caused by the blue tide. Figure 3B depicts an electropherogram of the Nishinomiya Harbor seabed water.

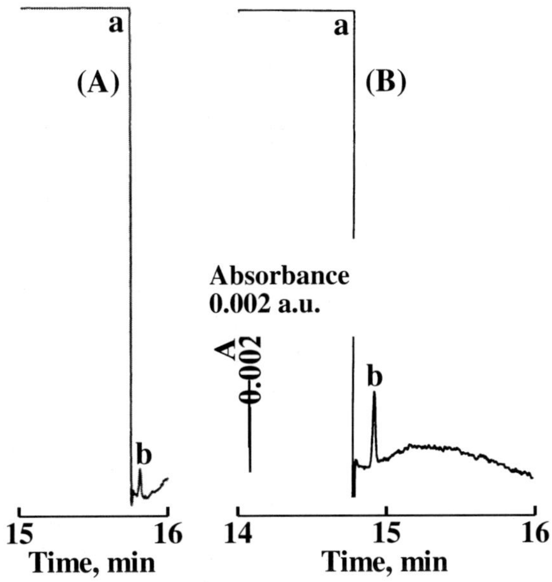

Figure 3. Electropherograms of MOOS-1 and seabed seawater from Nishinomiya Harbor. (A) Sample, MOOS-1; (B) Sample, seabed seawater from Nishinomiya Harbor. Electrophoretic conditions are identical to those of Table 4. Peaks: a = Cl^- and SO_4^{2-}, b = PO_4^{3-}, and c = NO_2^-. Reproduced with permission from Wiley-VCH Verlag GmbH & Co.

Table 6. Analytical results for phosphate in seawater

Sampling site[a]	Depth (m)	Temp. (°C)	pH	S[b]	DO[c] (mg/l)	PO_4^{3-}–P (mg/l) CZE[d]	MBS
Port of Kobe	0	27.5	7.78	30.2	2.70	0.078	0.078
Port of Kobe	4.0	26.9	7.81	30.4	2.16	0.059	0.068
Rokko Island	0	27.8	8.12	28.5	5.30	0.060	0.070
Rokko Island	11.0	25.8	7.70	31.2	0.02	0.117	0.118
Pond at KU[c,e]	0	26.7	7.92	27.4	3.16	0.058	0.054
Pond at KU[c,e]	4.5	26.1	7.77	30.5	0.01	0.117	0.116
Fukaehama	0	27.0	8.02	28.1	4.61	0.055	0.060
Fukaehama	5.5	25.7	7.69	31.1	0.04	0.161	0.153
Nishinomiya Harbor	0	27.0	7.78	28.8	1.90	0.104	0.114
Nishinomiya Harbor	2.5	25.9	7.60	30.7	0.01	0.200	0.195

[a]Sampling date: 11 September 2006. [b]S, salinity. [c]DO, dissolved oxygen. [d]Electrophoretic conditions are identical to those of Table 4; results obtained using peak area and working curve. Reproduced with permission from Wiley-VCH Verlag GmbH & Co.

The migration time for phosphate increased linearly with increasing salinity of the sample solutions of 13.6–34.0 (data not shown). A regression equation relating the migration time to salinity (x, 13.6–34.0) was $y = 1.69\times10^{-3}x^2 + 8.16\times10^{-2}x + 11.23$ ($r = 0.9992$). The salinity values calculated using this equation closely approximated those shown in Table 6 ($r = 0.9722$). In addition to the determination of phosphate in seawater samples, the sample salinity can be estimated using the proposed method.

5. Simultaneous Determination of Iodide and Iodate [10, 11]

5.1. Outline

We developed CZE with tITP as an on-line concentration procedure for the simultaneous determination of iodide and iodate in seawater. The effective mobility of iodide was decreased by the addition of 20 mM cetyltrimethylammonium chloride (CTAC) to the artificial seawater BGE so that tITP functioned for both the iodide and iodate. The LODs for iodide and iodate were 4.0 and 5.0 µg/l (as iodine) at S/N of three. The values of the RSD of peak area, peak height, and migration time for iodide and iodate were 2.9, 1.3, 1.0 and 2.3, 2.1, 1.0%, respectively. The proposed method was applied to the simultaneous determination of iodide and iodate in seawater collected around Osaka Bay. A sufficient recovery percentage was obtained for iodide (87–112%) and iodate (93–112%) in the standard addition experiments.

5.2. Procedure

Iodide and iodate in the seawater sample were determined using the following procedure. A seawater sample was filtered through a 0.45-µm membrane before analysis. No pretreatment procedure was necessary, except for filtration. The detection wavelength was set at 221 nm for CZE determination of iodide and iodate. The capillary was thermostated at 30°C. A new capillary was washed with 1 M sodium hydroxide for 40 min and then with water for 10 min. The capillary was filled with BGE (artificial seawater containing 20 mM CTAC, pH 7.9) by vacuum for 3 min. After a sample was vacuum injected into the CZE apparatus for 20 s (420 nl), the 2 M phosphate terminating ion solution was injected for 4 s (84 nl). Voltage of 8 kV was applied with the sample inlet side as the cathode.

5.3. Calibration Graphs

Standard solutions for iodide and iodate were prepared using artificial seawater containing 0.05 mg/l nitrite and 0.5 mg/l nitrate. Calibration graphs for iodide and iodate were linear using both the peak area and peak height. Regression equations relating area response to concentration for iodide (x, 0 – 0.1 mg/l) and iodate (x, 0 – 0.1 mg/l) were $y = 7.27\times10^4x + 4$ ($r = 0.9997$) and $y = 1.18\times10^4x + 344$ ($r = 0.9885$), respectively; those relating peak height were $y = 2.53 \times 10^4x - 6$ ($r = 0.9997$) and $y = 2.02\times10^4x + 247$ ($r = 0.9979$). Table 7 presents the values of the RSD and LOD for iodide and iodate.

Table 7. Precision and detection limits of determination of iodide and iodate[a]

	RSD (%, n = 7)			LOD (I^-, IO_3^-–I)
	Area	Height	Time	(μg/l, S/N = 3)
I^-	2.9	1.3	1.0	4.0
IO_3^-	2.3	2.1	1.0	5.0

[a]Electrophoretic conditions: capillary, $L_{tot.}$ = 72 cm, $L_{det.}$ = 50 cm, 75 μm i.d.×375 μm o.d.; BGE, artificial seawater containing 20 mM CTAC (pH 7.9); voltage, –8 kV; wavelength for detection, 221nm. Sample, artificial seawater containing 0.05 mg/l NO_2^-–N, 0.5 mg/l NO_3^-–N, 0.1 mg/l I^-, and 0.1 mg/l IO_3^-–I; vacuum (16.9 kPa) injection period, 20 s (420 nl). Terminating ion solution, 2 M phosphate; vacuum injection period, 4 s (84 nl). Reproduced with permission from Elsevier.

5.4. Analytical Results

The proposed method was applied to the determination of iodide and iodate in seawater samples taken from the surface and the seabed around coastal areas of Osaka Bay on 26 August 2003. The results are presented in Table 8. Duplicate analyses were performed and the average values were calculated. Iodide and iodate were detected in all samples. Total iodine concentrations ($I^- + IO_3^-$) in the surface seawater samples were 0.065–0.075 mg/l, which were higher than the total iodine concentrations (0.051 ± 0.004, 0.052 ± 0.002 mg/l) at the surface of the Seto Inland Sea reported by Ito et al. [12]. Iodate concentrations (0.038–0.061 mg/l) were higher than the iodide concentrations (0.010–0.032 mg/l), except for the seawater samples taken from the seabed of the Rokko Island and the Port of Kobe. The iodide concentrations (0.045 and 0.055 mg/l), however, were higher than iodate concentrations (0.030 and 0.027 mg/l) in the seabed samples from the Rokko Island and the Port of Kobe, where the DO values were low. The iodide concentrations in these samples were higher than those in other samples. Iodate concentrations in the Nishinomiya Harbor were much higher than those at other sampling sites; iodide concentrations here were much lower than those at other sampling sites. The ratios of iodate to the total iodine concentrations for the surface and the seabed waters were, respectively, 86% and 84%. The Nishinomiya Harbor was a shallow and closed area where a small river flowed into the area. The surface and seabed samples from the pond at KU and the Tarumi Harbor, with 0.03–0.05 mg/l iodide and iodate added, were analyzed. The recoveries for iodide and iodate were, respectively, 87–112% and 93–112%. Figure 4 depicts an electropherogram of seawater taken from the seabed in the Tarumi Harbor. The sharp peaks for iodide and iodate with baseline separation were detected within 11 min.

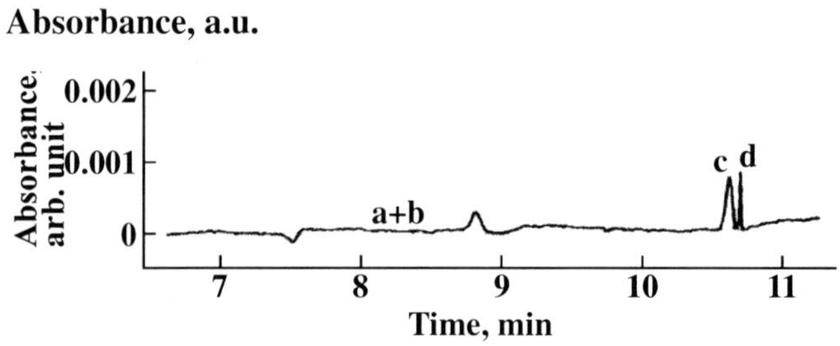

Figure 4. Electropherogram of seabed seawater from Tarumi Harbor. Electrophoretic conditions are identical to those of Table 7. Peaks: a = NO_2^-, b = NO_3^-, c = I^-, and d = IO_3^-. Reproduced with permission from Elsevier.

Table 8. Analytical results for iodide and iodate in seawater[a]

Sampling site[b]	Depth (m)	Temp. (°C)	pH	S[c]	DO[d] (mg/l)	I^- (mg/l)	IO_3^-–I (mg/l)	I^-+ IO_3^-–I (mg/l)
Port of Kobe	0	29.1	7.37	24.1	7.57	0.031	0.041	0.072
Port of Kobe	8.5	25.6	6.90	29.7	1.28	0.055	0.027	0.082
Rokko Island	0	28.8	7.75	18.5	7.97	0.022	0.043	0.065
Rokko Island	10.5	24.1	6.83	30.9	0.31	0.045	0.030	0.075
Pond at KU[e]	0	28.5	7.63	14.4	6.48	0.020	0.049	0.069
Pond at KU[e]	4.0	26.5	7.29	22.2	3.58	0.031	0.053	0.084
Nishinomiya Harbor	0	30.2	7.33	8.2	7.06	0.010	0.061	0.071
Nishinomiya Harbor	2.0	29.2	7.57	15.1	3.65	0.011	0.057	0.068
Tarumi Harbor	0	26.8	7.25	29.0	6.16	0.028	0.047	0.075
Tarumi Harbor	5.0	25.2	7.13	30.6	5.36	0.032	0.038	0.070

[a]Electrophoretic conditions are identical to those of Table 7; results for duplicate analyses using peak area. [b]Sampling date: 26 August 2003. [c]S, salinity. [d]DO, dissolved oxygen. [e]KU, Kobe University. Reproduced with permission from Elsevier.

6. DETERMINATION OF BROMATE [13]

6.1. Outline

We developed CZE with direct UV detection for the determination of bromate in highly saline samples such as seawater and salts using tITP as an on-line concentration procedure. The following optimum conditions were established: BGE, artificial seawater containing no bromide adjusted to pH 3.0; detection wavelength, 210 nm; vacuum injection period of sample, 18 s (378 nl); terminating ion solution, 600 mM sodium acetate; vacuum injection period of the terminating ion solution, 7 s (147 nl) for seawater and 12 s (252 nl) for salts; applied voltage of 7 kV with the sample inlet side as the cathode. The LOD for bromate was 30 μg/l (BrO_3^-–Br) with S/N of 3. The respective values of the RSD of the peak area, peak height, and migration time for bromate were 6.4, 1.5, and 0.51%. Seawater and salt samples, with bromate added, were analyzed using this method. The recovery of bromate in seawater samples was 84–104%. Linear regression equations relating area and height responses to the bromate concentration were obtained using the salt samples.

6.2. Procedure

Bromate contents in seawater and salt samples were ascertained using the following procedure. No pretreatment procedure was necessary except for filtration. The detection wavelength was set at 210 nm. The capillary was thermostated at 30°C. A new capillary was washed with 1 M sodium hydroxide for 40 min and then with water for 10 min. The capillary was filled with BGE (artificial seawater containing no bromide, adjusted to pH 3.0 with 1 M hydrochloric acid) by vacuum for 3 min. After a sample was vacuum-injected into the CZE apparatus for 18 s (378 nl), the terminating ion solution (600 mM acetate) was injected for 7 s (147 nl) for seawater samples and 12 s (252 nl) for salt samples. Voltage of 7 kV was applied with the sample inlet side as the cathode.

6.3. Calibration Graphs

Calibration graphs for bromate were linear, using both the peak area and peak height. Regression equations relating the area and height responses to the concentration for bromate (x, 0 – 1.0 mg/l) were, respectively, $y = 6.70\times10^3x + 1.25\times10^2$ ($r = 0.9987$) and $y = 4.75\times10^3x + 6.71 \times 10^2$ ($r = 0.9995$). The LOD for bromate (BrO_3^-–Br) was 30 μg/l (S/N = 3). The respective values of the RSD of the peak area, peak height, and migration time for bromate were 6.4, 1.5, and 0.51% (0.5 mg/l, $n = 8$).

Table 9. Analytical results for bromate in surface seawater[a]

Sampling site[b]	Temp (°C)	S[c]	BrO_3^-–Br Added (mg/l)	Found (mg/l)	Recovery (%)
Pond at KU	10.7	32.3	0	ND[d]	–
Pond at KU			0.25	0.21	84
Pond at KU			0.50	0.46	92
Pond at KU			0.75	0.74	99
Pond at KU			1.0	1.04	104

[a]Electrophoretic conditions: capillary, $L_{tot.}$ = 72 cm, $L_{det.}$ = 50 cm, 75 μm i.d.×375 μm o.d.; BGE, artificial seawater containing no bromide adjusted to pH 3.0 with 1 M hydrochloric acid; voltage, –7 kV; wavelength for detection, 210 nm. Sample, seawater added 0–1.0 mg/l BrO_3^-–Br; vacuum (16.9 kPa) injection period, 18 s (420 nl). Terminating ion solution, 600 mM acetate adjusted to pH 4.8; vacuum injection period, 15 s (315 nl). Results using peak area and working curve. [b]Sampling date: 13 February 2008. [c]S, salinity. Reproduced with permission from Wiley-VCH Verlag GmbH & Co.

6.4. Analytical Results

Seawater samples were taken from the surface at a small harbor at our university and from the surface close to an outlet of discharged water from a sewage treatment plant in September 2008. The seawater samples, with 0.25–1.0 mg/l of bromate (BrO_3^-–Br) added, were analyzed using the method. The recovery of bromate from seawater samples collected from the small harbor was 84–104%, as presented in Table 9. Figure 5 depicts an electropherogram of the seawater containing 0.75 mg/l bromate. A sharp bromate peak was detected within 19 min with a dip of unknown origin immediately after the bromate peak. Bromate peaks were not observed clearly without tITP. The recovery of bromate was 45–60% when the seawater sample was collected from the surface near the sewage treatment plant. The reason for the low recovery was presumed to be the low salinity of the sample that might be mixed with the treated water. The peak area, height, and migration time for bromate varied according to salinity of the sample solution because chloride in the sample solutions corresponds to the leading ion for tITP (data not shown).

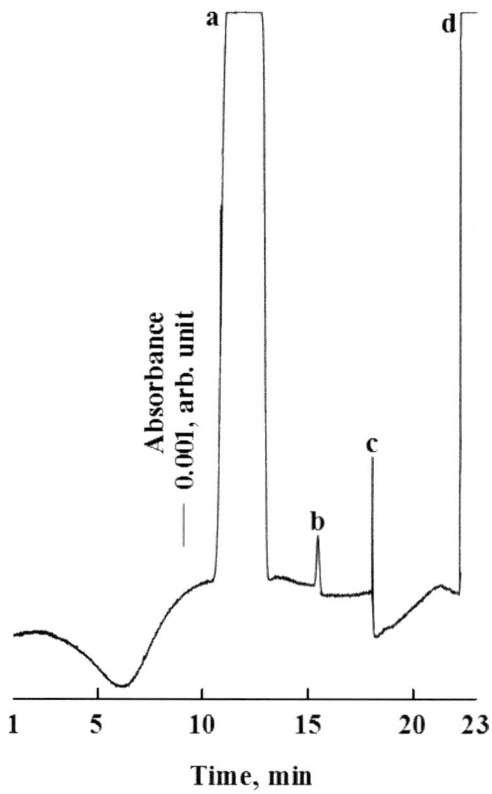

Figure 5. Electropherogram of surface seawater, 0.75 mg/l bromate added, from a small harbor at our university. Electrophoretic conditions are identical to those of Table 9. Peaks: a = Br^-, b = NO_3^-, c = BrO_3^-, and d = CH_3COO^-. Reproduced with permission from Wiley-VCH Verlag GmbH & Co.

The method was also applied to the determination of bromate in salt samples. Sea salts of two kinds, A and B, and a commercially available type of rock salt, C, were used. The chloride concentration in seawater is *ca.* 20,000 mg/l [14]. Therefore, 3.31 g of each salt was dissolved in 100 ml water to constitute 20,000 mg/l chloride in the solutions. Similarly, the salt samples, with 0.25–1.0 mg/l of bromate added, were analyzed using the method. Regression equations relating the area and height responses to concentration for bromate (x, 0–1.0 mg/l) were linear, as shown in Table 10. However, the regression equations' slopes differed depending on the kind of salts. Therefore, a standard addition method must be used to determine the bromate in salt samples using the proposed procedure.

Table 10. Regression equations for bromate added to salt samples[a]

Salt	Regression equation (peak area)	Regression equation (peak height)
A	$y = 1.43\times10^4x + 3.62\times10^2$ ($r = 0.9979$)	$y = 5.44\times10^3x - 1.37\times10^2$ ($r = 0.9859$)
B	$y = 1.38\times10^4x - 4.31\times10^2$ ($r = 0.9948$)	$y = 8.99\times10^3x - 1.79\times10^2$ ($r = 0.9974$)
C	$y = 1.44\times10^4x + 1.09\times10^2$ ($r = 0.9995$)	$y = 4.50\times10^3x + 2.26\times10^2$ ($r = 0.9945$)

[a]Electrophoretic conditions are identical to those in Table 9, except for the injection period for terminating ion solution: 12 s (252 nl). y: peak area or peak height for bromate. x: concentration for bromate (0–1.0 mg/l). r: correlation coefficient. Reproduced with permission from Wiley-VCH Verlag GmbH & Co.

Conclusion

This chapter presented a summary of the analytical procedures and results for the determination of trace inorganic anion, such as nitrite and nitrate, phosphate, iodide and iodate, and bromate in seawater (and salt). In addition to the beneficial points presented above, CZE presents the benefit that different analytes in different samples can be determined using only a single cheap capillary if a suitable BGE is prepared. Ion chromatography requires expensive columns of different kinds according to the analytes. In spite of the several benefits, CZE is apparently not used sufficiently for analyses of actual samples including environmental waters. Environmentally friendly CZE methods are anticipated for use extensively in actual analyses.

References

[1] Timerbaev, A. R. & Fukushi, K. (2003). Analysis of seawater and different highly saline natural waters by capillary zone electrophoresis. *Mar. Chem.,* 82, 221–238.

[2] Japanese Standards Association, Lubricants – Determination of Rust-Preventing Characteristics, *JIS K* 2510: 1998, Tokyo 1998, p. 8 (Japanese).

[3] Urbánek, M, Křivánková, L. & Boček, P. (2003). Stacking phenomena in electromigration: From basic principles to practical procedures. *Electrophoresis,* 24, 466–485.

[4] Fukushi K., Nakayama, Y. & Tsujimoto, J. (2003). Highly sensitive capillary zone electrophoresis with artificial seawater as the background electrolyte and transient isotachophoresis as the on-line concentration procedure for simultaneous determination of nitrite and nitrate in seawater. *J. Chromatogr.* A, 1005, 197–205.

[5] Melanson, J. E. & Lucy, C. A. (2000). Ultra-rapid analysis of nitrate and nitrite by capillary electrophoresis. *J. Chromatogr.* A, 884, 311–316.

[6] Japanese Standards Association, Testing Methods for Industrial Waste Water, *JIS K* 0102: 1998, Tokyo 1998, p. 153 (Japanese).

[7] Okamoto, T., Fukushi, K., Takeda, S. & Wakida, S. (2007). Determination of phosphate in seawater by CZE with on-line transient ITP. *Electrophoresis,* 28, 3447–3452.

[8] Japanese Standards Association, Testing Methods for Industrial Waste Water, *JIS K* 0102: 1998, Tokyo, 1998, p. 175 (Japanese).

[9] Belias, C., Dassenakis, M. & Scoullos, M. (2007). Study of the N, P and Si fluxes between fish farm sediment and seawater. Results of simulation experiments employing a benthic chamber under various redox conditions. *Mar. Chem.,* 103, 266–275.

[10] Yokota, K., Fukushi, K., Takeda, S. & Wakida, S. (2004). Simultaneous determination of iodide and iodate in seawater by transient isotachophoresis-capillary zone electrophoresis with artificial seawater as the background electrolyte. *J. Chromatogr.* A, 1035, 145–150.

[11] Yokota, K. & Fukushi, K. (2004). Simultaneous determination of iodide and iodate in seawater by capillary zone electrophoresis. *Bull. Soc. Sea Water Sci. Jpn.,* 58, 75–79 (Japanese).

[12] Ito, K., Shoto, E. & Sunahara, H. (1991). Ion chromatography of inorganic iodine species using C_{18} reversed-phase columns coated with cetyltrimethylammonium. *J. Chromatogr.* A, 549, 265–272.

[13] Fukushi, K., Yamazaki, R. & Yamane, T. (2009). Determination of bromate in highly saline samples using CZE with on-line transient ITP. *J. Sep. Sci.,* 32, 457–461.

[14] Isshiki, K. (2005). Chemistry of Seawater. In Fujinaga, T., Sorin, Y. & Isshiki, K. (Eds.), *The Chemistry of the Oceans and Lakes* (First ed., pp. 3–560). Kyoto: Kyoto University Press (Japanese).

In: Capillary Electrophoresis (CE)
Editor: Christian Reed
ISBN: 978-1-63483-122-2

Chapter 3

APPLICATIONS OF CAPILLARY ELECTROPHORESIS TO PHARMACEUTICAL AND BIOCHEMICAL ANALYSIS

S. Flor[1,2], M. Contin[2], M. Martinefski[2], C. Dobrecky[2], J. P. Cattalini[2], O. Boscolo[2], V. Tripodi[1,2] and S. Lucangioli[1,2]

[1]Department of Pharmaceutical Technology.
Faculty of Pharmacy and Biochemistry, University of Buenos Aires, Argentina
[2]National Council of Scientific and Technical Research (CONICET)

ABSTRACT

In the last decades, miniaturized separation techniques have rapidly gained popularity in different areas of analysis such as pharmaceutical, biopharmaceutical, clinical, biological, environmental, and forensics. The great advantages presented by the analytical miniaturized techniques, including high separation efficiency and resolution, rapid analysis and minimal consumption of reagents and samples, make them an attractive alternative to the conventional chromatographic methods.

In this sense, capillary electrophoresis (CE) is a family of related techniques that employs narrow-bore capillaries to perform highly efficient separations from large to small molecules. Different modes are applied in CE. Capillary zone electrophoresis (CZE) using a simple buffer as electrolyte, is widely used for the analysis of inorganic and organic ions. Another CE mode is electrokinetic chromatography (EKC), in which the separation principle is based on the differential partition between the analytes and a pseudostationary phase as well as the migration behavior of the analytes. Several nanostructures are used as pseudostationary phases like micelles, microemulsion droplets, and polymers, increasing the selectivity and versatility of the analytical system.

CE advantages with respect to other analytical techniques comprise very high resolution in short time of analysis, versatility, the possibility to analyze molecules without chromophore groups, simultaneous analysis of compounds with different hydrophobic characteristics, small sample volume, and low cost. Moreover, it is possible to adapt this technique to the analysis of numerous types of compounds like biological macromolecules, chiral compounds, inorganic ions, organic acids, DNA fragments and even whole cells and virus particles. An increasing number of CE applications are in progress in many clinical laboratories. As this technique employs small sample volumes, is ideal for the analysis of biological fluids in which the limited amount of sample represents a challenge. In pharmaceutical quality control, it is possible to determine active ingredients in the presence of related substances with different physicochemical characteristics, especially chiral impurities in the final products using the same analytical system with a relatively simple instrumental. Moreover, numerous applications are reported in the analysis of inorganic ions. Also, the determination of macromolecules such as polysaccharides, therapeutic proteins, and flavonoids present in plant extracts, biological and biopharmaceutical products may also be analyzed with this technique.

In summary, CE has become an important analytical tool in the field of research, clinics and pharmaceutical industry offering a large number of applications, in biological, natural and pharmaceutical samples, as an alternative or complementary option to traditional analytical techniques to implement in the routine laboratory.

1. Introduction

Capillary electrophoresis (CE) is a powerful analytical technique introduced in 1980, and since then it has been applied from the analysis of small molecules to biomolecules.

CE separation depends on the differential migration of the compounds in narrow-bore capillaries filled with a background electrolyte (BGE) when an electric field is applied.

The CE advantages include short analysis time, high efficiency, low cost of operation, reduction in solvent consumption and especially, the possibility of simultaneous analysis of compounds with different hydrophobicity in a simple run.

CE instrumentation consists of a power supply, electrodes, sample introduction systems and a detector. The instrumentation used for CE is simple compared to other chromatographic systems (Figure 1).

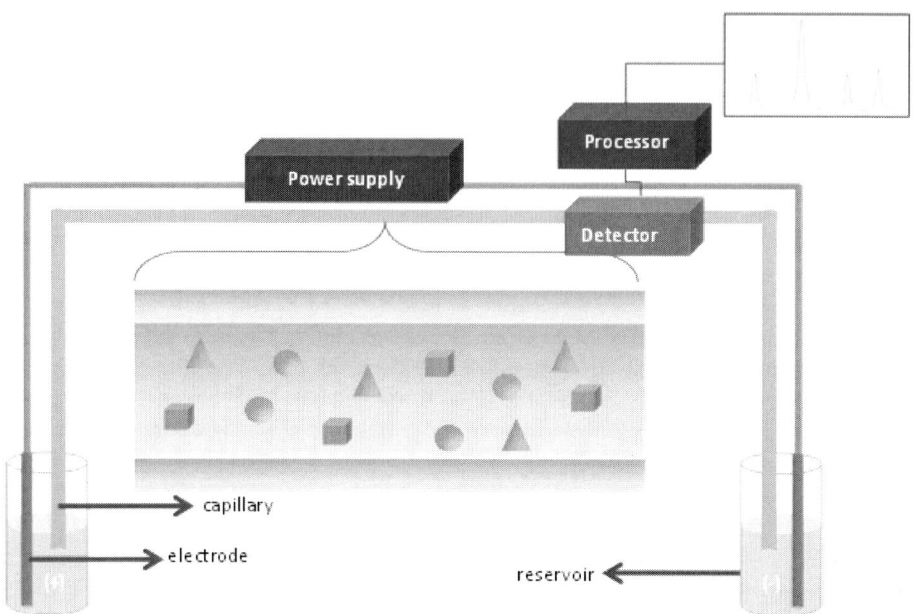

Figure 1. Scheme of capillary electrophoresis equipment.

The most use detector is the UV/diode–array, although spectrofluorometry, electrochemical detection and mass spectromety are applied [1-3].

CE has been employed in different modes, the most applied is capillary zone electrophoresis (CZE), where the separation of the analytes is based on the charge-to-mass ratio and involve the use of a simple buffer solution as BGE. Electrokinetic chromatography (EKC) is a CE mode where the mobile phase is generally an aqueous buffer with a pseudo-stationary phase (PSP) of micelles (MEKC), vesicles (VEKC) microdroplets (MEEKC) or polymers. EKC is based on the chromatographic partition of the analytes between the PSP and the aqueous buffer. The presence of the PSP increases versatility and selectivity of the analytical system. Other modes include capillary isoelectric focusing (CIEF) based on analyte isoelectric points, capillary gel electrophoresis (CGE) based on the analyte size/molecular weight ratio and capillary isotachophoresis (CITP) based on moving boundaries [2].

In summary, CE has become an important analytical tool in different areas such as clinical, natural products, biopharmaceutical, chiral, pharmaceutical and ions as an alternative or complementary to traditional analytical techniques.

2. Principles of Capillary Electrophoresis

In CE, especially CZE, analytes must be electrically charged, and the separation of them is based on their different migration velocities in a solution, when a voltage is applied. The electrophoretic mobility of the analyte depends on their properties (electrical charge, molecular size, and shape) and the characteristics of the running buffer (type and ionic strength of the electrolyte, pH, viscosity and type of the additives) in which the migration takes place [3-5].

The direction and velocity of the analyte (ionic) inside the capillary are determined by the sum of two components: the migration of the ionic solute, forced by the electric field towards the opposite end of the capillary, where the detector is placed, and the electroosmotic flow (EOF). EOF is a flow generated inside the capillary when an electric field is applied (Figure 2).

The migration velocity of the analyte is proportional to the applied electric field, allowing short analysis time and it depends on the size and charge of the ionic analyte. In CZE, low percentage of organic solvent can be used to improve the selectivity as well as other additives such as complex agents or EOF modifying agents.

Capillaries are narrow tubes of fused silica of about 25-100 um inner diameter and several centimeters in length. The voltage usually applied is 5 to 30 kV.

EKC systems are applied when the compounds are neutral at the pH of the BGE or if the electrophoretic mobilities of the compounds are similar. In EKC the separation is carried out by the partition of the analytes between the mobile phase and a pseudostationary phase (PSP). The mobile phase is normally an aqueous buffer at different concentration and pH values and the pseudo-sationary phase may be micelles (MEKC, micellar electrokinetic chromatography), vesicles (VEKC, vesicles electrokinetic chromatography), microdroplets (MEEKC, microemulsion electrokinetic chromatography)and, more recently, polymers. In all cases the PSP must be charged at the pH of the mobile phase. The differential interaction of the analyte and the PSP (micelle, microdroplets, polymers, etc) allows the separation of the analytes in the sample. Figure 3 shows the MEEKC separation principle [3, 6].

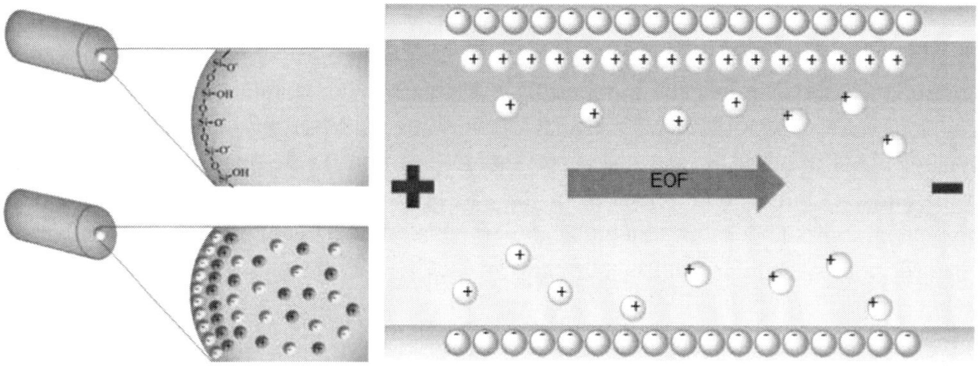

Figure 2. Schematic representation of the electroendosmotic flow (EOF).

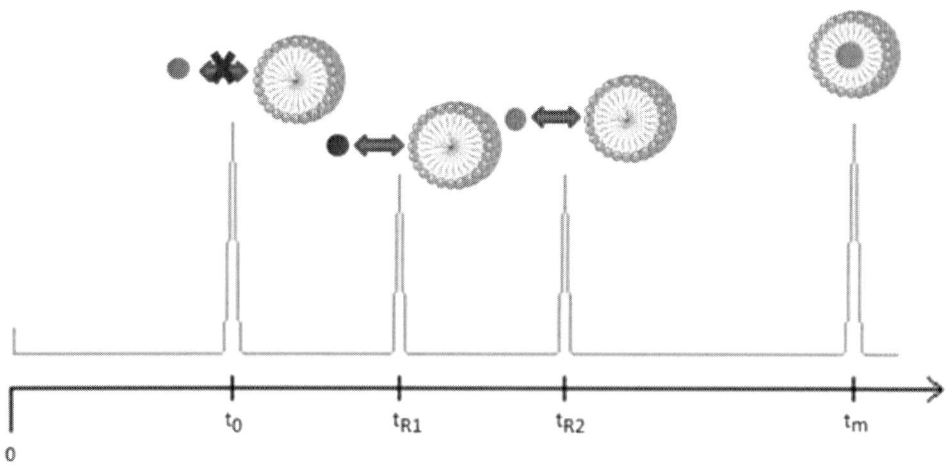

Figure 3. Schematic representation of the separation principle in MEEKC.

2.1. Micellar Electrokinetic Chromatography (MEKC)

MEKC was introduced by Terabe in 1984 as a CE system that allows resolution both neutral and ionic molecules [7]. Since its introduction, MEKC has been consolidated as a method of choice applied to the analysis of neutral as well charged compounds [8]. In the development of MEKC method the choice of surfactant is the key to achieve appropriate selectivity. One of the most useful tensioactive agent employed as PSP in MEKC is sodium dodecyl sulfate (SDS), a single-chain structure surfactant [9]. However, separations of some hydrophobic compounds using SDS may be unsuccessful [10]. Different strategies can be used to optimize the hydrophobicity and consequently the retention of the analytes in the PSP. As an example, two tails tensioactives like bis-2(ethylhexyl)sulfosuccinate (AOT) or biotensioactives like bile acids or the use of mixed surfactant agents.

Another strategy used to improve selectivity in MEKC is the employments of organic solvent in a low percentage, due to high percentages of them reduce the number of micelles. Moreover, the incorporation of polymeric micelles as PSP can resolve this problem [11-12].

2.2. Microemulsion Electrokinetic Chromatography (MEEKC)

MEEKC has been found to provide better separation efficiency than other EKC modes probably due to the improvement of the mass transfer process between the microdroplet and the aqueous buffer solution [13]. Moreover, the microemulsion system possess high stability, transparency, easy preparation and the ability to interact with a range compounds of different hydrophobicity in a single run [14-16].

In this regard, when the PSP is a polymer, the EKC system can be used to improve selectivity and sensibility by a complexation process [17].

2.3. Capillary Electrochromatography (CEC)

CEC is a technique derivate from CE and HPLC. In CEC, a capillary packing a stationary phase is submitted to a high electric field, which causes a EOF [18]. In general, there are three types of capillary columns used in CEC, open tubular (OT), packed columns (PC) and monolithic columns (MLC). OT columns the selector agent is covalent attached, coated or physically adsorbed to the internal surface of the capillary. PC, the stationary phase is packed into the capillary and retained through inlet and outlet frits. MLC the stationary phase is included in a porous continuous bed and they are the most promising ones [19-20].

3. Bioanalysis

3.1. Introduction

The first reproducible and useful separations by electrophoresis were published in 1937 by Arme Tiselius who achieved electrophoretic separation of blood serum [21]. Since that time, electrophoresis has evolved as the key technique for analysis and preparative separations of complex biological samples. Main advantages of capillary electrophoresis compared to other analytical techniques used for biological samples are listed:

1. Small sample requirement (few nanolitres), which is an outstanding advantage in the analysis of biological fluids or tissues, with different analytical techniques.
2. Minor sample treatment, just enough to avoid capillary clogging, because capillaries are rinsed after each run and apparently no irreversible retention is produced.
3. Ability to separate compounds in aqueous media. The main components of biological fluids display high polarity and water solubility which makes them difficult to retain on reversed-phase stationary phases, allowing it a suitable technique alternative to HPLC.

4. Provides additional information to the mass, because analytes migration is based on mass to charge ratio.
5. High efficiency
6. Multiple modes of CE can be applied to the analysis of the same sample, which provide more information.

Main disadvantages

1. Low sensitivity compared to traditional analytical methodology.
2. Robustness, shift in migration time.

3.2. Applications

An alternative to overcome the drawbacks of sensitivity and selectivity is immunoaffinity CE (IACE). IACE combines on-line coupling of highly selective antibody-capture agents with the high resolving power of CE. This powerful analytical tool is extremely useful for enrichment and quantification of ultra-low-abundance analytes in complex biological matrices. [22] This technology is applied to the analysis of multiple predictive biomarkers of disease [23-24]. An example is the development of a chip-based IACE to measure cytokines in serum and on dry blood spot with a LOD of approximately 0.5pg. [24-25]. The hallmark of CE immunoassay is the use of biological-related ligand to selectively bind and recognize a given analyte or a group of analytes. CE immunoassay can be classified in two main divisions: homogeneous immunoassays, in which all of the assay components are present in solution, and heterogeneous immunoassays, in which one or more of the assay components is used in an immobilized form (e.g., on a solid support). On the other hand, based on whether they involve a competition between an analyte and a labeled binding agent (for example, a competitive binding immunoassay) or whether binding by only the analyte is required for detection (as occurs in noncompetitive immunoassays or immunometric assays) [26]. Nowadays, both formats of CE, conventional or microchip, coupled to immunological platforms seem to be a key to solve some of the problems concerning the increased need to develop highly sensitive, fast and low cost methods of analysis in biological diagnostics [22]. In the field of endocrinology, these methods have been created to measure such analytes as insulin [7-8], glucagon [7-8], thyroxine [29], steroid hormones [30-31], vasopressin [32], follicle-stimulating hormone [33], and luteinizing hormone [34]. Finally, various applications of CE immunoassays have been reported in the field of cancer, with examples including the detection of cancer biomarkers, the assessment of DNA damage by potential carcinogens, and the determination of multi-drug resistance in cancer cells. An example of the lastone, is the use of a noncompetitive format and fluorescein as the label. The resulting method provided a LOD of 0.2nM for a multi-drug resistance associated protein and was utilized to examine the expression of this protein by various cancer cell lines [35].

On the other hand, CE-MS coupling offers the structural identification of both mass spectrometer and migration time relationship to structure. Almost all types of mass analyzers have been coupled to CE, but taking into account the narrow peaks resulting from CE separation, in addition to complexity of the sample require high mass accuracy and high resolution are required to resolve closely migrating components with similar nominal mases. For that reason CE-TOF/MS is used in most of the applications [36].

Versatility of using multiple modes ofCE, makes it a tool with huge possibilities for analyzing biological samples compared to other analytical techniques. In this way, CE was applied to biological sample fingerprinting. Hanna-Brown et al. showed how sulphated β-CD-modified MECK can be used to provide a useful tool for urine fingerprints, allowing for separations of more than 80 signals in fewer than 25 min [37]. The possibility of measuring the same samples in CE operating with different polarities and buffer systems affords an amazing opportunity, since all compounds in a given sample can then be resolved and detected in one or other mode.

Another area where CE has been successfully applied is the study of viruses and bacteria. An attractive advantage of CE is the ability to simultaneously identify microorganisms. For example, capillary IEF in pH gradient 3-10, with pressurized mobilization to detect zones by on-line UV detector at 280 nm was applied to the analytical separation of *Serratia rubidae*, *Pseudomona putida* and *Escherichia coli* in 18 min [38]. The study of multidrug resistant microorganisms isgrowing, there is a need for fast and unequivocal identification of suspect organisms to supplement existing techniques in the clinical laboratory, to quickly begin an appropriate and effective therapy. Recently, Fleurbaaij F et al., [29] developed a CE-ESI-MS/MS bottom-up proteomics workflow for sensitive and specific peptide analysis with the emphasis on the identification of β-lactamases in bacterial species. Moreover, separation and identification as well as the detection of their ability to form biofilm of phenotypically indistinguishable *Candida* species based on capillary isoelectric focusing (CIEF) and CZE with UV detection were applied. CZE narrow zones of the cells of *Candida* species were detected with sufficient resolution and migration time were obtained. The values of the isoelectric point and the migration velocities of the examined species were independent on the origin of the tested strains [40]. To prevent adsorption of microorganism onto the capillary wall appropriate buffer solution additives were needed, an example is the use of poly(ethylene glycol) [41].

Recently, many methods for analysis of a single cell have been developed. This analysis offers unique information about the differentiation, specialization, proliferation, senescence and cells death. In summary, direct injection of intact cells into the capillary, followed by lysis, allow electrophoretic separations of different components and their detection [42-43].

Taking into account the importance of the methodology used in the diagnosis and in the follow up of treatment in pathology, development of accuracy and sensitive methods suitable to be applied in routine laboratory are required. In this way, a CD-modified MEKC was employed to determine individual bile acid profiles (total of 15 free and most conjugated forms of bile acids) [44]. This system was applied in the study of cholestatic pathologies. On the other hand, it was demonstrated that the method proposed by Tripodi et al. provided more information compared to the enzymatic method commonly used in routine laboratory [45] (Figure 4). Another MECK system combined with polymeric micelles was developed to analyze, in a single run, nine steroid hormones in human urine [46] (Figure 5).

A new microemulsion as pseudostationary phase employing two combined surfactant agents like bis (2-ethylhexyl) sulfosuccinate (AOT) with cholic acid was applied in the determination of coenzyme Q10 in human plasma [47]. It is foreseen that MEEKC system will be useful to diagnose and follow up mitochondrial diseases after CoQ10 treatment.

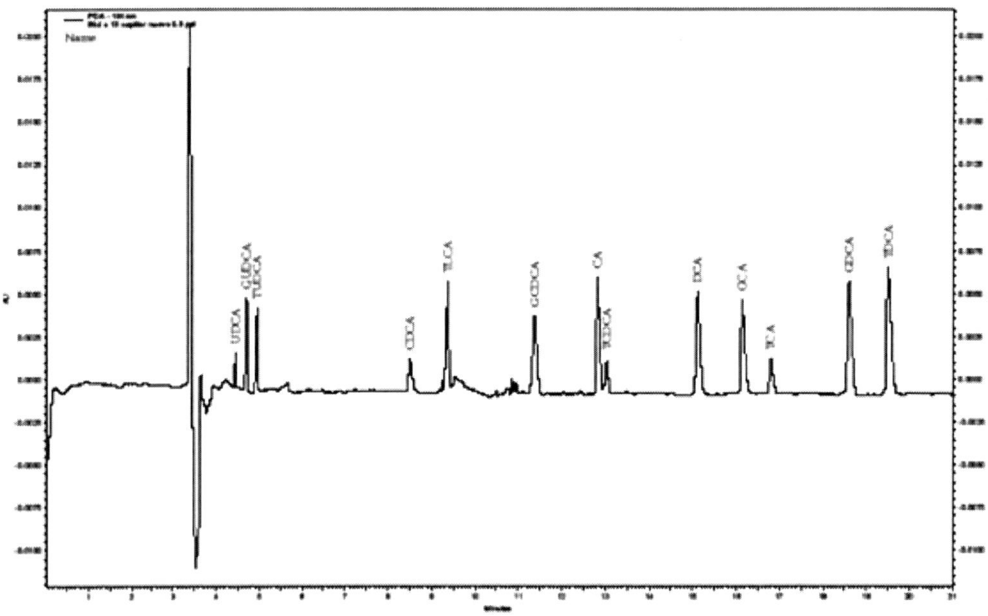

Figure 4. Electropherogram of 13 bile acids by MECK UDCA, GUDCA, TUDCA, CDCA, TLCA, GDCA, CA, TCDCA, DCA, GCA, TLCA, GDCA, TDCA. Electrophoretic system consisted of: 50 mM SDS, 0.5 M β-CD, 0.5 M β-OHpropylCD, 10 % acetonitrile in 10 mM borate-phosphate buffer (1:1).

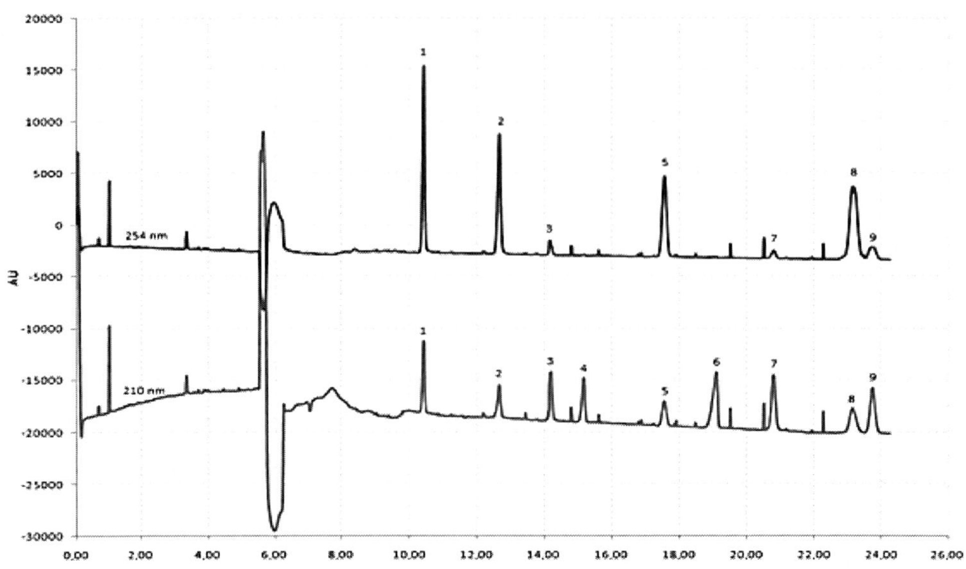

Figure 5. Electropherogram of nine steroids standard by MEKC-CA-SDS-poloxamine (50 mM CA, 10 mM SDS, 0.05% poloxamine in borate: phosphate buffer pH 8.0 with 2.5% ME, 2.5% THF. 1: Cort, 2:D4, 3: E3, 4: SDHEA, 5: To, 6: DHEA, 7: E1, 8: Pg, and 9: E2.

The most successful application of CE both scientifically and commercially is DNA sequencing [48]. CE is a completely automated alternative to other physical gel techniques employed to analyze sequences and fragments of DNA and proteins. DNA sequencers based on capillary array electrophoresis are the best selling products in the history of analytical instrumentation. The instrumentation and methodology for DNA sequencing resulted in a range of other applications, like clinical molecular diagnostics, especially mutation detection for prenatal diagnostics and genetic screening, genotyping, paternity testing, etc [42].

In the last time, CE has gained an important place in all areas as an alternative and complementary technique, especially in bioanalysis. It has demonstrated considerable advantages compared to traditional methodologies, due to the wide field of application, only some examples are mentioned here.

4. Capillary Electrophoresis in the Analysis of Flavonoids

4.1. Introduction

Flavonoids are ubiquitous secondary plant metabolites that are widely used because of their spasmolytic, antiphlogistic, antiallergic, antioxidant and diuretic properties. Their common structure is that of diphenylpropanes (C_6-C_3-C_6) and consist of two aromatic rings linked through three carbons that usually form an oxygenated heterocycle. Biogenetically, the A ring usually comes from a molecule of resorcinol or phloroglucinol synthesized in the acetate pathway, whereas B ring is derived from the shikimate pathway. They can be found as aglycones, but most commonly as glycoside derivatives. They are divided in different subgroups and they either occur as aglycones or as O- or C-glycosides. Figure 6 represents the basic structure of flavonoids.

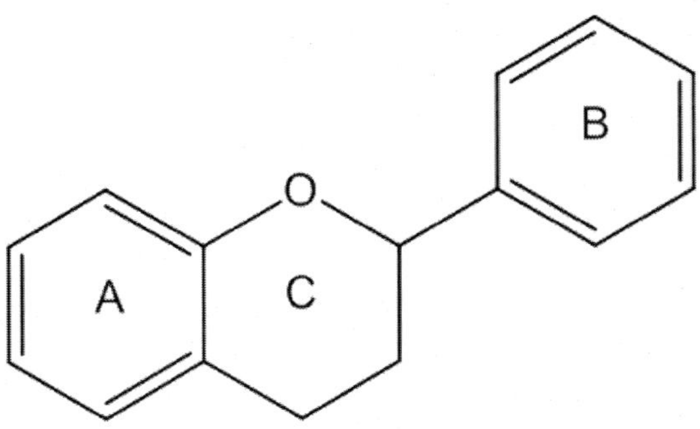

Figure 6. Basic structure of flavonoids.

Among the flavonoids, flavones (e.g., apigenin, luteolin, diosmetin), and flavonols (e.g., quercetin, myricetin, kaempferol), and their glycosides are the most common compounds [49].

4.2. Capillary Zone Electrophoresis (CZE)

Borate buffers have been widely applied to the analysis of flavonoids due to their ability to complex polyphenol aglycone and/or saccharides. Two hydroxyl groups of boric acid complex with cis-diol groups. Borate ions form charged and mobile five-membered-ring complexes (with 1,2-diols) and six-membered ring complexes (with 1,3-diols) and therefore, the separation selectivity is increased [50]. It is also confirmed by CE that an ion-dipole interaction takes place between flavonoids and borate. In this sense, quercetin, isorhamnetin, luteolin, luteolin-7-O-glycoside and apigenin are used as model compounds [51]. The complex formation reaction is a strongly-pH dependent equilibrium which is favored at higher pH values [52]. Early works by Morin, Seitz and McGhie show the use of borate buffers to study the effect of several parameters (pH, buffer concentration, structure, among others) on the electrophoretic mobility of selected flavonoids. This methodology is applied to separate flavonoid-O-glycosides in plant extracts [52-54]. Recent approaches based on the same methodology include 20 mM borate buffer pH 9.5 with UV detection for the analysis of hyperin, isoquercetrin, myricetin and quercetin-3-O-robionobioside [55]; 50mM borax pH 9,3 for studying flavonoids in *Apocynum venetum* [56]; 10 mg/ml sodium tetraborate for analyzing apigenin and apigenin-7-O-glucoside in *Chamomille* flowers [57]. Additional examples are listed [58-62].

Mixed phosphate-borate buffers have also been proven suitable for the separation of flavonoids in plant extracts. This approach is applied to the analysis of flavonoids in *Floslonicerae* [63], *Acanthopanaxsenticosus* [64], *Anaphalismargaritacea* [65].

4.3. Additives

The use of organic modifiers is sometimes necessary to achieve proper separation of critical compounds. For this purpose, methanol (5% – 40%) and acetonitrile (8 – 22%) are often employed [66]. Sodium hydrogen phosphate and sodium dihydrogen phosphate at pH 8 were also used for the analysis of kaempferol, rutin, quercetin, myricetin and apigenin from *Centellaasiatica, Rosahybridis* and *Chromolaenaodorata* but the addition of organic solvents as ACN (10%, v/v) and methanol (6%, v/v) are required to obtain adequate separation of critical compounds, such as kaempferol, quercetin and myricetin [67]. ACN is also used for the analysis of *Lamiophlomis rotate* [51] and *Epimediumspp* [69], whereas methanol is employed in *Achilleamillefolium* [68] and *Passifloraincarnata* [70].

Cyclodextrins are particularly attractive additives to provide chiral resolution for enantioseparation of bioactive compounds and improve resolution of analytes.

Native CDs have been successfully employed. β-CD is applied in *Chrysanthemum morifolium* for the analysis of apigenin, catechin, epicatechin, kaempferol, luteolin, quercetin [71] and γ-CD is used for separating catechin and epicatechin enantiomers [72] in plant food. Chemical modifications lead to a significant improvement in their physicochemical properties and chiral recognition abilities. Alone or in combination, DM-β-CD [73], HP-β-CD and HP-γ-CD [74-75] are common examples.

A novel electrophoretic tungstate buffer is proposed as complex-forming reagent for the analysis of flavonoids in *Hypericum perforatum.* Ionic liquids (such as 1-alkyl-3-methylimidazolium-based ionic liquid) are substances with melting points at or close to room temperature and have been explored as additives to borate running buffers for flavonoid separation. 1-butyl-3-methylimidazolium tetrafluoroborate has been found to be suitable in the separation of kaempferol, quercetin, isorhamnetin in *Hippophaerhamnoides* extracts and phytopharmaceutical preparations. A particular mode of CZE is non-aqueous capillary electrophoresis (NACE), which employs a non-aqueous buffer system. Use of organic solvents instead of water firstly helps in increasing the solubility of hydrophobic analytes but also improves selectivity [76].

4.4. Electrokinetic Capillary Electrophoresis (EKC)

MEKC (Micellarelectrokinetic capillary electrophoresis) with SDS has also been proposed as an alternative to traditional CZE [77, 78]. It has become the method of choice for the analysis of catechins in green tea extracts [79]. It has also been applied to *Amaranthus* spp. for the analysis of rutin and total quercetin [80]. RF-MEKC has been employed for catechins, which become neutral at acidic pH. Addition of CD to the MEKC system gives rise to the to the inclusion-complexationequilibria of the analytes into the cyclodextrin cavity, which occurs simultaneously to the partitioning into the SDS micelle [76].

An example of the combination of SDS and CD in reversed-flow is related to green tea infusion for the separation of catechins and catechin-gallates [82].

MEEKC (Microemulsionelectrokinetic chromatography) is another methodology that has also proven to be suitable for flavonoid analysis. 0.5% (w/v) ethyl acetate, 2.0% (w/v) SDS, 9 mM DTAC (Dodecyltrimethylammonium bromide), 4.0% (w/v), 1-butanol, 25 mM phosphoric acid (pH 2.0) have been successfully applied for the evaluation of flavonoids in *Astragalus*spp [81].

Modern approaches include the use of carbon nanotubes. They are made of a novel material with unique properties such as high electrical conductivity, large surface area, and chemical stability. Functionalized multiwalled carbon nanotubes as the additive in a microemulsion buffer [83] or pseudostationary phase [84] of MEEKC were used for separation of 13 analytes in Compound Xueshuantong capsule [83] and ten analytesinQishenyiqi dropping pills [84], respectively [85].

4.5. Detection

The UV spectra of flavonoids exhibit two major absorption bands in the region of 240 – 400 nm; the band at 300 – 380 nm is associated with the absorption of the cinnamoyl system, whereas the band at 240 – 280 nm is associated with the absorption of the benzoyl system [86]. Even though the concentration sensitivity is low in CE due to its shorter optical path length [85], it is not a challenging concern because of the relatively high content of bioactive flavonoids in medicinal plant extracts.

Since polyphenols can be electrochemically oxidized at a relatively moderate potential, electrochemical detection (ED) is a useful alternative detection approach. Borate running buffer is also suitable in CZE separation of flavonoids with ED detection; in general tetraborate in the concentration range 50–100 mM (pH 8.45–9.0) is used [86]. Examples of CZE of borate or borate-phosphate with ED are listed [87 – 90]. MEEKC with ED is also applied [91, 92]. Cao and coworkers propose a novel detection approach for flavonoids which is highly sensitive. They use a zwitterionic microemulsion electrokinetic chromatography (ZI-MEEKC) coupled with light-emitting-diode-induced fluorescence (LED-IF, 480 nm). The BGE consists of 92.9% (v/v) 5 mM sodium borate, 0.6% (w/v) ZI surfactant, 0.5% (w/v) ethyl acetate and 6.0% (w/v) 1-butanol. It has been successfully applied to hawthorn plant and food products [93]. ECL (electrochemiluminescence) is also applied to CE [94]. The increasing trend towards the use of MS (mass spectrometry) detection is mostly due to the recent availability of easy-to-use and robust CE-MS commercial equipment. An example of the application of MS associated to flavonoid analysis is the determination of naringenin from a phytomedicine by CE-ESI (electrospray interface)-MS [95]. It has been used in the characterization of secondary metabolites in *Genistatenera* [96].

4.6. Quality Control and Fingerprinting

Due to the chemical complexity of herbal drugs, an effective quality control poses a challenge. It comprises the identification and assay of known active ingredient/s, the chemical profiling of known and unknown compounds and the identification or detection of adulterants [97]. Considering that herbal drugs are complex matrices in which no single constituent is responsible for the overall efficacy, instead of analyzing particular marker compounds, fingerprinting assures herbal quality by constructing specific patterns of recognition (based on analytical data).

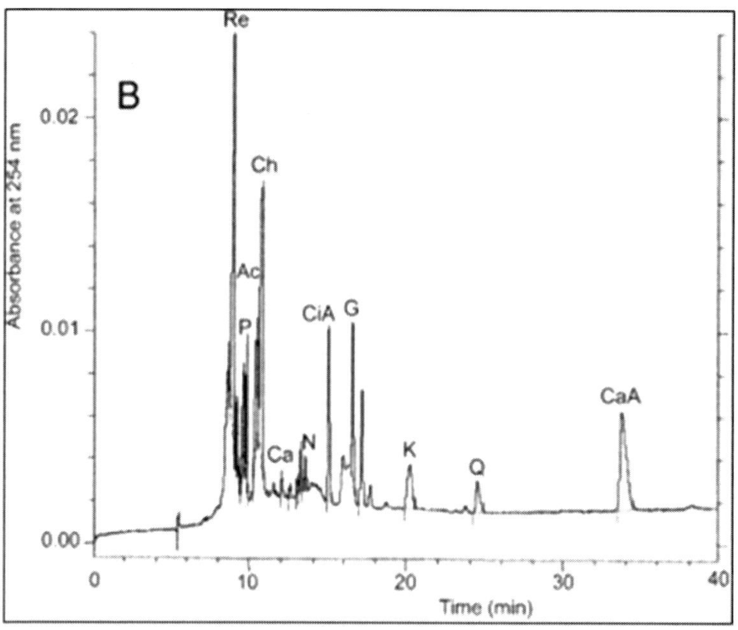

Figure 7. Propolis fingerprint. CZE electropherograms of propolisethanolic extract. Separation and analysis were carried out on an uncoated fused-silica capillary tube (50 mm ID, 70 cm total length, and 50 cm from the injection point to the detector) at 25°C. The operatingbuffer was constituted by sodium tetraborate, 30mM, pH 9.0, UV detector set at 254 nm. Abreviations: Re (resveratrol), P (pinocembrin), Ac (acacetin), Ch (chrysin), Ca (catechin), N (naringenin), CiA (cinnamic acid), G (galangin), K (kaempferol), Q (quercetin), CaA (caffeic acid). From (102), with permissions.

In this way, sets of ratios of detectable compounds can be evaluated as well. This represents a more comprehensive approach, enabling authentication, quality assurance and stability studies [66]. This strategy is applied to *Ginkgo biloba* extracts [98], *Polentyphae*

[99], *Frucusaurantii* [100], *Rhizoma Smilacis Glabra* (101), *Radix Scutellariae* [102] (Figure 7).

4.7. Conclusion

Since its beginning, CE has evolved to become a versatile separation technique. Apart from its many advantages (such as outstanding separation efficiency, high speed and low cost analysis, low solvent and sample consumption, rapid method development and superior specificity) the researcher is able to choose from several separation modes. Taking into account the hyphenation of CE with LED-IF, ECL and MS, apart from the traditional UV-diode array, it allows to further widen the applications of CE in the field of natural product analysis making it an attractive tool for the challenging quality control of herbal drugs.

5. Biopharmaceutical Analysis

5.1. Introduction

The use of peptides, proteins and complex carbohydrates as therapeutic compounds has been increased during the last decades due to the great improvement in protein chemistry and biotechnological procedures. This growing in the production field has been accompanied inexorably with the demand to advanced analytical techniques in order to use them in the quality control of these products to ensure their safety and efficiency [103].

Capillary electrophoresis (CE) is a powerful technique for the analysis of macromolecules due to the high separation efficiency, high resolution and small sample required. However, there are some drawbacks that have to be taken into account like the possibility of adsorption of molecules onto the inner wall of capillary or a lower sensitivity, although there are some solutions to deal with.

CE methods applied in the analysis of macromolecules include the use of CZE, CGE, isotacoforesis CEC, EKC and CIEF.

5.2. CGE Applications

In CGE, the separation is based on molecular size differences by the restricted migration of the molecules through porous gel media. In the cases where the slab gel is the sodium dodecyl-sulfate-polyacrilamide gel, the SDS-protein complexes in free solution migrate with the same mobility independent on the mass. The development of CGE was based on the experience gained from the traditional physical gel electrophoresis, which was used (and is still used) in biology and genetics, for the analysis of proteins, DNA and RNA. This technique was developed with the idea of avoiding the thermal diffusion due to the fact that the separations are performed in anti-convective media. The narrow capillary used in CE makes unnnecessary the use of an anti-convective media which lead finally in modern CE [104].

CGE has been employed for the purity analysis of proteins as well as for the analysis of DNA. The principal advantages of CGE over the traditional physical SDS-PAGE are the possibility of automatization, the great resolution and short separation time.

The quality control of biopharmaceuticals includes the PEGylated conjugated of INF-α; the pattern of oligosaccharides of rituximab, trastuzumab and palivizumab and several antibodies. [105-106].

The best resolution for the analysis of proteins and peptides compared to other CE modes has been harvested to analyze the charge heterogeneity of biopharmaceuticals. This CE mode is vitally important to study the charge pattern and has been successfully employed for the study of degraded products of myeloprotein, and also the study and separation of the different forms of erythropoietin (EPO) which can be carried out in just six minutes, and the method is also able to discriminate little differences between glycosilation patterns from different sources of EPO. [107-108].

5.3. CEC Applications

There are also reports of methods based on CEC, where different stationary phases were used. One of the most innovative stationary phase is based on the molecular imprinting technique. In this case, the stationary phase consists of a polymer synthesized in the presence of a template. After polymerisation, the template molecules are removed making the binding sites and the cavities (which are complementary to the template in size, shape and functionality), accessible. The stationary phase possesses a molecular "memory", and thus, it is able to specifically recognize the target molecule (Paper imprinting). Using a tetra peptide as template (which form part of oxytocin), it was possible to develop a stationary phase able to analyze it in the presence of other proteins without any interference [109].

Despite CZE is the simplest and the first developed CE mode, it has been employed in the analysis of biopharmaceuticals including proteins, therapeutic peptides hormones and also to study electrophoretic mobility and to quantify net charge [110].

5.4. EKC Methods

There are few reports of EKC methods in the field of protein assay, most of them based on SDS micelles as pseudo-stationary phases [111]. However, employing a polymeric electrokinetic system it was possible to perform the determination of contaminants and impurities of heparin samples. This analytical method reported the possibility to assay and quantify related compounds even in the final product, not only in the raw material, due to the incorporation of polymeric β-CD which increases the sensitivity of the system [112]. Figure 8 represents the electropherograms of heparin and related compounds, notice the increase in the response when β-CD polymeric is added.

At the moment this chapter is being written, CE has been accepted as an official method for the analysis of recombinant human erythropoietin in the European Pharmacopeia. A detailed review by *E. Taminzi* resumes the principal applications of CE in the analysis of biopharmaceuticals, including peptides and proteins.

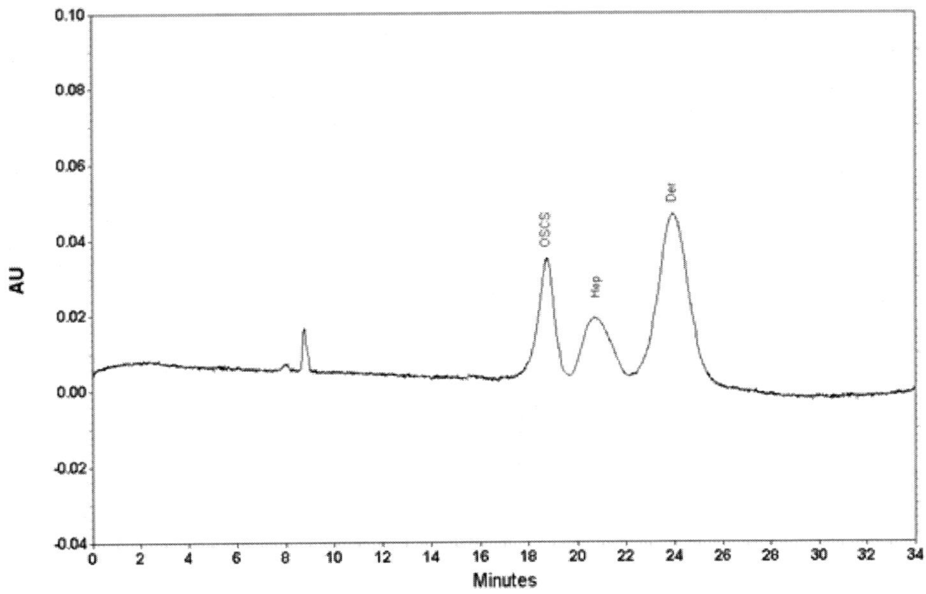

Figure 8. Electropherograms of Hep (0.1 mg/mL), OSCS (10 ug/mL) and Der (10 ug/mL) in BGE with 0.5% w/v polymeric-β-CD and 0.4% w/v Tetronics 1107.

6. Chiral Analysis

6.1. Introduction

Among the active ingredients used in pharmacological therapies, a high percentage of them possess asymmetric centers responsible for the optical activity. These stereochemical differences determine their pharmacodynamics and pharmacokinetic profiles. From pharmacodynamic point of view, we can describe four possible scenaries. First, only one of the enantiomers owns the pharmacological activity, while the other one can be inactive or even toxic. Many examples are described like, rivastigmine, citalopram [113-114]. Other possibility is that both enantiomers possess the same pharmacological activity. In the third group we can find enantiomers which have the same pharmacological activity but different potency. Finally, the enantiomers could have completely different pharmacological activities. To pharmacodynamic differences, we should add pharmacokinetics, for instance, stereospecific metabolization in omeprazole.

For these reasons, in the last twenty years, the discussion about the administration of a single enantiomer versus a racemic formulation has been in the spotlight, and given place to which has been called racemic switch, that has affected not only the pharmaceutical industry but also the requirements for chiral compounds by regulatory authorities. As a consequence, in 2009, 12 drugs of the top 20 products according to their sales are single enantiomer drugs, while 4 drugs are achiral compounds, 3 products are marketed as racemates and one product is a combination of a chiral and a racemic drug. In fact, the top 3 products are single enantiomer drugs [115]. This fact has led to the necessity of developing new technologies which enable the production of a single stereoisomer but also new methodologies to quality control [114]. These new methods should be sensitive and show a good resolution for each enantiomer, in order to be suitable for the analysis of enantiomeric purity, chiral separations in pharmaceutical dosage forms and intermediate products, and enantioseparations of drugs and metabolites in biological matrices [115].

Among all the chromatographic methods applied to enantioseparation, CE has displayed real advantages over the others, and has recently become a complementary technique to HPLC [116]. CE parameters like sensitivity, simplicity, high resolution power and versatility have settled a CE at this level. The main advantage of CE is the possibility to use a single or combined chiral selectors added to the BGE instead of the chiral chromatographic columns, giving countless possibilities and therefore a variety of pseudostationary phases [117].

Enantioseparation in CE has been conceived in two modes: indirect and direct. In the indirect approach, the compound must have a functional group that can be derivatized and the derivatization reagent has to be of high enantiomeric purity. Due to the fact that derivatization increases time analysis and the risk of racemization under the reaction conditions, it is rarely used. Direct mode is the most frequently used in enantiomeric separation by CE. In this mode a chiral selector is added to the BGE as a pseudoestationary phase [116]. Chiral selectors are molecules which interact with the enantiomers forming temporary diastereomers complexes. Intramolecular interactions (electrostatic ion-ion, ion-dipole, and dipole-dipole, hydrogen bonds, π-π) explain the action mechanism. There exist a variety of chiral selectors which have been widely used in CE and the quantity is continuously growing given multiple current developments, although cyclodextrins (CDs) are still the most commonly used [118-119].

6.2. Chiral Selectors

6.2.1. Cyclodextrins

CDsare cyclic oligosaccharides compounds by 6, 7 or 8 units of D-glucose(α, β and γ-CDs respectively) and the recently introduced δ-CD with nine units. CDs own several profitable properties for CE analysis such as UV transparency, availability, wide application range (polar, nonpolar, charged, uncharged analytes), and reasonable solubility in water plus broad selectivity spectra [116, 119]. It has been proposed that in the numerous chiral centers (e.g., 35 chiral centers in β-CD) resides the ability of CDs to form chiral complexes with a large number of substances. Juvancz et al., state that with the help of CDs, there are almost no restrictions in the structure of the analyte. CDs are able to separate enantiomers not only with central chirality but also with planar or axial chirality, chiral centers provided by heteroatoms such as sulfur, silicon, phosphorus, and nitrogen as well as regioisomers [120].

The basis of the mechanism of separation in chiral CE carried out with CDs could be explained by two phenomena. One of them can be considered as chromatographic, based on the inclusion of the analyte (guest) or at least its hydrophobic part into the relatively hydrophobic cavity of CD (host), and the other one electrophoretic [116, 120] (Figure 9).

Through derivatization of hydroxyl groups in native CDs several new CDs with different properties have been developed. So they can be classified as neutral, charged and polymeric CDs and they can also be classified as native, randomly substituted or specifically substituted CDs.

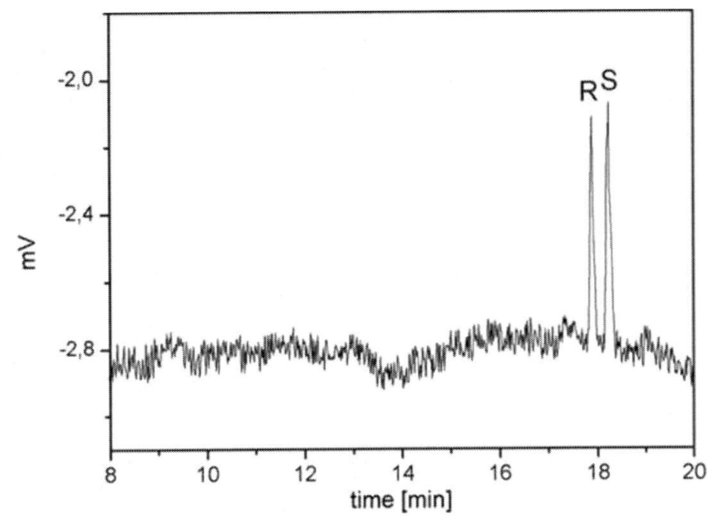

Figure 9. Electropherogram of *S*- and *R*-enantiomers of rivastigmine. BGE, triethanolammonium phosphate, pH 2.5, 75 mM; β-CD, 7.5 mM. From [118], with permissions.

6.2.2. Crown Ethers

Chiral crown ethers are cyclic polyethers that form stereoselectively inclusion complexes with primary amines [116]. Crown ethers have been widely used in liquid chromatography (LC) as bonded stationary phases. They were applied as chiral selectors in CE for the first time for chiral separation of amino acids by Kuhn et al. Complexation mechanism involves the inclusion of the hydrophilic part of the analyte into the cavity, unlike CDs-analyte complexes [116]. The chiral recognition mechanism is based on the formation of hydrogen bonds between the three amine hydrogens and the oxygens of the macrocyclic ether.

This crown ether was shown to be applicable also to the chiral separation of dipeptides, sympathomimetics, and various other drugs containing primary amino groups by CE [121-122].

Main limitation of the use of crown ethers is that BGE must be potassium-ammonium ions free, because they compete with the enantiomers for the cavity.

6.2.3. Polysaccharides

The use of polysaccharides as chiral selectors is based on their higher molecular structures, which define helical hydrophobic cavities and pores of polymeric network. Electrostatic interactions provide additional stereoselectivity effects in case of ionic polysacharides [116].

A variety of linear neutral and charged polysaccharides such as heparin, chondroitin sulfates, dextran sulfate, carrageenan, dextran, dextrin, laminaran, and pullulan have been successfully employed in CE [123].

Polimaltodextrin was the first polisaccharide that have been used for enantioseparations in CE. It was described in 1992 by D'Hulst and Verbeke for the chiral separation of nonsteroidal anti-inflammatory drugs [123].

The positively charged linear polysaccharides have not been used as chiral separation recently due to their low solubility and their adsorption to the inner surface of the capillary wall. Heparin is a natural occurring, linear, polydisperse, polyanionic glycosaminoglycan. Its electrophoretic mobility is a consequence of its high anionic character along with the chirality inherent to the constituent monosaccharide, making ita useful chiral selector. Agyei et al., applied Heparin for the analysis of chloroquine and chlorpheniramine [118].

6.2.4. Proteins

Proteins provide strong and highly selective interactions, influenced by their tertiary structure. Different analytes can selectively bind to different binding sites of proteins. Proteins can be used, depending on their ionization state, for neutral, basic, or acidic analytes [116].

Up to now, the most used chiral selectors in CE are 1-acid glycoprotein, cellullase, ovoglicoprotein, casein, cellobiohydrolase, avidin, human serum transferrin and serum albumins of different species such as HSA or BSA [124-125]. However, there are several limitations associated with the use of proteins in CE enantioseparations, mainly, low separation efficiency obtainable, adsorption of proteins to the capillary wall, high UV absorption.

6.2.5. Antibiotics

Various classes of antibiotics have been reported, and these include ansamycins, macrolides, lincosamides, aminoglycosides and β-lactams [123]. Macrocyclic antibiotics were introduced for the first time as chiral selectors by Armstrong et al., Their use is based on the existence of several asymmetric centers and defined structures (like hydrophobic pocket) making multiple possible interactions. The primary interaction of macrocyclic antibiotics is carried out by charge to charge or ionic interactions. The secondary interactions are hydrogen bonding, steric repulsion, hydrophobic dipole-dipole and, π-π [118].

Since these compounds have strong UV absorption, applications in this field were accomplished by antibiotics immobilized in CEC stationary phases.

6.2.6. Ligand-Exchange Selectors

Ligand-exchange chromatography was firstly applied in HPLC and TLC an lately, Gassmann et al. applied this principle to CE for the separation of 12 dansylated amino acids. Enantioseparation mechanism is based on the differential thermodynamic stability of diastereomeric complexes (selector ligand, metal and analyte) [126]. For example, histidine-copper(II) hemicomplex, could interact properly whith analytes containing aminocarboxyl or hydroxycarboxyl groups like amino acids and hydroxyl [116].

Several compounds have been used as ligands like amino acids, oligopeptides, hydroxy acids or amino alcohols.

6.4. Chiral MEKC and MEEKC

Chiral recognition of analytes in MEKC and MEEKC is based on differences in their partition coefficients between the chiral drops and micelles, and the electrolyte bulk phase. Mikus et al., have amply reviewed the application of many amphiphilic molecules, charged and also neutral such as bile salts, long-chain N-alkyl-L-aminoacids, n-alkanoyl-L-amino acids, N-dodecoxycarbonylaminoacids, alkylglycosides, alkylglucosides, as chiral selectors [116]. The use of polymer micelles as chiral selectors has also been widely investigated given their advantages over the conventional surfactants. Poly(sodium N-undecanoyll-leucylvalinate)(poly-l-SULV) and Poly-N-undecenoyl-l-amino acid-sulfate (poly-l-SUCAAS) have been successfully applied to the separation of amino acid enantiomers. Sugar-based chiral surfactants have been synthesized and applied to the MEKC enantioseparation of amino acids by *Kitagawa et. al.* who have demonstrated that the carbohydrate head groups, including their anomeric configurations, have significant effects on the enantiomeric separation and the migration behavior [127-128]. Another example is the resolution of sertraline cis-trans isomers by MEKC with two cyclodextrins and cholate as tensiactive agent (Figure 10) [129].

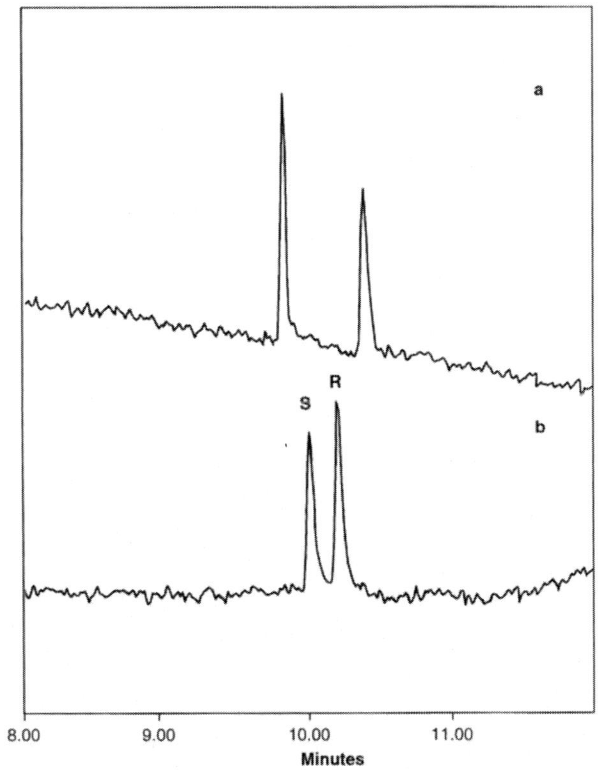

Figure 10. Electropherogram showing the enantioseparation of (a) racemic *trans* isomer (50.0 mg/ml) (b) *cis*-(1*S*,4*S*) enantiomer (sertraline hydrochloride) (25.0 mg/ml) and *cis*-(1*R*,4*R*) enantiomer (30.0 mg/ml) From reference 129, with permissions.

On the other hand, Foley et al. have made a great contribution to the study and development of chiral microemulsions and their application as pseudostationary phases in MEEKC. Dodecoxycarbonylvaline (DDCV) was the first chiral surfactant applied in MEEKC. Chiral microemulsions have been applied to the analysis of epinephrine, arterenol, ephedrine, atenolol, and methylephedrine. Poly-D-SUV, also used in MEKC, was applied for the analysis of binaphthyl derivatives and (±)-barbiturates. Dual-chirality MEs, have been developed where chiral surfactant DDCV and the chiral co-surfactant S-2-hexanol were employed together to analyze N-methyl ephedrine and pseudoephedrine. Another combination of DDCV and a chiral oil (diethyl tartrate) were also used to separate atenolol, metoprolol, N-Methyl ephedrine, ephedrine, synephrine, and pseudoephedrine [128-130].

7. Pharmaceutical Analysis

7.1. Introduction

Capillary electrophoresis (CE) is a very important analytical method, which is applied to drug discovery, in qualitative and quantitative analysis in purity tests, in chiral separation, impurity profiles, pharmaceutical quality control, so as to provide efficacy and safety for patients [131], due to the high resolution and efficiency in a short analysis time, the ability to mix multiple detection techniques, automated analytical equipment, low-reagents consumption and the development of rapid methods [132].

Different CE techniques offer numerous possibilities in pharmaceutical analysis depending on the complexity of the sample, the nature of its components, application and nature of the analytes; each of these techniques provide several advantages for the separation and detection of different pharmaceuticals.

An important advantage of CE is the availability of various techniques based on different separation principles. The most commonly used in pharmaceutical analysis are capillary zone electrophoresis (CZE) and electrokinetic chromatography (EKC), specially MEKC, microemulsion electrokinetic chromatography (MEEKC), capillary electrochromatography (CEC) and non-aqueous capillary electrophoresis (NACE) are increasing in order to improve the separation efficiency, selectivity, sensitivity and the flexibility of CE. [131-132].

7.2. Capillary Zone Electrophoresis (CZE)

The CZE is the simplest form of CE, used in the separation of ionizable analytes. It Is used in the separation of catecholamines [133], in the analysis of capsules, tablets and injectable alendronate [134], atenolol tablets, fenoterol, metoprolol, propranolol, terbutaline, clenbuterol [135], isoniazid and p-aminosalicylic acid [136], ethambutol [137]; Pharmaceutical formulations of low molecular weight heparins [138]; neomycin ointments [139]; depot tablets of salbutamol [140] and pharmaceutical formulations of a large amount of antibiotics such as fluoroquinolones (ciprofloxacin) [141], aminoglycosides (kanamycin and related compounds) [142], tobramycin [143], amikacin and its impurities [144] and sulfonamides (Sulfamethoxazole) [145].

7.3. Micellarelectrokinetic Chromatography (MEKC)

MEKC Is usually applied to the simultaneous separation of complex mixtures of pharmaceuticals with similar physicochemical and structural features. As example it can be mentioned the determination of drugs in samples having a high protein content (biological samples), in the chiral separation of optically active pharmaceuticals and profiles of impurities of similar physicochemical and structural features with the active substance [132]. Since they tend to have ratios of charge / mass similar to the active ingredient they can be separated with improved selectivity by a combination of charge / mass, hydrophobicity and charge/ mass interactions on the surface of the micelles [133].

Used MEKC purity analysis of drugs such as paclitaxel, cisplatin, carboplatin [133]; Pharmaceutical formulations of β-lactam antibiotics (biapenem) [146], penicillin [147], cephalosporins [148], macrolide [149], tetracycline [150], sulfonamides [151]; standards fluoroquinolones (ciprofloxacin and norfloxacin) [152]; synthetic analogues and insulin [153]; paracetamol tablets, 4-aminophenol [154]; oseltamivir capsules [155]; omeprazole tablets [156]; antifungals [157]; barbiturates [158]; benzodiazepines [159]; phenothiazines [160]; Tricyclic antidepressants [161] and xanthine [162].

MEKC is the most popular capillary electrophoresis techniques because of its high resolving power and ability to separate ionic and neutral analytes, provides a high selectivity to a large number of compounds and is considered the method of separation of choice for the analysis of pharmaceuticals. One difficulty that MEKC presents is reproducibility in quantitative analysis, including the migration time and peak height or area [132] (Figure 11).

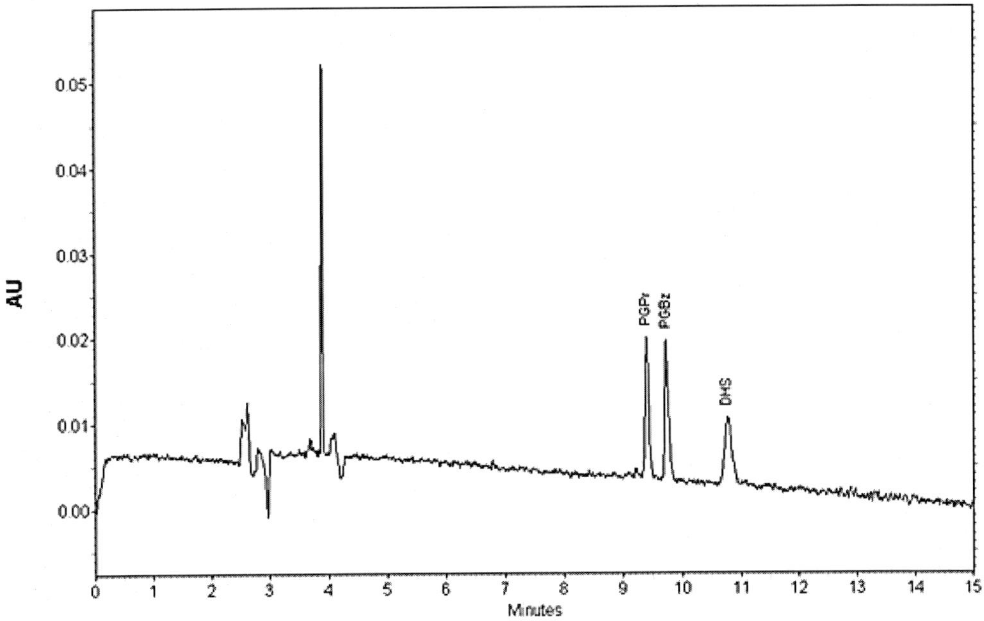

Figure 11. Electropherogram of a standard solution of DHS, PGPr and PGBz at BGE final conditions.

7.4. Microemulsionelectrokinetic Chromatography (MEEKC)

MEEKC is a mode of CE where the microemulsion droplets are used as pseudo-stationary phases [131] for the separation of both hydrophilic and hydrophobic particles as well asthe determination of physicochemical properties [163]. Based on studies where CE systems (MEKC Y MEEKC) were compared with respect to the retention characteristics for a given number of lipophilic drugs (betamethasone and its derivatives) it was observed that the phosphatidylcholine MEEKC systems are apparently the best models for estimating hydrophobicity of betamethasone and its derivatives, which may also be better models in drug delivery studies thereof when applied to pharmaceutical microemulsions [163].

MEEKC applications in drug analysis include: betamethasone and its derivatives [163], pharmaceutical formulations of hidrosoluble vitamins and liposoluble [164], tablets and injectable suspensions and synthetic estrogens [165] and paracetamol suppositories and impurities [166].

7.5. Capillary Electrochromatography (CEC)

It is a hybrid technique that combines the characteristics of liquid chromatography (high selectivity) and capillary electrophoresis (most efficient). CEC is very suitable for the separation of α, β and δ-tocopherols and α-tocopherol acetate in pharmaceutical preparations (167).

7.6. Non-Aqueous Capillary Electrophoresis (NACE)

NACE is used for the separation of hydrophobic compound wich cannot be analyzed in aqueous media. [133]. It Is uses in the analysis of bromazepan, nicotine, fluoxetine and tamoxifen [168].

In recent years, different CE techniques have improved very quickly and have led to many advances and applications in pharmaceutical products, drug tests, determination of impurities of the active substance, determination of physicochemical properties, chiral separations. Therefore, CE is a very valuable tool for the pharmaceutical industry that allows drug discovery, development and quality control.

8. Ion Analysis

Since capillary electrophoresis (CE) was introduced into the domain of inorganic ion analysis more than twenty years ago, the separation and quantification of metal ions is one of the areas in which CE is being used to an increasing extent (Timerbaev & Shpigun 2000). Numerous publications on this regard reveal that CE is a more suitable method of choice than HPLC in terms of superior resolution and separation efficiency, which are achievable in a shorter time, simpler instrumentation, lower consumption of reagents and sample, and versatility. The advantages of CE along with a broad range of applications to inorganic ion determinations in various matrices have been evaluated and many works have been reported in the literature [169]. Inorganic ions are routinely monitored in a variety of samples that are important to several industries such as pharmaceutical and metal plating companies, and the drinking and waste water industries. Given that inorganic ions have little UV absorbance or no chromophore groups in their chemical structure, conventional methods such as HPLC are not possible to employ for their quantification and more sophisticated techniques as ion chromatography (IC) [170-174], flame atomic absorption spectrometry (FAAS) [175-176] and flame atomic emission spectrometry (FAES) [175-176] are usually required. Nevertheless, the mentioned techniques are highly complex and very expensive. In this context, CE has emerged as a suitable alternative and has gained importance as a highly efficient separation method for the analysis of ions present in different matrices [177-183]. In CE, the most applied method for the determination of inorganic ions is based on the use of an indirect UV detection mode where ions can be simply analysed [180, 184-188]. In this kind of methods, the composition of the background electrolyte (BGE) is important in terms of UV-absorbing properties, pH and the presence of complexing agents to aid in the separation of the inorganic ions [189]. Additionally, inorganic ions can be determined and quantified by using a direct UV detection mode where complexing agents are added to the BGE in order to form an UV-absorbing compound *in situ* [190-192]. When anionic ion analysis is considered, the capillary surface is usually coated with long-chain alkyl ammonium salts (such as cetyltrimethylammonium bromide -CTAB-, tetradecyltrimethylammoniumbromide -TTAB-), or polycations (such as hexadimetrine -HDB-) to reverse the electroosmotic flow (EOF) and produce the ion migration, and the analysis is achieved by switching the polarity of the power supply from positive to negative (193). On the other hand, conductometric detection is an alternative to UV detection for the analysis of inorganic ions [189].

The determination of inorganic ions by CE is important for quality control of pharmaceutical entities and products, among qualitative and quantitative analyses, purity testing and stoichiometric determination. In this regard, several works based on CE methods for quantifying inorganic counter ions, inorganic ionic impurities and inorganic ions as therapeutic agents being part of different pharmaceutical formulations or drug delivery systems have been published.

Many drugs are prepared in salt forms to improve their solubility or stability. Sulphate and chloride are common counter ions for basic drugs, and cationic metals as sodium, potassium, magnesium and calcium are also used as drug counter ions [193]. For quality purposes, types and contents of these ions must be determined [193] and an extended review on this matter was reported in the literature by de L´Escaille and Falmagne [181]. To mention some of them, the sulphate counter ion was quantified by an indirect UV CE method in aminoglycoside antibiotics by using chromic acid as a component of the BGE [194]. In another work, an indirect UV method by CE was performed for the determination of sulphate in indinavir sulphate raw material using a BGE containing ammonium molybdate as an UV-absorbing agent [195]. The method showed short analysis time and good linearity, precision and sensitivity. In addition, an indirect UV-CE method was performed for the quantification of sulphate as an oxidation product of metabisulphite present in pharmaceutical formulations [196]. Similarly, the sulphate ion was quantified as one of the inorganic degradation products in the topiramate antiepileptic drug in final product and raw material [197]. Also, a quantitative analysis of anions such as chloride, sulphate, nitrate and phosphate from a prenatal vitamin formulation based on UV indirect mode was reported [181]. Other UV indirect methods for the quantification of anionic and cationic drug counter ions were developed [198], where acceptable method performance in terms of linearity, accuracy, precision and migration times were demonstrated. Furthermore, a CE method for the determination of sodium levels in the sodium salt of an acidic drug (cephalosporin and cephalotin) has been developed based on an indirect UV mode of detection using imidazole as an UV-absorbing probe in the BGE [199]. In another report, a CE technique with conductivity detection has been evaluated as a method for determining potassium counter ion content and for screening of inorganic impurities in pharmaceutical drug substances [200]. Besides, a different detection technique for inorganic ions such as the conductivity detection (particularly the contactless mode) has been applied in a simple CE method for the determination of potassium, sodium, calcium and magnesium in parenteral nutrition formulations [182].

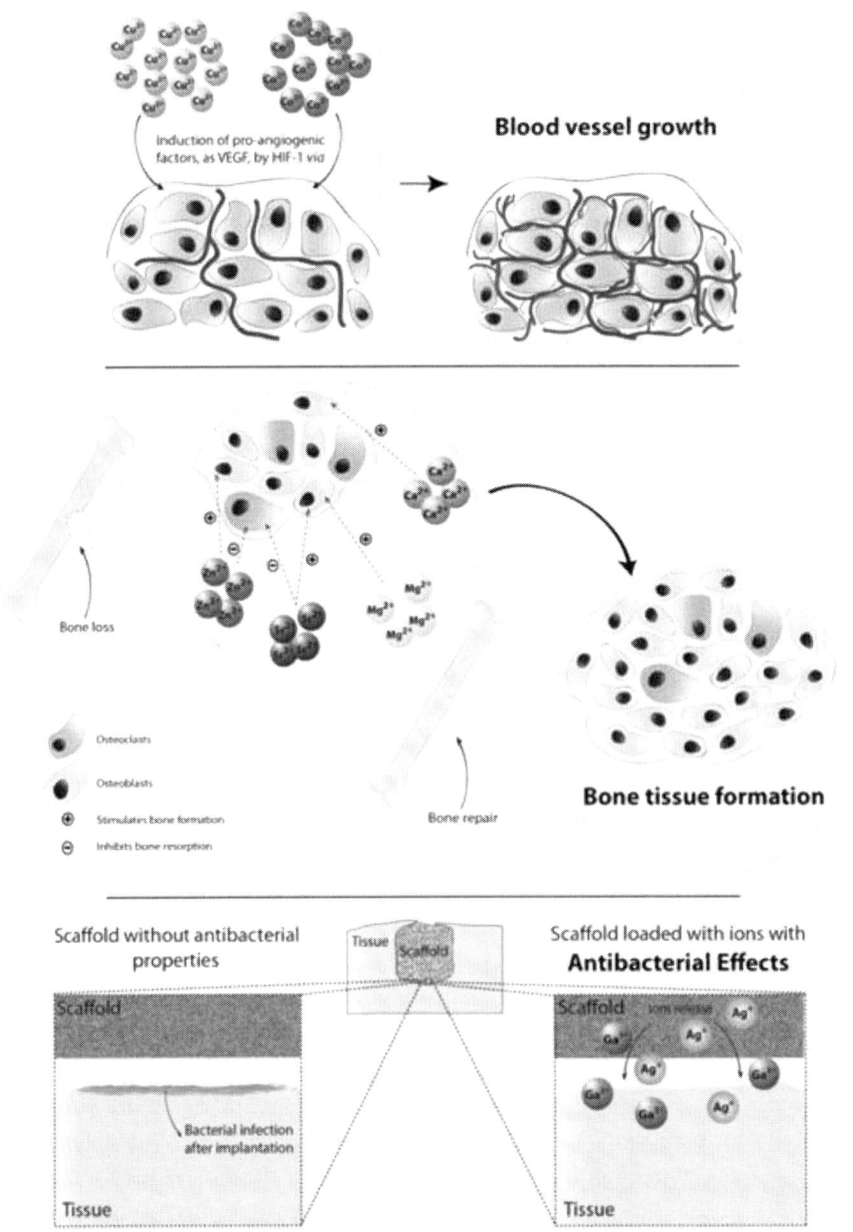

Figure 12. This figure shows the effects of relevant inorganic ions in their role of therapeutic agents in tissue engineering applications from reference 203 with permission.

On the other hand, as it was previously mentioned, drug impurities with non-cromophore groups can be analyze by CE using indirect UV detection. For instance, a method using a BGE consisting of potassium chromate as an UV-absorbing probe has been developed for the determination of phosphyte and phosphate impurities in ibandronate [201]. Similarly, an indirect UV CE method was developed and validated to quantify the same impurities from zoledronate, where a BGE containing phthalic acid and TRIS was used [202]. Both methods offered good sensitivity and resolution for phosphyte and phosphate impurities and can be used to evaluate the quality of regular production samples of ibandronate and zoledronate.

Lately, an interesting approach is the use of inorganic ions as therapeutic agents in the field of tissue engineering and regenerative medicine. Ions such as copper, calcium, cobalt, iron, gallium, magnesium, strontium and zinc can be considered in this regard; and most of them are essential cofactors of enzymes in the organism [203]. In tissue engineering, dissolution products from scaffolds have an important role in the integration between biomaterial and the host tissue, and produce different intra and extracellular responses [204-206]. These considerations incited the incorporation of inorganic ions into different releasing systems, e.g., scaffolds, intended for therapeutic applications (Figure 12) [203-204, 207-208].

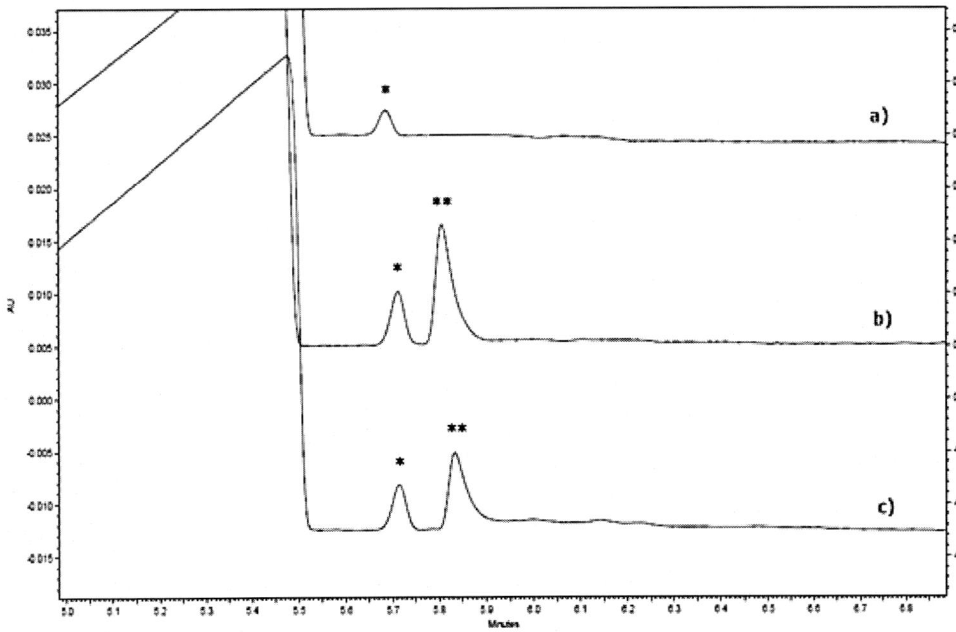

Figure 13. Electropherograms obtained from the method for quantifying calcium ions are presented in this figure, which includes an electropherogram of ammonium phosphate buffer, used as blank (a); an electropherogram of calcium standard solution of 5 μg mL^{-1} (b); and an electropherogram of a sample from release study of the matrixes containing calcium ions (c). Ca^{2+} peak is pointed as **, and Na^{+} peak (present in the sample) is pointed as * from reference 211 with permission.

Due to the fact that ions incorporated into the scaffolds are released in small amounts - ppm or ppb-, there is a need to develop new, efficient and sensitive methods to quantify the release of those ions from the matrices, and this context CE techniques are considered as highly efficient and simple separation methods for the analysis of ions in different matrices. In relation to this, calcium ions which are involved in bone formation, bone metabolism and bone mineralization [209-210] were incorporated within biomaterials for bone tissue engineering, and the ion release from these matrixes has been quantified for the first time by a novel CE indirect UV method developed and validated by Cattalini et al. [211]. In this work, a BGE containing imidazole as UV-absorbing molecule at 214 nm, α-hydroxyisobutyric acid, 1,4,7,10,13,16-hexaoxacyclo-octadecane and methanol was utilized, and the method showed good results in terms of sensitivity achieving LOD and LOQ values of 0.03 and 0.1 $\mu g\ mL^{-1}$, respectively (Figure 13). Furthermore, copper ions have also been incorporated in matrices for tissue engineering due to their important effects on blood vessels formation [201, 212-213], and their release was quantified by a CE direct UV method based on the complexation of copper ions with EDTA, which was added to the BGE to aid in the *in situ* complex formation. The proposed method was suitable for the quantification of copper ions released from biomaterials with LOD and LOQ values of 0.05 and 0.16 $\mu g\ mL^{-1}$, respectively.

Final Conclusion and Perspectives

In the last years CE has consolidated as powerful analytical technique which has been applied in the analysis of a wide range of compounds from inorganic ions to macromolecules, in different matrices, such as biological, pharmaceutical and environmental samples. Their application in clinical and biomolecular analysis, has gained an important place as an alternative and complementary technique, and is continuously growing.

Moreover, in pharmaceutical quality control, CE has become an important tool for determination of active ingredients, the analysis of related substances and principally to the determination of enantiomeric purity in chiral analysis, which has led to their implementation in the official pharmacopoeias.

Today the eye is set into the potential of CE for the quality control of biopharmaceuticals, especially therapeutic peptides and proteins. Their intrinsic characteristics as high resolution, reduced analysis time, and small sample volume plus its feasibility to couple to a variety of highly sensitive detectors, make it a reliable technique to quality control in finished products in terms of activity, heterogeneity, stability, and purity.

In conclusion, CE has demonstrated considerable advantages compared with traditional methodologies, here we mentioned some relevant examples because the field of application is wide.

References

[1] Capillary Electrophoresis: an alternative to chromatography for analysis of natural organic binders. Mazanek M, Kaml I, Kenndler E. *Studies in conservation.* 2006, 51: 139-141.

[2] Recent advances of capillary electrophoresis in pharmaceutical analysis. Suntornsuk L. *Anal. Bioanal. Chem.* 2010, 398, 29-52.

[3] Tripodi V, Lucangioli S. Microemulsions as pseudostationary phases in electrokineticchromatoraphy: I. Estimation of physicochemical parameters. II. Analysis of drugs in pharmaceutical and biofluidics matrices. Progress in Colloid and Interface

Science Series, 144: Microemulsion: Properties and Applications. 2008. Chap. 19, 501-525. Ed Taylor and Francis. New York.

[4] Khaledi M. High Performance Capillary Electrophoresis. 1998. Ed. John Wiley and Song, New York.

[5] Neubert R, Ruttinger H. Affinity Capillary Electrophoresis in Pharmaceutics and Biopharmaceutics. 2003. Ed Marcel Dekker, New York.

[6] Lucangioli S, Tripodi V. Capillary Electrophoresis: Fundamentals, Techniques and Applications. Applications of capillary electrophoresis to the clinical and pharmaceutical analysis. 2012. Chap.6, 187-209. Ed Nova Science Publishers Inc. ISBN: 978-1-62081-867-1.

[7] Electrokinetic separations with micellar solutions and open-tubular capillaries. Terabe, S., Otsuka K., Ichikawa K., Tsuchiya A., Ando T. *Anal. Chem.* 1984, 56: 111-113.

[8] Characterization of polymeric-mixed micelles applied as pseudo-stationary phase in MEKC. Flor S., Tripodi V., Chiappetta D., Iglesias Rando M., Lucangioli S. *Trends in chromatography*. 2013,8: 43-54.

[9] Micellar electrokinetic chromatography: methodological and instrumental advances focused on practical aspects. Silva M. *Electrophoresis.* 2009, 30: 50-64.

[10] Comparison of three different anionic surfactants for the separation of hydrophobic compounds by nonaqueous capillary electrophoresis. Lin J., Nakagawa M., Uchiyama K., Hobo T. *Electrophoresis.* 2002, 23: 421-425.

[11] Influence of organic solvents on the separation selectivity in capillary electrophoresis. Kenndler E, Sarmini K. *J. Chromatogr A.* 1997, 792: 3-11.

[12] Electrokinetic chromatography with micelles polymeric and monomeric additives with similary chemical functionality as pseudo-stationary phases. Gogová K, Maichel B, Gaš B, Kenndler E. *J. Chromatogr A.* 2001, 916: 79-87.

[13] Recent advances in the development and application of microemulsion EKC. McEvoy E, Marsh A, Altria K, Donegan S, Power J. *Electrophoresis.* 2007, 28: 193-207.

[14] Recent advances in microemulsion electrokinetic chromatography. Marsh A, Clark B, Broderik M, Power J, Donegan S, Altria K. *Electrophoresis.* 2004, 25: 3970-3980.

[15] Background and operating parameters in microemulsion electrokinetic chromatography. Altria K, Mahuzier P, Clark B. *Electrophoresis.* 2003, 24: 315-324.

[16] Recent applications of microemulsion electrokinetic chromatography. Hui C. *Electrophoresis.* 2006, 27: 60-75.

[17] Recent progress in the use of ionic polymers as pseudostationary phases for electrokinetic chromatography. Palmer C. *Electrophoresis*. 2009, 30: 163-168.

[18] Monolithic stationary phases in classic and chiral pharmaceutical analysis with CEC and pCEC. Tanret I, Mangelings D, Vander Heyden Y. *J. Chromatogr Sci.* 2009, 47: 407-417.

[19] Enantiomeric separation by CEC using chiral stationary phases. Kato M, Toyooca T. *Chromatography.* 2001, 22: 159-170.

[20] Enantiomeric separation by capillary electrochromatography on a sulfated poly b-cyclodextrin modified silica-based monolith. Yuan R, Wang Y, Ding G. *Analytical Sciences*. 2010, 26: 943.

[21] A new apparatus for electrophoretic analysis of colloidal mixtures. Tiselius A. 1937, *Trans. Farad Soc*, 524-531.

[22] Immunoaffinity capillary electrophoresis: a new versatile tool for determining protein biomarkers in inflammatory process. Guzman N, Phillips T. *Electrophoresis.* 2011, 32: 1565-1578.
[23] Coupling Immunoaffinity Capillary Electrophoresis with Sensitive Detectors for the Characterization of Protein Biomarkers in Biological Fluids and Cells. Guzman N. *Am. Laborat.* 2000, 40: 18-29.
[24] Rapid analysis of inflammatory cytokines in cerebrospinal fluid using chip-based immunoaffinity electrophoresis. Phillips T. *Electrophoresis.* 2004, 25: 1652-1659.
[25] Validity and Reliability of Perinatal Biomarkers after Storage as Dry Blood Spots on Paper. Mihalopoulos N, Phillips T, Slater H, Thomson J, Varner M, Moyer-Mileur L. *Am. J. Hum. Biol.* 2011, 23(5): 717-719.
[26] Clinical applications of capillary electrophoresis based immunoassays. Moser A, Willicott C, Hage D. *Electrophoresis.* 2014, 35: 937–955.
[27] Rapid simultaneous determination of glucagon and insulin by capillary electrophoresis immunoassays. German I, Kennedy R. *J. Chromatogr B Biomed. Sci. Appl.* 2000, 742(2): 353-62.
[28] Simultaneous capillary electrophoresis competitive immunoassay for insulin, glucagon, and islet amyloid polypeptide secretion from mouse islets of Langerhans. Guillo C, Truong T, Roper M. *J. Chromatogr A.* 2011, 1218(26): 4059-4064.
[29] Zhao S, Liu Y, Kalish, H. Phillips. Clinical Applications of Capillary Electrophoresis: Methods and Protocols. 2013, Chap. 8. Ed. Humana, New York.
[30] Direct immunoassay of estriol in pregnancy serum by capillary electrophoresis with laser-induced fluorescence detector. Su P, Zhang X, Wang Y, Chang W. *Talanta.* 2003, 60: 969–975.
[31] Capillary electrophoresis-based immunoassay for cortisol in serum. Schmalzing D, Nashabeh W, Yao X, Mhatre R, Regnier F, Afeyan N, Fuchs M. *Anal. Chem.* 1995, 67:606–612.
[32] Analysis of vasopressin using capillary electrophoresis with laser-induced fluorescence detector based on competitive immunoassay. Han K, Ban E, Yoo Y. *J. Chromatogr A.* 2003, 1013(1-2):215-20.
[33] Non competitive immunoassay for follicle-stimulating hormone in human serum using capillary electrophoresis with chemiluminescence detection. Zhang Y, Zheng H. *J. Immunoassay Immunochem.* 2010, 31(3): 193-204.
[34] Non-competitive immunoassay for luteinizing hormone in human serum using capillary electrophoresis with chemiluminescence detection. Zhang Y, Zhang Z, Yang F. *Luminescence.* 2007, 22(6):511-517.
[35] Rapid determination of multidrug resistance-associated protein in cancer cells by capillary electrophoresis immunoassay. Mbuna J, Kaneta T, Imasaka T. *J. Chromatogr A.* 2011, 1218(25):3923-3927.
[36] Electrophoresis as a metabolomics tool for non-targeted fingerprinting of biological samples. Barbas C, Moraes EP, Villaseñor A. *J. Pharm Biomed. Anal.* 2011, 55(4):823-831.
[37] Micellar electrokinetic capillary chromatography and data alignment analysis: a new tool in urine profiling. Guillo C, Barlow D, Perrett D, Hanna-Brown M. *J. Chromatogr A.* 2004, 1027(1-2): 203-12.

[38] Separating microbes in the manner of molecules. Capillary electrokinetic approaches. Armstrong D, Schulte G, Schneider heinze J, et al. *Anal. Chem.* 1999, 71: 5465-5469.
[39] Capillary-electrophoresis mass spectrometry for the detection of carbapenem ases in (multi-)drug-resistant Gram-negative bacteria. Fleurbaaij F, Heemskerk A, Russcher A, Klychnikov O, Deelder A, Mayboroda O, Kuijper E, Van Leeuwen H, Hensbergen P. *Anal. Chem.* 2014, 86(18):9154.
[40] Separation of phenotypically indistinguishable Candida species, C. orthopsilosis, C. metapsilosis and C. parapsilosis, by capillary electromigration techniques. Horká M, Růžička F, Kubesová A, Němcová E, Slais K. *J. Chromatogr A.* 2011, 1218(25):3900-3907.
[41] CE separation of proteins and yeasts dynamically modified by PEG pyrenebutanoate with fluorescence detection. Horká M, Růzicka F, Holá V, Slais K. *Electrophoresis.* 2007, 28(13):2300-2307.
[42] DNA diagnostics by capillary electrophoresis. Kleparnik K, Bovek P. *Chem. Rev.*2007,107: 5279-5317.
[43] Electrically assisted sampling across membranes with electrophoresis in nanometer inner diameter capillaries. Woods LA, Gandhi PU, Ewing AG. *Anal. Chem.* 2005, 77(6):1819-1823.
[44] Simultaneous determination of free and conjugated bile acids in serum by cyclodextrin-modified micellar electrokinetic chromatography. Tripodi V, Lucangioli S, Scioscia S, Carducci C. *J. Chromatogr B Analyt. Technol. Biomed. Life Sci.* 2003, 785(1): 147-55.
[45] The importance of analytical methodology in accurate diagnosis and monitoring of intrahepatic cholestasis of pregnancy. Martinefski M, Contin M, Samassa P, Di Carlo MB, Lucangioli S, Tripodi V. *Int. J. Gynaecol. Obstet.* 2013, 123(1):78-9.
[46] Simultaneous determination of nine endogenous steroids in human urine by polymeric-mixed micelle capillary electrophoresis. Flor S, Lucangioli S, Contin M, Tripodi V. *Electrophoresis.* 2010, 31(19):3305-3313.
[47] A capillary electrophoretic system based on a novel microemulsion for the analysis of coenzyme Q10 in human plasma by electrokinetic chromatography. Lucangioli S, Flor S, Contin M, Tripodi V. *Electrophoresis.* 2009, 30(11):1899-905.
[48] DNA sequencing by CE. Karger B, Guttman A. *Electrophoresis.*2009, 30: 196-202.
[49] Polyphenols: Chemistry, Dietary Sources, Metabolism, and Nutritional Significance. Bravo L. *Nutrition Reviews.* 1998, 56(11).
[50] Influence of Borate Buffers on the Electrophoretic Behavior of Humic Substances in Capillary Zone Electrophoresis. Schmitt-Koppling P, Hertkorn N, Garrison A, Freitag D, Kettrup A. *Anal. Chem.* 1998, 70: 3798 – 3808.
[51] Separation and determination of flavonoids in Lamiophlomisrotata by capillary electrophoresis using borate as electrolyte. Luo M, Lu H, Ma H, Zhao L, Liu X, Jiang S. *J. Pharm. Biomed. Anal.* 2007, 44: 881–886.
[52] Borate complexation of flavonoid-O-glycosies in capillary electrophoresis. II. Separation of flavonoid-3-O-glycosides differing in their sugar moiety. Morin Ph, Villard F, Dreux M. *J. Chromat.* 1993, 628: 161-169.
[53] Capillary electrophoresis analysis of flavonoids. Seitz U, Oefner P, Nathakarnkitkool S, Popp M, Bonn G. *Electrophoresis.* 1992,13: 35-38.

[54] Separation of flavonols by capillary electrophoresis: the effect of structure on electrophoretic mobility. McGhie T, Markham K. Phytochemical Analysis. 1994, 5: 121-126.
[55] Simultaneous Determination of the Active Ingredients in Abelmoschusmanihot (L.) Medicus by CZE. Liu Y, Li W, Ling X, Lai X, Li Y, Zhang Q, Zhao Y. Chromatographia. 2008, 67: 819–823.
[56] CZE determination of five flavonoids in *Apocynum venetum* from different regions. Meng M, Zeng H, Li J, Yao S, Sun W. *Chin. Pharm J.* 2010, 3: 19-25.
[57] Use of capillary electrophoresis for estimating the quality of chamomile flowers. Gavrilin M, Senchenko S, *Pharm. Chem. J.* (2009), 43, 582–584.
[58] Quantification of main phenolic compounds in sweet and bitter orange peel using CE–MS/MS. Sawalha S, Arráez-Román D, Segura-Carretero A, Fernández-Gutiérrez A. *Food Chem.* 2009, 116: 567–574.
[59] Differentiation of Resina Draconis from Sanguis Draconis by CE. Gong W, Cao Y, Wang Y. *Chromatographia*. 2007, 66: 767–771.
[60] Separation and Determination of Flavonoids in Ixeridiumgracile by Capillary Electrophoresis. Zhang Y, Zhao L, Shi Y. *J. Chromatogr Sci.* 2007, 45: 600–604.
[61] Capillary electrophoretic analysis of flavonoids in single-styled hawthorn (Crategusmonogyna Jacq.) ethanolic extracts. Urbonaviciute A, Jakstas V, Kornysova O, Janulis V, Maruska A. *J. Chromatogr A.* 2006, 1112: 339–344.
[62] Determination of flavonoids by high-performance liquid chromatography and capillary electrophoresis. Wang S, Huang K. *J. Chromatogr A.* 2004,1032 (1-2): 273-279.
[63] Qualitative and quantitative analysis of active flavonoids in Floslonicerae by capillary zone electrophoresis coupled with solid-phase extraction. Chen J, Li S, Li P, Song Y, Chai X, Ma D. *J. Sep. Sci.* 2005,(28)365–372.
[64] Simultaneous Determination of Rutin and Hyperin in *Acanthopanaxsenticosus* (Rupr. et Maxim) Leaves by High-Performance Capillary Electrophoresis. Zheng C, Liu S, Li H, Shi Z, Ping W. *Chin. Pharm. J.* 2008, 43: 783–785.
[65] Simultaneous determination of nine flavonoids in Anaphalismargaritacea by capillary zone electrophoresis. Ren Z, Zhang Y, Shi Y. *Talanta.* 2009, 78: 959–963.
[66] Quality control of herbal medicines by capillary electrophoresis: Potential, requirements and applications. Ganzera M. *Electrophoresis.*2008,28:3489-3503.
[67] Precision improvement for the analysis of flavonoids in selected Thai plants by capillary zone electrophoresis. Suntornsuk L, Anurukvorakun O. *Electrophoresis.*2005,26: 648–660.
[68] Comparison of capillary electrophoresis and high performance liquid chromatography for determination of flavonoids in Achilleamillefolium. Kocevar N, Glavac I, Injac R, Kreft S, *J. Pharm. Biomed. Anal.* 2008, 46: 609–614.
[69] Effect of stability of internal standard on quantification of 15 flavonoids in *Epimedium* using CZE. Chen X, Tu P, Jiang Y, Wang Y, Li S. *J. Sep. Sci.* 2009, 32: 275–281.
[70] Quantification of the flavonoid glycosides in Passiflora incarnata by capillary electrophoresis. Machart E, Krenn L, Kopp B. *Planta Med.* 2003, 69: 452–456.
[71] Simultaneous determination of flavonoids in chrysanthemum by capillary zone electrophoresis with running buffer modifiers. Zhang S, Dong S, Chi L, He P, Wang Q, Fang Y. *Talanta*. 2008, 76: 780–784.

[72] Enantioseparation of catechin and epicatechin in plant food by chiral capillary electrophoresis. Kofink M, Papagiannopoulos M, Galensa R. *European Food Research and Technology*. 2007, 225(3-4): 569-577.

[73] Fast determination of flavonoids in Hippophaerhamnoides and its medicinal preparation by capillary zone electrophoresis using dimethyl-β-cyclodextrin as modifier. Yue M, Jiang T, Shi Y. *Talanta*. 2004, 62(4): 695 – 699.

[74] An enantiomeric assay for the flavonoids medicarpin and vestitone using capillary electrophoresis. Allen D, Gray J, Paiva N, Smith J. *Electrophoresis*. 2000, 21(10): 2051-2057.

[75] Separation of flavanone enantiomers and flavanone glucoside diastereomers from Balanophora involucrate Hook. f. by capillary electrophoresis and reversed-phase high-performance liquid chromatography on a C18 column. Pan J, Zhang S, Yan L, Tai J, Xiao Q, Zou K, Zhou Y, Wu J, Pan J, Zhang S, Yan L, Tai J, Xiao Q, Zou K, Zhou Y, Wu J. *J. Chromatogr A*. 2008,1185: 117–129.

[76] Capillary electrophoresis of phytochemical substances in herbal drugs and medicinal plants. *Gotti R. J. Pharm. Biomed. Anal.* 2011, 55: 775 – 801.

[77] Application of micellar electrokinetic capillary chromatography to the determination of flavonoid drugs. Pietta P, Mauri P, Rava A, Sabbatini G. *J. Chromatogr.*1991, 549: 367 – 373.

[78] Influence of structure on the behavior of flavonoids in capillary electrophoresis. Pietta P, Mauri P, Bruno A, Gardana C. *Electrophoresis*. 1994, 15: 1326 – 1331.

[79] Analysis of green tea extract dietary supplements by micellar electrokinetic chromatography. Weiss D, Austria E, Anderton C, Hompesch R, Jander A. *J. Chromatogr A*.2006, 1117: 103–108.

[80] Rutin and total quercetin content in amaranth (Amaranthus spp.). Kallinova J, Dadakova E. *Plant Food Hum. Nutr*. 2009, 64: 68–74.

[81] Separation and sweeping of flavonoids by microemulsion electrokinetic chromatography using mixed anionic and cationic surfactants. Cao J, Dun W. *Talanta*. 2011, 84: 155–159.

[82] Determination of catechins in green tea infusions by reduced flow micellar electrokinetic chromatography. Peres R, Tonin F, Tavares M, Rodríguez Amaya D. *Food Chem*. 2011, 127: 651–655.

[83] Enhanced separation of Compound Xueshuantong capsule using functionalized carbon nanotubes with cationic surfactant solutions in MEEKC. Cao J, Li P, Chen J, Tan T, Dai H. *Electrophoresis*. 2013, 34: 324–330.

[84] Ionic liquids coated multi-walled carbon nanotubes as a novel pseudostationary phase in electrokinetic chromatography. Cao J, Li P, Yi L. *J. Chromatogr A*. 2011, 1218: 9428–9434.

[85] Advance of CE and CEC in phytochemical analysis (2012–2013). Zhao J, Hu D, Lao K, Yang Z, Li S. *Electrophoresis*. 2014, 35: 205 – 214.

[86] Harborne J, Mabry T, Mabry H. The Flavonoids, first ed., Chapman and Hall, London, (1975).

[87] Quantitative and qualitative analysis of flavonoid markers in Frucusaurantii of different geographical origin by capillary electrophoresis with electrochemical detection. Peng Y, Liu F, Ye J. *J. Chromatogr B*. 2006, 830: 224–230.

[88] Simultaneous determination of flavones and phenolic acids in the leaves of Ricinuscommunis Linn. by capillary electrophoresis with amperometric detection. Chen Z, Zhang J, Chen G. *J. Chromatogr B.* 2008, 863: 101–106.

[89] Determination of flavonoids in Houttuyniacordata Thunb. and Saururuschinensis (Lour.) Bail. by capillary electrophoresis with electrochemical detection. Xu X, Ye H, Wang W, Yu L, *Chen G. Talanta.* 2006, 68: 759–764.

[90] An improved method for the quantitation of flavonoids in herba Leonuri by capillary electrophoresis. Xu X, Ye H, Wang W, Chen G. *J. Agric. Food Chem.* 2005, 53: 5853–5857.

[91] Separation and determination of flavonoids using microemulsion EKC with electrochemical detection. Yu L, Xu X, Huang L, Ling J, Chen G. *Electrophoresis.* 2008, 29: 726 – 733.

[92] Separation and determination of flavonoids using microemulsion EKC with electrochemical detection. Yu L, Xu X, Huang L, Ling J, Chen G. *Electrophoresis.* 2008, 29: 726–733.

[93] Highly sensitive analysis of flavonoids by zwitterionic microemulsion electrokinetic chromatography coupled with light-emitting diode-induced fluorescence detection. Cao W, Hu S, Li X, Pang X, Cao J, Ye L, Dai H, Liu X, Da J, Chu C. *J. Chromatogr A.* 2014, 1358: 277 – 284.

[94] Capillary electrophoresis and microchip capillary electrophoresis with electrochemical and electrochemiluminescence detection. Du Y, Wang E. *J. Sep. Sci.* 2007, 30: 875 – 890.

[95] Analysis of three flavonoids by CE-UV and CE-ESI-MS. Determination of naringenin from a phytomedicine. Huck C, Stecher G, Ahrer W, Stöggl W, Buchberger W, Bonn G. *J. Sep. Sci.* 2002, 25: 904–908.

[96] Capillary electrophoresis-mass spectrometry characterization of secondary metabolites from the antihyperglycaemic plant Genistatenera. Edwards E, Rodrigues J, Ferreira J, Goodall D, Rauter A, Justino J, Thomas-Oates J. *Electrophoresis.* 2006, 27: 2164 – 2170.

[97] Capillary electrophoresis of natural products: 2011–2012. Tubaon R, Rabanes H, Haddad P, Quirino J. Electrophoresis. 2014, 35: 190–204.

[98] Sequential uniform designs for fingerprints development of Ginkgo biloba extracts by capillary electrophoresis. Ji Y, Alaerts G, Xu C, Hu Y, VanderHeyden Y. *J. Chromatog A.*2006, 1128: 273–281.

[99] Simultaneous chromatographic fingerprinting and quantitative analysis of flavonoids in Pollen Typhae by high-performance capillary electrophoresis. Han L, Liu X, Yang N, Li J, Cai B, Cheng S. *Acta Pharmaceutica Sinica B.* 2012, 2: 602 – 609.

[100] Quantitative and qualitative analysis of flavonoid markers in Frucusaurantii of different geographical origin by capillary electrophoresis with electrochemical detection. Peng Y, Liu F, Ye J. *J. Chromatogr B.* 2006, 830: 224 – 230.

[101] Development of capillary electrophoresis fingerprint for quality control of Rhizoma Smilacis Glabrae. Zhang Q, Cheung H. *Phytochemical Analysis.* 2011, 22: 18 – 25.

[102] Separation of flavonoids and phenolic acids from propolis by capillary zone electrophoresis. Volpi N. *Electrophoresis.* 2004, 25: 1872 – 1878.

[103] Bioanalysis: Heri, hodie, cras. Pier Giorgio Righetti. *Electrophoresis.* 2013, 34: 1442-1451.

[104] Guzman N. Capillary electrophoresis technology.1993. Ed Marcel Dekker INC, Nueva York.

[105] Capillary electrophoresis with laser-induced fluorescence detection for detailed studies on *N*-linked oligosaccharide profile of therapeutic recombinant monoclonal antibodies. Komoda S, Ishikawa R, Kakei K. J. Chromatogr A. 2006, 1133: 332-339.

[106] Characterization of Glycoprotein Biopharmaceutical Products by Caliper LC90 CE-SDS Gel Technology. Han M, Phan D, Nighlinge, N, Taylor L, Jankhah S, Woodruff B, Yates Z, Freeman S, Guo A, Balland A, Pettit D. *Chromatographia.* 2006, 64: 335-342.

[107] Validatibility of a capillary isoelectric focusing method for impurity quantitation. Lasdun A, Kurumbali R, Leimgruber N, Rathore A. *J. Chromatogr A.* 2001, 917: 147-158.

[108] Rapid and high-resolution glycoform profiling of recombinant human erythropoietin by capillary isoelectric focusing with whole column imaging detection. Dou P, Liu Z, He J, Xu J, Chen H. *J. Chromatogr A.* 2008, 1190: 372-376.

[109] Recognition of oxytocin by capillary electrochromatography with monolithic tetrapeptide-imprinted polymer used as the stationary phase. Zheng C, Liu Z, Gao R, Zhang L, Zhang Y. *Anal. Bioanal. Chem.* 2007, 388: 1137-1145.

[110] Capillary electrophoresis and isoelectric focusing in peptide and protein analysis. Righetti P, Sebastiano R, Citerrio A. *Proteomics.* 2013, 13: 325.340.

[111] The potential of the capillary electrophoresis techniques for quality control of biopharmaceuticals–A review. Tamiz E, Jouyban A. *Electrophoresis.* 2015, 36: 831-858.

[112] Novel and highly sensitive mixed-polymeric electrokinetic chromatography system for determination of contaminants and impurities of heparin samples. Tripodi V, Flor S, Dobrecky C, Contin M, Lucangioli S. *Electrophoresis.* 2012, 31(21) 3606–3612.

[113] Van Eeckhaut A, Michotte Y. Chiral Separations by Capillary Electrophoresis, *Chromatographic science series.* 2009, Vol 100 Ed CRC Press, United States.

[114] The chiral switch: The development of single enantiomer drugs from racemates. Hutt A, Valentová J. *Acta Facult Pharm Univ. Comenianae.* 2003, 50:7-23.

[115] Fundamental aspects of chiral electromigration techniques and application in pharmaceutical and biomedical analysis. Scriba. *Journal of Pharmaceutical and Biomedical Analysis.* 2011, 55: 688–701.

[116] Advanced CE for chiral analysis of drugs, metabolites, and biomarkers in biological samples. Mikus P, Markova K. *Electrophoresis* 2009, 30: 2773–2802.

[117] Chiral selectors in CE: Recent developments and applications. Tsioupi R, Stefan-van Staden R, Kapnissi-Christodoulou C. *Electrophoresis* 2013, 34: 178–204.

[118] Enantioselective separation of rivastigmine by capillary electrophoresis with cyclodextrines. Lucangioli S, Tripodi V, Masrian E, Scioscia S, Carducci C, Kenndler E. *J. Chromatogr A.* 2005, 1081: 31–35.

[119] Cyclodextrins in Capillary Electrophoresis: Recent developments and new trends. Escuder-Gilabert L., Martın-Biosca M, Medina-Hernandez J, SagradoS. DOI: http://dx.doi.org/doi:10.1016/j.chroma.2014.05.074.

[120] Application of cyclodextrins in chiral capillary electrophoresis. Řezanka P, Navrátilová K, Řezanka M, Král V, Sýkora D. *Electrophoresis.* doi: 10.1002/elps.201400145.

[121] Use of Chiral Crown Ethers in Capillary Electrophoresis. Schmid M, Gübitz G. *Chiral Separations Methods in Molecular Biology.* 2004, 243: 317-321.

[122] Enantioseparation of amino acids on an 18-crown-6-tetracarboxylic acid-bonded silica by capillary electrochromatography. Leea T, Leeb W, Hyunc M, ParkaJ. *J. of Chromatogr A*. 2010, 1217: 1425–1428.
[123] Chiral selectors in CE: Recent developments and applications (2012-mid 2014). Stavrou I, Mavroudi M, Kapnissi-Christodoulou C. *Electrophoresis* 2015, 36: 101–123.
[124] Separation of basic drug enantiomers by capillary electrophoresis using ovoglycoprotein as a chiral selector: Comparison of chiral resolution ability of ovoglycoprotein and completely deglycosylated ovoglycoprotein. Matsunaga H, Haginaka J. *Electrophoresis*. 2001, 22: 3251-3256.
[125] Chiral resolution of tryptophan derivatives by CE using canine serum albumin and bovine serum albumin as chiral selectors. Hödl H, Koidl J, Schmid M, Gübitz G. Electrophoresis. 2006, 27:4755–4762.
[126] Chiral metal-ion complexes for enantioseparation by capillary electrophoresis and capillary electrochromatography: A selective review. SchmidM. *J. of Chromatogr A*. 2012,126: 710– 716.
[127] Recent progress in capillary electrophoretic analysis of amino acid enantiomers. Kitagawa F, Otsuka K. *J. Chromatogr B,* 2011, 879: 3078– 3095.
[128] Differentiation of Enantiomers by Capillary Electrophoresis. ScribaG. *Top Curr Chem*. 2013. DOI: 10.1007/128_2013_438.
[129] Analysis of *cis–trans* isomers and enantiomers of sertraline by cyclodextrin-modified micellar electrokinetic chromatography. Lucangioli S, Hermida L, Tripodi V, Rodrıguez V, Lopez E, Rouge P, Carducci N. *J. Chromatogra A*. 2000, 871: 207–215.
[130] Twenty-one years of microemulsion electrokinetic chromatography (1991–2012): A powerful analytical tool. YangH, DingY, CaoJ, LiP. *Electrophoresis*.2013, 34: 1273–1294.
[131] Recent advances of capillary electrophoresis in pharmaceutical analysis. SuntornsukL. *Anal. Bioanal. Chem*. 2010,398:29–52.
[132] Principles of Micellar Electrokinetic Capillary Chromatography Applied in Pharmaceutical Analysis, HancuG, SimonB, RusuA, MirciaE, GyéresiÁ. *Advanced Pharmaceutical Bulletin*.2013, 3(1): 1-8.
[133] Capillary electrophoresis application in the analysis of the anti-cancer drugs impurities, ZalewskaM, WilkK, Milnerowicz H. *Acta Poloniae Pharmaceutica drug research*. 2013, 70(2): 171 -180.
[134] Rapid analysis of alkylphosphonate drugs by capillary zone electrophoresis using indirect ultraviolet detection. Prutthiwanasan B, Suntornsuk L, *J. Sep. Sci*. 2010, 33(2):228-34.
[135] Simultaneous separation of eight beta-adrenergic drugs using titanium dioxide nanoparticles as additive in capillary electrophoresis, Zhou S, Wang Y, De Beer T, Baeyens WR, Fei GT, Dilinuer M, Ouyang J, *Electrophoresis*. 2008 Jun; 29(11):2321-9.
[136] Simultaneous determination of isoniazid and p-aminosalicylic acid by capillary electrophoresis using chemiluminescence detection. Zhang X, Xuan Y, Sun A, Lv Y, Hou X. *Luminescence*. 2009, 24(4):243-249.
[137] Optimization of an electrolyte system for analysis of ethambutol in pharmaceutical formulations by capillary zone electrophoresis using complexation with copper(II).

Faria A, de Souza M, Bruns R, de Oliveira M, *J. Chromatogr A.* 2008, 1202(2):224-228.

[138] A capillary electrophoretic method for fingerprinting low molecular weight heparins. King J, Desai U, *Anal. Biochem.* 2008, 380(2):229-234.

[139] Rapid analytical procedure for neomycin determination in ointments by CE with direct UV detection. Huidobro A, García A, Barbas C, *J. Pharm. Biomed. Anal.* 2009, 49(5):1303-1307.

[140] Quantitative determination of salbutamol in tablets by multiple-injection capillary zone electrophoresis, Lodén H, Pettersson C, Arvidsson T, Amini A, *J. Chromatogr A.* 2008, 1207(1-2):181-185.

[141] Development and implication of a capillary electrophoresis methodology for ciprofloxacin, paracetamol and diclofenac sodium in pharmaceutical formulations and simultaneously in human urine samples. Solangi A, Memon S, Mallah A, Memon N, Khuhawar M, Bhanger M, *Pak. J. Pharm. Sci.* 2011, 24(4):539-44.

[142] Capacitively coupled contactless conductivity detection as an alternative detection mode in CE for the analysis of kanamycin sulphate and its related substances. El-Attug M, Adams E, Hoogmartens J, Van Schepdael A. *J. Sep. Sci.* 2011, 34(18):2448-54.

[143] Optimization of capillary electrophoresis method with contactless conductivity detection for the analysis of tobramycin and its related substances. El-Attug M, Hoogmartens J, Adams E, Van Schepdael A. *J. Pharm. Biomed. Anal.* 2012, 58:49-57.

[144] Development and validation of a capillary electrophoresis method with capacitively coupled contactless conductivity detection (CE-C(4) D) for the analysis of amikacin and its related substances. El-Attug M, Adams E, Van Schepdael A. *Electrophoresis.* 2012, 33(17):2777-2782.

[145] Fast simultaneous determination of trimethoprim and sulfamethoxazole by capillary zone electrophoresis with capacitively coupled contactless conductivity detection. da Silva I, Vidal DT, do Lago C, Angnes L. *J. Sep. Sci.* 2013, 36(8):1405-1409.

[146] Determination of biapenem in a medicinal product by micellar electrokinetic chromatography with sweeping in an enhanced electric field. Michalska K, Cielecka-Piontek J, Pajchel G, Tyski S. *J. Chromatogr A.* 2013, 1282:153-60.

[147] Development of a screening method for analytical control of antibiotic residues by micellar electrokinetic capillary chromatography. Nozal L, Arce L, Rios A, Valcarcel M. *Anal. Chim. Acta.*2004, 523(1): 21-28.

[148] Adaptation of capillary electrophoresis to the determination of selected cephalosporins for injection. Pajchel G, Tyski S. *J. Chromatogr A.*2000. 895(1-2): 27-31.

[149] Micellar electrokinetic capillary chromatography of macrolide antibiotics. Separation of tylosin, erythromycin and their related substances. Tobback K, Li YM, Pizarro NA, De Smedt I, Smeets T, Van Schepdael A, et al. *J. Chromatogr A.*1999, 857(1-2): 313-20.

[150] Migration behavior and separation of tetracycline antibiotics by micellar electrokinetic chromatography. Chen Y, Lin C. *J. Chromatogr A.*1998, 802(1): 95-105.

[151] Rapid determination of sulfonamides in milk using micellarelectrokinetic chromatography with fluorescence detection. Lamba S, Sanghi S, Asthana A, Shelke M. *Anal. Chim. Acta.*2005, 552(1): 110-115.

[152] Simultaneous Separation of Ciprofloxacin, Norfloxacin and Ofloxacin by Micellar Electrokinetic Chromatography. Hancu G, RusuA, Simon B, Boia G, Gyéresi Á. *J. Braz. Chem. Soc.* 2012, 23(10): 1889-1894.

[153] Capillary electrokinetic chromatography of insulin and related synthetic analogues. Ortner K, Buchberger W, Himmelsbach M. *J. Chromatogr A.* 2009, 1216(14):2953-2957.
[154] Determination of paracetamol and its main impurity 4-aminophenol in analgesic preparations by micellar electrokinetic chromatography. Németh T, Jankovics P, Németh-Palotás J, Koszegi-Szalai H. *J. Pharm. Biomed. Anal.* 2008, 47(4-5):746-749.
[155] Analysis of oseltamivir in Tamiflu capsules using micellar electrokinetic chromatography. Jabbaribar F, Mortazavi A, Jalali-Milani R, Jouyban A. *Chem. Pharm. Bull.* 2008, 56(12):1639-1644.
[156] Determination of omeprazole, hydroxyomeprazole and omeprazole sulfonesing automated solid phase extraction and micellarelectrokineticcapillary chromatography. Pérez-Ruiz T, Martínez-Lozano C, Sanz A, Bravo E, Galera R. *J. Pharm. Biomed. Anal.* 2006, 42(1):100-106
[157] Simultaneous determination of triazole antifungal drugs in human plasma by sweeping-micellar electrokinetic chromatography. Lin SC, Liu HY, Lin SW, Yao M, Wu UI, Kuo HP, et al. *Bioanal. Chem.* 2012, 404(1): 217-28.
[158] Application of micellar electrokinetic capillary chromatography to forensic analysis of barbiturates in biological fluids. Ferslew K, Hagardorn A, McCormick W. *J. Forensic. Sci.*1995, 40(2):245-249.
[159] Micellar electrokinetic capillary chromatography of benzodiazepines in human urine, Schafroth M, Thormann W, Allemann D. *Electrophoresis.*1994, 15(1): 72-78.
[160] Enantioseparation of phenothiazines in cyclodextrin-modified micellar electrokinetic chromatography. Lin C, Chen K, Hsiao Y, Liao W, Chen C. *J. Chromatogr A.*2002, 971(1-2): 261-266.
[161] Determination of tricyclic antidepressants in human plasma by micellar electrokinetic capillary chromatography. Lee K, Lee J, Moon D. *J. Chromatogr.* 1993, 616(1): 135-43.
[162] Electrophoretic behaviour of pharmacologically active alkylxanthines. Blanco M, Valverde I. *J. Chromatogr A.*2002, 950(1-2): 293-299.
[163] Comparison ofthe retention characteristics of different pseudostationary phases for microemulsion and micellar electrokinetic chromatography of betamethasone and derivatives, Lucangioli S, Carducci C, Scioscia S, Carlucci A, Bregni C, Kenndler E. *Electrophoresis.* 2003, 24(6):984-991.
[164] Rapid determination of water- and fat-soluble vitamins with microemulsion electrokinetic chromatography. Yin C, Cao Y, Ding S, Wang Y. *J. Chromatogr A. 2008*, 1193(1-2):172-177.
[165] Simultaneous determination of natural and synthetic estrogens by EKC using a novel microemulsion. Tripodi V, Flor S, Carlucci A, Lucangioli S. *Electrophoresis.* 2006, 27(22):4431-4438.
[166] Application of MELC and MEEKC for the Analysis of Paracetamol and Related Impurities in Suppositories. McEvoy E, Donegan S, Power J, Altria K. *Chromatographia.* 2008, 68(1-2): 49-56.
[167] Separation of alpha-, beta-, gamma-, delta-tocopherols and alpha-tocopherol acetate on a pentaerythritoldiacrylatemonostearate-ethylene dimethacrylate monolith by capillary electrochromatography. Chaisuwan P, Nacapricha D, Wilairat P, Jiang Z, Smith N. *Electrophoresis.* 2008, 29(11):2301-2309.

[168] Nonaqueous capillary electrophoresis in pharmaceutical analysis. Geiser L, Veuthey J. *Electrophoresis*. 2007, 28(1-2):45-57.

[169] Recent progress in capillary electrophoresis of metal ions. Timerbaev A, Shpigun O. *Electrophoresis*.2000, 21:4179-91.

[170] Simultaneous determination of common inorganic anions, magnesium and calcium ions in various environmental waters by indirect UV-photometric detection ion chromatography using trimellitic acid-EDTA as eluent. Ohta K, Tanaka K. *Anal. Chim. Acta*.1998, 373:189-195.

[171] Eith C, Kolb M, Seubert A, Viehweger K. Practical ion chromatography. An introduction.2001. Ed Metrohm Ltd, Switzerland.

[172] Simultaneous separation and detection of anions and cations in ion chromatography. Nesterenko P. *Trends Analyt. Chem*. 2001, 20:311-319.

[173] Ion-exchange chromatographic analysis of soluble cations, anions and sugars in milk whey. Cataldi T, Angelotti M, D'Erchia L, Altieri G, Di Renzo G. *Eur. Food Res. Technol*. 2003, 216:75-82.

[174] Determination of common inorganic anions and cations by non-suppressed ion chromatography with column switching. Amin M, Lim L, Takeuchi T. *J. Chromatogr A*. 2008, 1182:169-75.

[175] Atomic absorption and atomic emission spectrometry for the determination of the trace element content of selected fruits consumed in the United States. Miller-Ihli N. *J. Food Compost Anal*.1996, 9:301-311.

[176] Capillary electrophoresis in pharmaceutical analysis: a survey on recent applications. Suntornsuk L. *J. Chromatogr Sci*. 2007, 45:559-577.

[177] Comparison of ion chromatography and capillary electrophoresis for the determination of inorganic ions. Haddad P. *J. Chromatogr A*. 1997, 770:281-90.

[178] Metal ion capillary zone electrophoresis with direct UV detection: determination of transition metals using an 8-hydroxyquinoline- 5-sulphonic acid chelating system. Timerbaev A, Buchberger W, Semenova O, Bonn G. *J. Chromatogr A*. 1993, 630:379-389.

[179] Recent advances and trends in capillary electrophoresis of inorganic ions. Timerbaev A. *Electrophoresis*. 2002, 23:3884-3906.

[180] Enhancement of detection sensitivity for indirect photometric detection of anions and cations in capillary electrophoresis. Johns C, Macka M, Haddad PR. *Electrophoresis,* 2003;24:2150-67.

[181] 13 ion analysis using capillary electrophoresis. del´Escaille F, Falmagne *J. Sep. Sci. Technol*. 2008, 9:317-355.

[182] Determination of potassium, sodium, calcium and magnesium in total parenteral nutrition formulations by capillary electrophoresis with contactless conductivity detection. Nussbaumer S, Fleury-Souverain S, Bouchoud L, Rudaz S, Bonnabry P, Veuthey J. *J. Pharm. Biomed. Anal.* 2010, 53:130-136.

[183] Application of capillary electrophoresis to determine metal cations, anions, organic acids, and carbohydrates in some Pinot Noir red wines. Rovio S, Sirén K, Sirén H. *Food Chemistry*.2011, 124:1194-1200.

[184] Metal ion capillary zone electrophoresis with direct UV detection: determination of transition metals using an 8-hydroxyquinoline- 5-sulphonic acid chelating system.

Timerbaev A, Buchberger W, Semenova O, Bonn G. *J. Chromatogr A*. 1993, 630:379-89.
[185] Determination of inorganic cations by capillary ion electrophoresis in Ilex paraguariensis (St. H.), a plant used to prepare tea in South America. Carducci C, Dabas P, Muse J. *J AOAC Int*. 2000, 83:1167-1173.
[186] Capillary electrophoresis for the analysis of small-molecule pharmaceuticals. Altria K, Marsh A, Sänger-van de Griend C. *Electrophoresis. 2006*, 27;2263-2282.
[187] Capillary electrophoresis of small ions using complex formation and indirect detection. Foret F. *Electrophoresis*.2009, 30:34-39.
[188] Separation and determination of cations in beverage products by capillary zone electrophoresis. Fung Y, Lau K. *J. Chromatogr A*. 2006, 1118:144-150.
[189] Simultaneous determination of inorganic and organic anions, alkali, alkaline earth and transition metal cations by capillary electrophoresis with contactless conductometric detection. Kubán P, Kubán P, Kubán V. *Electrophoresis*. 2002, 23:3725-3734.
[190] Simultaneous determination of Cr(III), Fe(III), Cu(II) and Pb(II) as UV-absorbing EDTA complexes by capillary zone electrophoresis. Baraj B, Martínez M, Sastre A, Aguilar M. *J. Chromatogr A*. 1995, 695:103-111.
[191] Simultaneous determination of chloride complexes of Pt(IV) and Pd(II) by capillary zone electrophoresis with direct UV absorbance detection. Baraj B, Sastre A, Martínez M, Spahiu K. *Anal. Chim. Acta*. 1996, 319:191-197.
[192] On-column complexation of metal ions using 2,6-pyridinedicarboxylic acid and separation of their anionic complexes by capillary electrophoresis with direct UV detection. Chen Z, Naidu R. *J. Chromatogr A*. 2002, 966:245-251.
[193] Capillary electrophoresis in pharmaceutical analysis: a survey on recent applications. Suntornsuk L. *J. Chromatogr Sci*. 2007, 45:559-577.
[194] Determination of sulfate in aminoglycoside antibiotics by capillary electrophoresis with indirect UV detection. Liu H, Sunderland B. *J Liq Chromatogr Relat Technol*. 2004, 24:677-687.
[195] Development and validation of a capillary electrophoresis method for the determination of sulfate in indinavir sulfate raw material. Pereira E, Micke G, Tavares M. *J. Brazilian Chem. Soc*. 2006, 17:251-256.
[196] Simultaneous analysis of metabisulfite and sulfate by CE with indirect UV detection. Application to and validation for a pharmaceutical formulation. Geiser L, Varesio E, Veuthey J. *J. Pharm Biomed. Anal*.2003, 31:1059-1064.
[197] Determination of the inorganic degradation products sulfate and sulfamate in the antiepileptic drug topiramate by capillary electrophoresis. Klockow-Beck A, Nick A, Geisshuesler S, Schaufelberger D. *J. Chromatogr B*. 1998, 720:141-151.
[198] Inter-company cross validation exercise on capillary electrophoresis. Quantitative determination of drug counter-ion level. Altria K, Clayton N, Harden R, Makwana J, Portsmouth M. *Chromatographia*. 1995, 40:47-50.
[199] Quantitative determination of drug counter-ion stoichiometry by capillary eletrophoresis. Altria KD, Goodall DM, Rogan MM. *Chromatographia* 1994, 38:637-642.
[200] Analysis of potassium counter ion and inorganic cation impurities in pharmaceutical drug substance by capillary electrophoresis with conductivity detection. Williams R, Boucher R. *J. Pharm. Biomed.* Anal.2000, 22:115-122.

[201] Validation of a capillary electrophoresis method for the analysis of ibandronate related impurities. Rodríguez J, Desimone M, Iglesias S, Giorgieri S, Diaz L. *J. Pharm. Biomed. Anal.*2007, 44:305-308.

[202] Zoledronate and related impurities analysis by capillary zone electrophoresis. Alvarez G, Foglia M, Camporotondi D, Giorgieri S, Desimone M, Díaz L. *Curr. Anal. Chem.* 2014, 10:231-234.

[203] Metallic ions as therapeutic agents in tissue engineering scaffolds: an overview of their biological applications and strategies for new developments. Mouriño V, Cattalini J, Boccaccini A. *J. R. Soc. Interface.* 2012, 9:401-419.

[204] Angiogenesis in calcium phosphate scaffolds by inorganic copper ion release. Barralet J, Gbureck U, Habibovic P, Vorndran E, Gerard C, Doillon CJ. *Tissue Eng. Part A.* 2009, 15:1601-1609.

[205] Bioactive glass and glass-ceramic scaffolds for bone tissue engineering. Gerhardt L, Boccaccini A. *Materials.*2010, 3:3867-3910

[206] A review of the biological response to ionic dissolution products from bioactive glasses and glass-ceramics. Hoppe A, Güldal N, Boccaccini A. *Biomaterials.* 2011, 32:2757-2774.

[207] Physicochemical, biological and drug-release properties of gallium crosslinked alginate/nanoparticulate bioactive glass composite films. Mouriño V, Newby P, Pishbin F, Cattalini J, Lucangioli S, Boccaccini A. *Soft Matter.* 2011, 7:6705-1

[208] Copper-releasing, boron-containing bioactive glass-based scaffolds coated with alginate for bone tissue engineering. Erol M, Mouriño V, Newby P, Chatzistavrou X, Roether J, Hupa L, Boccaccini A. *Acta Biomater.* 2012, 8:792-801.

[209] The effect of calcium ion concentration on osteoblast viability, proliferation and differentiation in monolayer and 3D culture. Maeno S, Niki Y, Matsumoto H, Morioka H, Yatabe T, Funayama A, Toyama Y, Taguchi T, Tanaka J. *Biomaterials.*2005, 26:4847-4855.

[210] The calcium-sensing receptor in bone cells: A potential therapeutic target in osteoporosis. Marie P. *Bone.* 2010, 46:571-576.

[211] Development and validation of a capillary zone electrophoresis method for the determination of calcium in composite biomaterials. Cattalini J, García J, Mouriño V, Lucangioli S. *Curr. Anal. Chem.* 2014, 10:465-472.

[212] The stimulation of angiogenesis and collagen deposition by copper. Gérard C, Bordeleau L, Barralet J, Doillon C. *Biomaterials.* 2010, 31:824-831.

[213] Copper-containing mesoporous bioactive glass scaffolds with multifunctional properties of angiogenesis capacity, osteostimulation and antibacterial activity. Wu C, Zhou Y, Xu M, Han P, Chen L, Chang J, Xiao Y. *Biomaterials.* 2013, 34:422-433.

In: Capillary Electrophoresis (CE)
Editor: Christian Reed

ISBN: 978-1-63483-122-2

Chapter 4

ON-LINE ELECTROPHORETIC-BASED PRECONCENTRATION METHODS IN CAPILLARY ZONE ELECTROPHORESIS: PRINCIPLES AND RELEVANT APPLICATIONS

Oscar Núñez*

Department of Analytical Chemistry, University of Barcelona,
Martí i Franquès, Barcelona, Spain
Serra Húnter Fellow, Generalitat de Catalunya, Spain

ABSTRACT

Capillary electrophoresis (CE) comprises a family of related separation techniques in which an electric field is used to achieve the separation of components in a mixture. Electrophoresis in a capillary is differentiated from other forms of electrophoresis in that separation is carried out within the confines of narrow-bore capillaries, from 20 to 200 μm inner diameter (i.d.), which are usually filled only with a solution containing electrolytes (typically, although not always necessary, a buffer solution). One of the key features of CE is the simplicity of the instrumentation required, and today this technique allows working in various modes of operation. Among them, capillary zone electrophoresis (CZE) is the most widely used due to its simplicity of operation and its versatility. The use of high electric fields results in short analysis times and high efficiency and resolution. In addition, the minimal sample volume requirement (in general few nanoliters), the on-capillary detection, the potential for both qualitative and quantitative analysis, the automation, and the possibility of hyphenation with other techniques such as mass spectrometry (MS) is allowing CZE to become one of the premier separation techniques in multiple fields, such as bio-analysis, food safety and environmental applications.

However, one of CZE handicaps is sensitivity due to the short path length (capillary inner diameter) when on-capillary detection is carried out, and the low amount of samples injected. For these reasons, many CZE applications will require of off-line and/or on-line preconcentration methods in order to improve limits of detection (LOD). Many different

* Corresponding author: oscar.nunez@ub.edu.

techniques have been developed to improve LODs in CZE. Among them, on-line electrophoretic-based preconcentration techniques are becoming very popular because no special requirement but a CE instrument is necessary for their application. These on-line preconcentration methods are designed to compress analyte bands within the capillary, thereby increasing the volume of sample that can be injected without losing separation efficiency. So, these methods are based on the principle of stacking analytes in a narrow band between two separate zones in the capillary where the compounds have different electrophoretic mobilities (for instance at the boundary of two buffers with different resistivities).

This chapter will address the principles of on-line electrophoretic-based preconcentration methods in capillary zone electrophoresis. Coverage of all kind of on-line electrophoretic-based preconcentration methods is beyond the scope of the present contribution, so we will focus on the most frequently used in CZE such as sample stacking, large-volume sample stacking (LVSS), field-amplified sample injection (FASI), pH-mediated sample stacking, and electrokinetic supercharging (EKS). Relevant applications of these preconcentration methods in several fields (bio-analysis, food safety, environmental analysis) will also be presented.

1. Introduction

1.1. Capillary Electrophoresis

Electrophoresis is the movement of dispersed particles and molecules relative to a fluid under the influence of a spatially uniform electric field. This electrokinetic phenomenon was observed for the first time in 1807 by Ferdinand Frederic Reuss who noticed that the application of a constant electric field caused clay particles dispersed in water to migrate [1]. It is ultimately caused by the presence of a charged interface between the particle surface and the surrounding fluid. Capillary electrophoresis (CE) is electrophoresis performed in a capillary tube [2, 3]. Electrophoretic separation of molecules in a glass tube and subsequent detection of the separated compounds by ultraviolet absorption was first described by Hjerten in the late 1960s [4, 5]. In these first works, separation of serum proteins, inorganic and organic ions, peptides, nucleic acids, viruses, and bacteria was described. The transformation of conventional electrophoresis to modern CE was spurred by the production of inexpensive narrow-bore capillaries for gas chromatography (GC) and the development of highly sensitive on-line detection methods for high performance liquid chromatography (HPLC). So, electrophoresis in a capillary is differentiated from other forms of electrophoresis in that it is carried out within the confines of narrow-bore capillaries, from 20 to 200 μm inner diameter (i.d.), which are usually filled only with a solution containing electrolytes (typically, although not always necessary, a buffer solution). The use of capillaries has numerous advantages, particularly with respect to the detrimental effects of Joule heating when applying an electric field through a fluid. Capillaries were introduced into electrophoresis as an anti-convective and heat controlling innovation. The high electrical resistance of capillaries enables the application of very high electrical fields (100 to 500 V/cm) with only minimal heat generation. In wide tubes thermal gradients caused band mixing and loss or resolution, however, with narrow-bore capillaries the large surface area-to-volume ratio efficiently dissipates the heat that is generated. However, electrophoresis in a tube did not become popular until early 1980s, when Jorgenson and Lucaks demonstrated the high resolution power of capillary zone electrophoresis [6, 7]. In fact, the introduction of 1981 of 75 μm i.d. capillary tubes by Jorgensen and Lukacs was the beginning of what is today known as modern high-performance CE (HPCE).

Capillary electrophoresis comprises today a family of related separation techniques in which an electric field is used to achieve the separation of components in a mixture. This family of techniques includes capillary zone electrophoresis (CZE), micellar electrokinetic capillary chromatography (MECC), capillary gel electrophoresis (CGE), capillary isoelectric focusing (CIEF), capillary isotachophoresis (CITP), none-aqueous capillary electrophoresis (NACE), and capillary electrochromatography (CEC), among others. In addition to this numerous separation modes in CE which offer various separation mechanisms and selectivities, the minimal sample volume requirements in all these techniques (in general few nanoliters, nL), the on-capillary detection, the potential for both qualitative and quantitative analysis, the automation, and the possibility of hyphenation with other techniques such as mass spectrometry (MS), is allowing CE to become one of the premier separation techniques in multiple fields. Thus, today capillary electrophoresis is emerging as an alternative technique for multiple application fields.

One of the most important features of CE is the simplicity of the instrumentation configuration used. Figure 1 shows a scheme of the components of a typical CE instrument. The basic instrumentation set-up consists of a controllable high voltage power supply (0 to 30 kV), a narrow-bore fused silica capillary with an optical viewing window, two electrolyte reservoirs, two electrodes, and an on-column detection system (typically an ultraviolet (UV) detector). Both ends of the capillary are placed in the electrolyte reservoirs that contain, in general, the same electrolyte solution that is filling the capillary. The electrodes used to make electrical contact between the high voltage power supply and the capillary are also immersed in the electrolyte reservoirs.

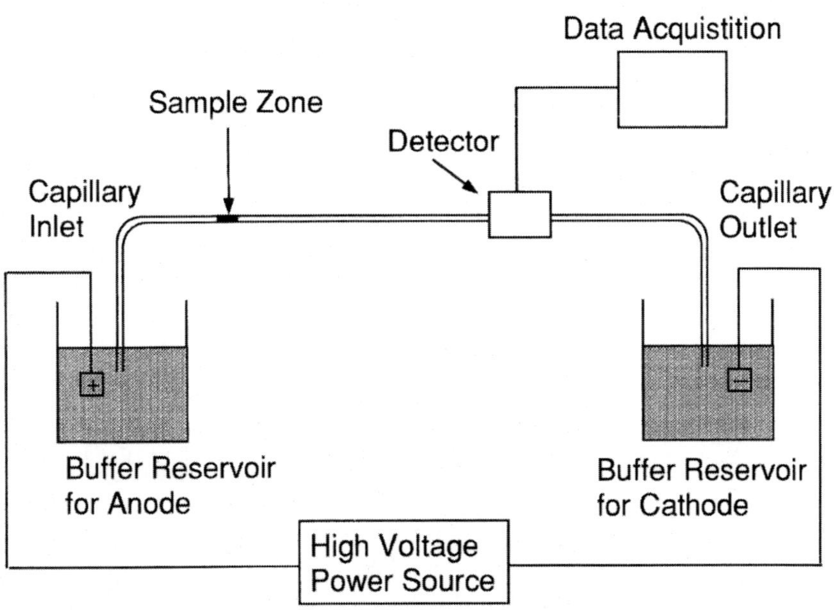

Figure 1. Scheme of the main components of a typical CE instrument.

After filling the capillary with the electrolyte solution, sample injection is accomplished by temporarily replacing the inlet electrolyte reservoir with a sample vial. A specific amount of sample, typically few nL as previously commented, is then introduced into the capillary by controlling either the injection pressure or the injection voltage. The optical window viewing included in the capillary is aligned with the detector, and on-column detection is carried out directly through the capillary close to the outlet capillary end.

Since the appearance in the marked of the first commercial CE instrument in the late 1980s, many advances and applications have taken place using this technique with tremendous impact on the progress of science. The characteristics of CE resemble a cross between traditional polyacrylamide gel electrophoresis and modern HPLC. Among them we can find:

- electrophoretic separations are performed in narrow-bore fused silica capillaries
- utilization of very high electric voltages (10 to 30 kV) generating high electric field strengths, often higher than 500 V/cm
- the high resistance of the capillary limits the current generated and the internal heating
- high efficiency (theoretical plates $N > 10^5$ to 10^6) on the order of capillary gas chromatography or even greater, with, in general, short analysis times
- relatively small sample requirement (1 to 50 nL injected) under conventional conditions
- easy automation for precise quantitative analysis and very easy to use
- limited consumption of reagents and, in general, operates in aqueous media
- presents numerous modes of operation to change selectivity and is applicable to a wider selection of analytes compared to other analytical separation techniques
- simple method development and automated instrumentation
- possibility of using different CE modes with the same commercial instrumental set-up
- compatible with multiple detection systems
- hyphenation to other techniques, such as mass spectrometry (CE-MS)

All these features make today CE one of the most promising separation techniques and it is being used in multiple application fields, such as in bio-analysis [8-12], cosmetic industry [13-17], food control and safety [18-23], or environmental applications [24-26] among others.

2. Capillary Zone Electrophoresis

Capillary zone electrophoresis (CZE) is one of the most widely used modes in CE due to its simplicity of operation and its versatility. In CZE the capillary is only filled with a simple electrolyte solution (usually a buffer solution), and the separation mechanism is based on the differences in the charge-to-mass ratio of the analytes. This mode is characterized for the homogeneity of the buffer solution and the constant field strength generated throughout the length of the capillary. After the sample is introduced into the capillary and the application of a capillary voltage, the analytes of a mixture separate into discrete zones at different velocities. Separation of both anionic and cationic species is possible by CZE; however neutral analytes cannot be separated. The principles of CZE separation are explained bellow.

2.1. Fundamentals of Capillary Zone Electrophoresis

It has long been known that molecules can be either positively or negatively electrically charged. When the numbers of positive and negative charges are the same, the charges cancel, creating a neutral (uncharged) molecule. Charged molecules in a solution will move under the effect of an electric field seeking the electrode with opposite charge. Cations (positively charged ions) move toward the cathode (negatively charged electrode), and anions (negatively charged ions) move toward the anode (positively charged electrode). This is the main fundamental of CZE separations, which is based on the different velocity experimented by charged analytes through the capillary under the application of an electric field. There are few significant differences between the nomenclature used in chromatography and capillary electrophoresis. For instance, a basic term in chromatography is the retention time (t_r). In CZE, under ideal conditions, nothing is retained, so the analogous term becomes migration time (t_m). Thus, the migration time is the time it takes a molecule to move from the beginning of the capillary to the detection window (point in the capillary where the on-column detection is carried out). If CZE is hyphenated with other techniques such as MS (CE-MS), the migration time will be the time it takes this molecule to move from the beginning to the end of the capillary. This migration time will depend on the migration velocity (v_m) of an analyte under the application of an electric field of intensity E.

In CZE, the migration velocity of a given analyte is determined by (i) the electrophoretic mobility of the analyte and (ii) the electroosmotic mobility of the electrolyte solution inside the capillary.

(i) Electrophoretic Mobility

The electrophoretic mobility of an analyte (μ_{ep}) depends on the characteristics of the analyte (electric charge, molecular size and shape) and those of the background electrolyte (BGE) solution in which the migration takes place (type and ionic strength of the electrolyte, pH, viscosity and presence of additives). Thus, the electrophoretic velocity (v_{ep}) of an analyte, assuming a spherical shape, is given by the next equation:

$$v_{ep} = \mu_{ep} \times E = \left(\frac{q}{6\pi\eta r}\right) \times \left(\frac{V}{L}\right)$$

where q is the effective charge of the analyte, η is the viscosity of the electrolyte solution, r is the Stoke's radius of the analyte, V is the applied voltage, and L is the total length of the capillary. So, the electrophoretic mobility is a constant value characteristic of the ion in a given medium. Small and highly charged molecules will have higher electrophoretic mobilities than large and minimally charged species.

In addition to the capillary voltage and the capillary length that directly influence the ion velocity, other physical parameters can also play an important role in the electrophoretic separation by indirectly modifying the electrophoretic mobility of a given ion, and the analysts can play an important role in creating media that exploit these differences between the molecules of a mixture to achieve their separation. For instance, the BGE pH will play an important role on the separation of molecules with acid-base properties. The temperature at which the separation will be performed can also affect the electrophoretic mobility of a given ion because it will directly modify the viscosity of the BGE.

(ii) Electroosmotic Mobility

WHEN an electric field is applied through a capillary filled with a BGE, a flow of solvent is generated inside the capillary. This phenomenon is known as electroosmotic, electroendoosmotic flow, or simply electroosmotic flow (EOF). This phenomenon occurs whenever the liquid near a charged surface is placed in an electrical field resulting in the bulk movement of fluid near that surface.

Generally, fused silica capillaries are employed in CZE, as well as other CE techniques. These capillaries have ionizable silanol groups in contact with the BGE solution within the capillary. The isoelectric point of fused silica capillaries is difficult to determine although it is considered to be close to 1.5, so its degree of ionization will be mainly controlled by the BGE pH. So, at a certain pH value the surface of a fused silica capillary can be hydrolyzed to yield a negatively charged surface as described in Figure 2. The negatively charged wall will attract positively charged ions that are hydrated from the BGE solution up near the surface to maintain charge balance, creating an electrical double layer, and consequently a potential difference very close to the wall, known as zeta potential (ξ). When an electric field is generated through the capillary by the application of a voltage, these hydrated cations forming the called diffuse double-layer migrate toward the cathode, pulling water along and creating a pumping action. The result is a bulk flow of BGE solution through the capillary towards the cathode (EOF). Because the surface to volume ratio is very high inside a capillary, EOF becomes a very significant factor in CZE.

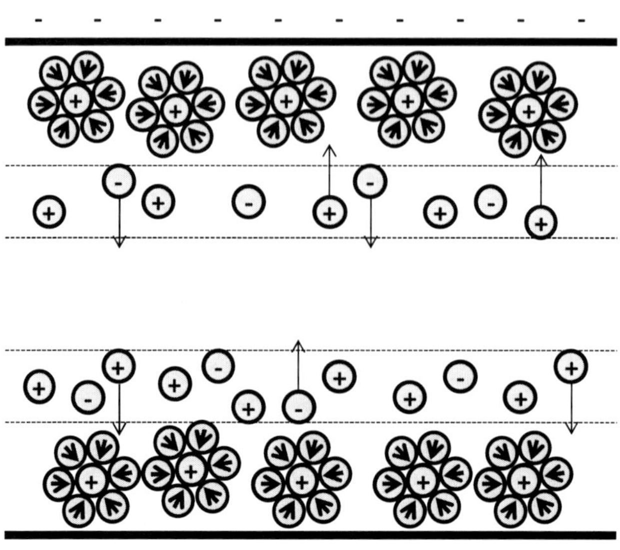

Figure 2. Scheme of the electric double layer generated within the capillary in CZE.

The velocity of the electroosmotic flow depends on the electroosmotic mobility (μ_{eo}) which in turns depends on the charge density on the capillary internal wall and the BGE characteristics. The electroosmotic velocity (v_{eo}) is given by the Smoluchowski [27] equation:

$$v_{eo} = \mu_{eo} \times E = \left(\frac{\varepsilon\xi}{4\pi\eta}\right) \times \left(\frac{V}{L}\right)$$

where ε is the dielectric constant of the BGE solution and ξ is the zeta potential generated in the capillary wall surface. The zeta potential is related to the inverse of the charge per unit surface area, the number of valence electrons, and the square root concentration of the electrolyte. Since this is an inverse relationship, increasing the concentration of the electrolyte, that is increasing the ionic strength of the BGE, results in a double-layer compression, a decrease in zeta potential and a reduction in EOF velocity.

In fused silica capillaries, the charge density on the capillary surface will change with the BGE pH. At high pH values, where silanol groups are predominantly deprotonated, the EOF is significantly greater than at low pH values where they become protonated. Depending on the conditions, it is possible to achieve EOF variations by more than one order of magnitude between pH value of 2 and 12 (Figure 3). Although it is difficult to completely suppressed EOF in fused silica capillaries, it can be considered close to zero at pH values bellow 2-2.5.

It should be mention that EOF direction is always toward the electrode that has the same charge as the capillary wall. For this reason, when using fused silica capillaries a cathodic EOF is generated. But other types of positively charged or even non-charged capillaries are available.

The electrophoretic mobility of the analyte and the electroosmotic mobility may act in the same direction or in opposite directions, depending on the charge of the analyte. In what is usually known as normal capillary zone electrophoresis, anions will migrate in the opposite direction to the electroosmotic flow and their velocities will be smaller than the electroosmotic velocity. Cations will migrate in the same direction as the electroosmotic flow and their velocities will be greater than the electroosmotic velocity. Under conditions in which there is a fast electroosmotic velocity (for instance when using BGE solutions with high pH values, Figure 3) with respect to the electrophoretic mobility of the solutes, both cations and anions can be detected in the same run.

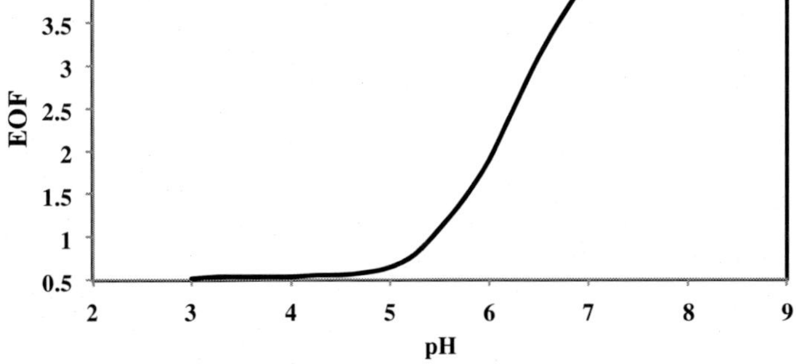

Figure 3. Effect of pH on electroosmotic flow in capillary zone electrophoresis.

Thus, the migration time (t_m) of an analyte is given by the equation:

$$t_m = \frac{l}{v_{ep} + v_{eo}} = \frac{l \times L}{(\mu_{ep} + \mu_{eo}) \times V}$$

where l is the distance from the injection end of the capillary to the detection point (capillary effective length). It is important to see that there are two different capillary lengths, the capillary effective length (l) and the total capillary length (L), and both must be controlled since the migration time and mobility are defined by the effective length, whereas the electric field is a function of the total capillary length. In general, when on-capillary detection is performed in CZE, the effective length is typically 5 to 10 cm (depending on the instrument set up) shorter than the total length. In contrast, when off-column detection is used such as in the case of CE-MS techniques the two lengths are equivalent.

So, the combination of both electrophoretic mobility and EOF will determine the migration time and, consequently, the migration order in capillary zone electrophoresis, as it is represented in Figure 4.

As can be seen in the figure, when working under a cathodic voltage (cathode in the outlet end of the capillary), the cations will migrate first from the capillary with a migration velocity $v_m = v_{ep} + v_{eo}$ with both factors contributing in the cathodic direction. The cation migration order will then depend on their specific electrophoretic mobility and, as previously described, this will be higher for lower ion size and higher ion charge. Then, neutral species will migrate from the capillary but they will not be separated because all of them will be moving at the EOF velocity ($v_m = v_{eo}$, for any non-charged molecule). Finally, the anions will migrate with a migration velocity $v_m = v_{eo} - v_{ep}$ (their electrophoretic velocity will be in the opposite direction than EOF). But only if $v_{eo} > v_{ep}$ the ions will be detected under a cathodic separation such as the one described in Figure 4 (mode known as counter-electroosmotic flow separation for the analysis of anions). Again, the migration order of these anions will depend on the magnitude of the electrophoretic mobility. In this case, anions with lower ion charge and higher ion size will migrate first from the capillary because their electrophoretic mobility will be opposing in a lower magnitude to the EOF than anions with higher ion charge and lower ion size.

Figure 4. Schematic representation of the migration order of cations (C), anions (A) and neutral (N) analytes in capillary zone electrophoresis when working with electroosmotic flow and under cathodic voltage conditions.

After the introduction of the sample into the capillary (that will be commented on the next section) each analyte ion of the sample migrates within the BGE as an independent zone, according to its electrophoretic mobility. Zone dispersion, that is the spreading of each solute band, results from different phenomena. Under ideal conditions the sole contribution to the solute-zone broadening is molecular diffusion of the solute along the capillary (longitudinal diffusion). In this ideal case, the efficiency of the analyte, expressed as the number of theoretical plates (*N*), is given by the next equation:

$$N = \frac{(\mu_{ep} + \mu_{eo}) \times V \times l}{2 \times D_m \times L}$$

where D_m is de molecular diffusion coefficient of the analyte in the BGE solution. However, in practice, other phenomena such as heat dissipation, sample adsorption onto the capillary wall, mismatched conductivity between sample and BGE solution, length of the injection plug, detector cell size and unleveled BGE reservoirs can also significantly contribute to band dispersion.

The use of high voltages will also provide for the greatest efficiency by decreasing the separation time. Today, the practical voltage limit in commercially available CE instruments is about 30 kV. Nevertheless, the practical limit of the field strength (very short capillaries can be used to generate high field strengths) is Joule heating. Joule heating is a consequence of the resistance of the BGE solution to the flow of current.

So, separation between two analytes can be obtained either by modifying the electrophoretic mobility of the analytes, the electroosmotic mobility induced in the capillary and by increasing the efficiency for the band of each analytes. The resolution (R_s) in CZE between two analytes can be calculated by the next equation:

$$R_s = \frac{\sqrt{N}(\mu_{ep,b} - \mu_{ep,a})}{4(\overline{\mu_{ep}} + \mu_{eo})}$$

where $\mu_{ep,a}$ and $\mu_{ep,b}$ are the electrophoretic mobilities of the two separated analytes and $\overline{\mu_{ep}}$ is the mean electrophoretic mobility of the two analytes:

$$\overline{\mu_{ep}} = \frac{(\mu_{ep,a} - \mu_{ep,b})}{2}$$

2.2. Sample Injection in Capillary Zone Electrophoresis

A very important operational aspect in CZE is the introduction of the sample into the capillary. The dimensions employed in CE are much smaller than those with which most chemical analysts are accustomed to work. Capillary internal diameters are typically in the micron scale and as such the entire volume of a capillary is usually a few microliters. Thus, in any conventional CE mode of operation only small volumes of the sample are loaded into the capillary (few nanoliters). However, because of the small volume of the capillaries used in CZE, the injection plug length is a more critical parameter than the sample volume in order to prevent sample overloading. If injection lengths longer than the diffusion controlled zone width are used, peak width broadening will be proportionally observed. Moreover, increasing injection lengths resulted in distorted peak shapes caused by mismatched conductivity between the BGE solution and the sample zones. So, sample overloading will have a detrimental effect on the resolution [28]. For this reason, as a rule, sample plug lengths lower than 1 to 2% of the total length of the capillary are typically used, which means an injection length of a few millimeters (sample volumes of 1 to 50 nL), depending on the capillary length and inner diameter.

Although this is sometimes considered one of the main advantages of CZE, because small amounts of samples are required since 5 μL of sample will be enough to perform several injections, for some applications will be one of the most important handicaps of this technique because of the decrease in sensitivity. So, for numerous CZE applications off-column and on-column preconcentration methods will be required.

Generally, sample introduction into the capillary in CZE can be accomplished by two main methods: (i) hydrodynamic injection and (ii) electrokinetic injection.

(i) Hydrodynamic Injection

Hydrodynamic injection is the most widely used method to accomplish the introduction of a small sample volume into the capillary. Although three basic strategies are available for that purpose: (i) the application of a positive pressure at the injection end of the capillary (inlet vial); (ii) the application of a vacuum at the exit end of the capillary (outlet vial); and (iii) by gravity (or siphoning action) obtained by inserting the inlet end of the capillary into the sample vial and raising the vial and capillary relative to the outlet end, in most of the cases hydrodynamic injections are accomplished by the application of a pressure difference between the two ends of the capillary. The amount of sample injected can be calculated by using the Hagen-Poiseuille equation:

$$V_c = \frac{\Delta P \pi d^4 t}{128 \eta L}$$

where V_c is the calculated injection volume, ΔP is the pressure difference between the two ends of the capillary, d is the inner diameter of the capillary, t is the injection time, η is the sample viscosity, and L is the total length of the capillary. Entering the Hagen-Poiseuille equation into a spreadsheet program simplifies fluid delivery calculations such as these. A free-available computer program called "CE expert" from Beckman Coulter Inc., can easily help to perform these calculations [29].

(ii) Electrokinetic Injection

Electrokinetic injection does not conform to the Hagen-Poiseuille equation. Usually it is performed by inserting the inlet end of the capillary into the sample vial and the outlet end into a BGE vial, and turning on the capillary voltage for a certain period of time. Through a combination of both electrophoretic migration and the pumping action of the EOF, sample is drawn into the capillary. Usually, field strengths 3 to 5 times lower than the one used for separation are employed.

In contrast to hydrodynamic injection, in electrokinetic injection the quantity of analyte loaded into the capillary is dependent on the electrophoretic mobility of the individual compound, so discrimination occurs for ionic species since compounds that migrate more rapidly in the electrical field will be over-represented in the sample introduced into the capillary compared to slower moving components [30]. The quantity of a given analyte injected into the capillary by electrokinetic injection, Q, can be calculated using the next equation:

$$Q = \frac{(\mu_{ep} + \mu_{eo})V\pi r^2 Ct}{L}$$

where μ_{ep} is the electrophoretic mobility of the analyte, μ_{eo} is the electroosmotic flow mobility, V is the capillary voltage applied during injection, r is the capillary radius, C is the analyte concentration in the sample, t is the injection time, and L is the total length of the capillary.

As described by this equation, sample loading is dependent of the EOF, the sample concentration, and analyte mobility. Variations in conductivity, which can be due to matrix effects such as the presence of ions (sodium, chloride…) could result in differences in the voltage drop and the quantity loaded [31]. Because of this, electrokinetic injection is generally not as reproducible as hydrodynamic injection. Despite quantitative limitations, electrokinetic injection is very simple, requires no additional instrumentation, and is advantageous when viscous media, or gels, are employed in the capillary, and when hydrodynamic injection is ineffective, for instance in order to increase sensitivity although it will always be analyte dependent.

2.3. Detection in Capillary Zone Electrophoresis

Most of the CZE detection is carried out on-capillary which means that a section of the capillary is linked to the detector and the capillary itself is the detection cell. Obviously, it is also possible to couple CE systems to detectors that are outside of the separation capillary or other systems such as mass spectrometry (CE-MS).

Absorbance-based detectors are the most commonly used in CE instruments. They rely on the absorbance of light energy by the analytes. This absorbance creates a shadow as the analytes pass between the light source and the light detector. The intensity of the shadow is proportional to the amount of analyte present. For absorptive detectors the absorbance of an analyte (A) is described by the Beer's law:

$$A = \varepsilon bC$$

where ε is the molar absorptivity of the analyte, b is the path length (capillary inner diameter in CZE), and C is the analyte concentration.

Detection through the capillary is complicated due to the curvature of the capillary itself [32]. The capillary and the fluid it contains make up a complex cylindrical lens. The curvature of this lens must be accounted for in order to gather the maximum amount of light and thereby maximize the signal-to-noise ratio. Generally, the effective length of the light path through the capillary is about 63.5% the stated capillary internal diameter. Thus, a 50 μm i.d. capillary has an effective path length of only 32 μm. In comparison to typical HPLC UV detectors that have light paths in the 5-10 nm range, the absorbance signal obtained in CE systems is very small, and a peak with an absorbance of 0.002 AU is a significant peak.

Photodiode array (PDA) detectors are a good alternative to single wavelength detection. These detectors consist on an achromatic lens system to focus the entire spectrum of light available from the source lamp into the capillary window. The light passing the capillary is diffracted into a spectrum that is projected on a linear array of photodiodes. An array consists of numerous diodes each of which is dedicated to measuring a narrow-band spectrum. In this manner it is possible to record the entire absorbance spectrum of analytes as they pass by the detection window. One of the advantages of using this kind of detector is that allows confirming the identity of analytes by using the spectral signature. Moreover, by comparing the change in spectra signature across the electrophoretic peak on an analyte it is also possible to estimate the analyte peak purity.

3. On-Line Electrophoretic-Based Preconcentration Methods in Capillary Zone Electrophoresis

Today, the benefits from the high number of theoretical plates obtained with CZE have been overshadowed by the poor sensitivity, and consequently the low limits of detection, achieved with this technique when UV detection is employed. As previously commented, due to the small dimensions of CZE capillaries, typically 20-200 μm I.D. and capillary lengths (40-80 cm in most of the applications), only very small sample volumes may be loaded into the capillary. For instance, for conventional 50 μm I.D. x 50 cm total length capillary only 1.18 nL/s of sample are introduced into the capillary when hydrodynamic injection (0.5 psi) is performed. Additionally, for the most common optical detection techniques, CZE suffers from a drastically reduced path length as compared to LC techniques. Since absorbance is directly proportional to path length and concentration, the concentration of the samples must be dramatically increased to obtain the same signal-to-noise ratio as would result from a typical LC analysis. Thus, for trace analysis applications, the amount of analytes injected into the capillary or the detector sensitivity must be increased. The latter aspect may be accomplished by utilizing light paths in connection with UV detection, or alternatively by using more sensitive detectors such as laser-induced fluorescence (LIF) detection [33-36] or even CE-MS techniques [18, 37-40]. Both bubble cells and z-shaped cells have been utilized as extended light paths for UV detection, which typically provide an enhancement of the signal-to-noise ratio by a factor of 3-6 [41], although this is not enough for some specific applications requiring trace analysis. High mass sensitivity has been reported with LIF detection but one of its handicaps is that is only applicable for some analytes (those with fluorescence properties) and the number of wavelengths available with commercial LIF detectors is limited. Regarding MS, CE-MS techniques are frequently used to increase sensitivity, but this coupling will require specialized interfaces and will not be addressed in this chapter.

The most convenient approach to improve sensitivity in CZE is to increase the amount of analyte injected into the capillary. This approach does not require any special instrument set-up configuration and, for this reason, is one of the most frequently proposed to improve sensitivity in CZE when UV detection is used.

A number of techniques have been developed to preconcentrate samples and to increase the amount of analytes that can be loaded onto the column without degrading the separation. Obviously, this may be accomplished by analyte enrichment during the sample preparation step, by using off-line extraction and/or preconcentration methods such as liquid-liquid extraction (LLE) [42] or solid phase extraction (SPE) [43-45]. Another approach involves increasing the amount of analyte introduced into the capillary tube by on-column electrophoretic-based preconcentration methods. An interesting review describing recent applications of on-line sample preconcentration techniques in capillary electrophoresis have been recently published [46]. These methods involve manipulating the electrophoretic velocity of the analyte during injection and separation on CZE, and include techniques such as normal sample stacking, large-volume sample stacking (LVSS), field-amplified sample injection (FASI), pH-mediated sample stacking, and electrokinetic supercharging (EKS). The principles of each method and some relevant applications will be discussed in the next sections.

3.1. Normal Sample Stacking

Most of the on-line electrophoretic-based preconcentration methods in CZE are based on the principle of stacking analytes in a narrow band between two separate zones in the capillary where the compounds have different electrophoretic velocities [47]. The simplest way of achieving this is by normal sample stacking, which is based on the differences of conductivity between the sample region and the BGE solution. For that purpose, sample is prepared in a matrix with lower ionic strength (and then lower conductivity) than that of the BGE solution. When a voltage is applied between both ends of the capillary, the field strength generated in the sample region will be higher than the one on the BGE region and, consequently, the analytes will have a higher electrophoretic velocity in the sample region than in the BGE region. The result is that analytes will stack-up in the boundary between the sample zone and the BGE due to their huge decrease in electrophoretic velocity when going from the sample region to the BGE region, as it is illustrated in Figure 5.

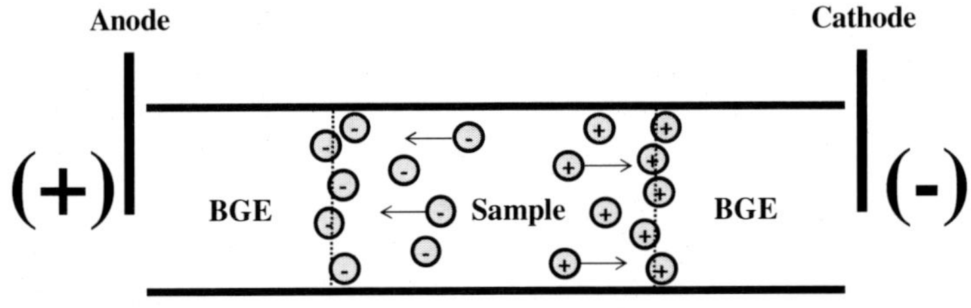

Figure 5. Scheme of normal sample stacking in CZE. Sample constituted of a matrix of lower ionic strength than that of BGE solution.

In the most simple form of normal sample stacking, the sample matrix is usually water or an electrolyte solution (buffer solution) of the same nature than the BGE used for the separation but at lower concentration, although it must be commented that the use of organic solvents can also help in improving sensitivity. For instance, acetonitrile is also frequently used because its resistivity helps in the stacking process (phenomenon known as acetonitrile stacking). For example, it was reported that the use of acetonitrile as sample solvent allowed a 10-fold improvement in sensitivity on the determination of procainamide and its metabolite n-acetylprocainamide in serum samples by CZE [48].

This simple on-line electrophoretic-based preconcentration method allow to increase the sample volume introduced into the capillary up to filling 10 to 20% of the total capillary volume without losing efficiency. However, the volume of sample cannot be increased indiscriminately because although an improvement on sensitivity will be achieved, peak resolution will be negatively affected not only due to the peak width broadening previously commented (see section 2.2) but also for the reduction on the effective length of the capillary used for the separation.

As an example of normal sample stacking, Heller et al. [49] developed a sample screening method for the authenticity control of whiskey using CZE with several on-line preconcentration methods. Figure 6 shows the electropherograms of a standard solution of aldehydes obtained by normal sample stacking.

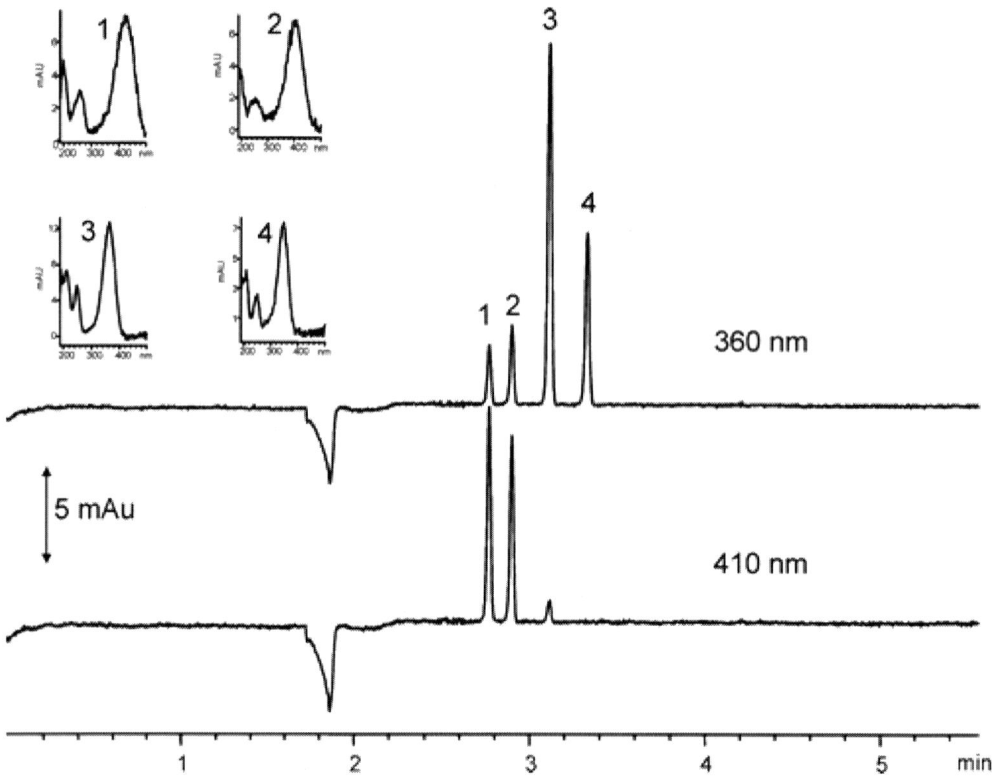

Figure 6. Electropherograms of a standard solution (aldehydes at 2 mg/L in deionized water with 40% (*v/v*) ethanol) using normal sample stacking. Peak identification: 1, sinapaldehyde; 2, coniferaldehyde; 3, syringaldehyde; 4, vanillin. Experimental conditions: fused silica capillary (48.5 cm (40 cm effective length) x 75 μm I.D.); BGE: 20 mM borate buffer and 10% methanol (pH 9.3); capillary voltage: +25 kV (positive polarity in the injection side); capillary temperature: 25 °C; hydrodynamic injection; 9 s, 50 mbar. UV spectra of each compound are also included in the figure. Reprinted with permission from reference [49]. Copyright (2011) American Chemical Society.

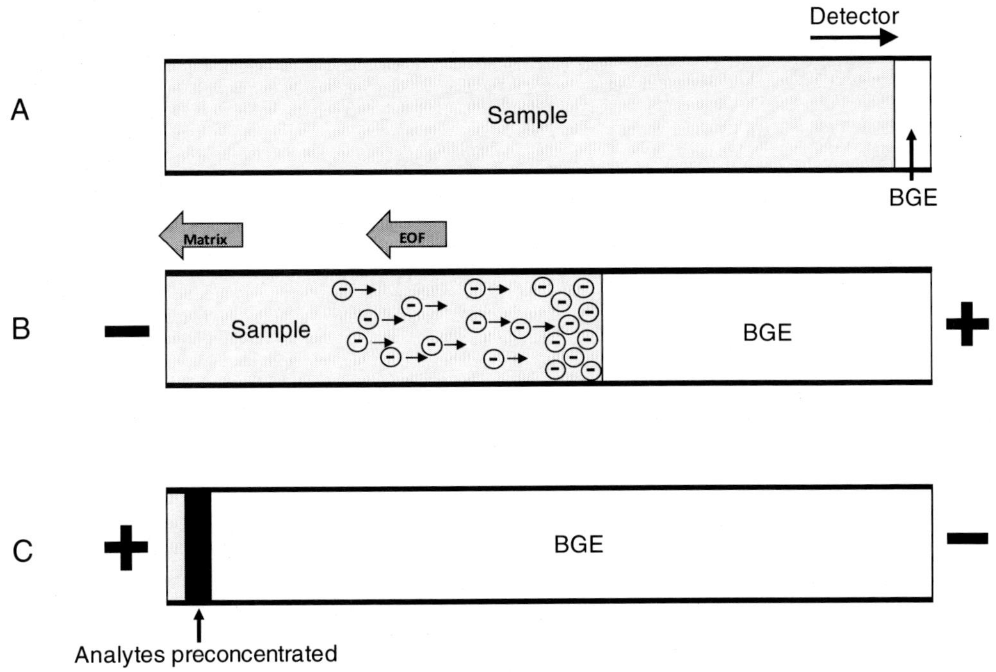

Figure 7. Scheme of large-volume sample stacking in CZE using uncoated fused-silica capillaries. Sample constituted of a matrix of lower ionic strength than that of BGE solution.

Aldehydes standard solution was prepared in deionized water with 40% (*v/v*) ethanol content. It can be observed that a good resolution was achieved for the analytes in a relatively short analysis time (less than 4 min). Limits of detection (LODs) in the range 30-100 μg/L were reported for the analyzed aldehydes by employing normal sample stacking.

3.2. Large-Volume Sample Stacking

Large-volume sample stacking (LVSS), also known as sample stacking with matrix removal, is a preconcentration technique designed by Chien and Burgi [50] that, similar to normal sample stacking, is performed by dissolving the sample in a matrix with lower ionic strength (lower conductivity) than the BGE and hydrodynamically filling between 30 to 50% of the capillary volume with the sample, as it is illustrated in Figure 7. After sample injection (Figure 7A), a reverse polarity (anode in the detection point of the capillary) is applied. Under these conditions, the analytes with negative charges stack-up at the boundary between the sample zone and the BGE due to stacking effect while the large volume of sample previously introduced into the capillary by hydrodynamic injection is pushed out of the capillary by the EOF (Figure 7B). When almost all the sample matrix is removed from the capillary, polarity is switched to normal (cathode in the detection point of the capillary) and separation takes place under counter-EOF conditions (Figure 7C).

Table 1. Selection of LVSS-CZE methods in environmental, food and bio-analytical applications

Compounds	Samples	LVSS conditions	CZE conditions	Detection	Analysis time	LODs	Ref.
Quaternary ammonium herbicides	Drinking water	Sample matrix: drinking water diluted 1:4 with Milli-Q water Hydrodynamic injection: 0.25 min, 137.9 kPa	Uncoated fused silica capillary of 57 cm (50 cm effective length) x 50 µm I.D. *BGE:* 50 mM acetic acid-ammonium acetate (pH 4.0) with 5% (*v/v*) methanol and 0.8 mM CTA *Capillary voltage:* +20 kV (sample matrix removal), -20 kV (electrophoretic separation)	UV: 220 and 255 nm	23 min	18-154 µg/L	[52]
15 Naphthalene- and benzene-sulfonates	Water samples	Sample matrix: real water sample Hydrodynamic injection: 15 nL, 50 mbar	Uncoated fused silica bubble-cell capillary of 64.5 cm (56 cm effective length) x 50 µm I.D. with an extended path length of 150 µm I.D. *BGE:* 20 mM borate buffer *Capillary voltage:* -30 kV (sample matrix removal), +30 kV (electrophoretic separation)	UV: 197-152 nm	16 min	20 µg/L	[54]
4 Linear alkylbenzene-sulfonates	Standards	Sample matrix: Milli-Q water Hydrodynamic injection: 90 s, 6.9 kPa	Uncoated fused silica capillary of 60 cm (50 cm effective length) x 50 µm I.D. *BGE:* 20 mM borate buffer with 30% acetonitrile at pH 9.0 *Capillary voltage:* -15 kV (sample matrix removal), +20 kV (electrophoretic separation)	UV: 200 nm	13 min	2-10 µg/kg (LOQs)	[55]
Acrylamide	Food products (biscuits, crisp breads, cereals, potato crisps, snacks, coffee)	Sample matrix: Milli-Q water (pH 10) Hydrodynamic injection: 20 s, 140 kPa	Uncoated fused silica capillary of 60 cm (50 cm effective length) x 75 µm I.D. *BGE:* 40 mM phosphate buffer (pH 8.5) *Capillary voltage:* -25 kV (sample matrix removal), +25 kV (electrophoretic separation)	UV: 210 nm	10 min	20 µg/kg	[56]
9 Sulfonamides	Meat and ground water	Sample matrix: 10 mM imidazol solution (pH 9.8) with 10% methanol Hydrodynamic injection: 0.5 min, 7 bar	Uncoated fused silica bubble-cell capillary of 64.5 cm (56 cm effective length) x 75 µm I.D. with an extended path length of 200 µm I.D. *BGE:* 45 mM phosphate buffer (pH 7.3)	UV: 265 nm	22 min	2.5-23 µg/L	[57]
Degradation products of metribuzin	Environmental water and soil samples	Sample matrix: Milli-Q water Hydrodynamic injection: 200 s, 50 bar	*Capillary voltage:* -28 kV (sample matrix removal), +25 kV (electrophoretic separation) Uncoated fused silica bubble-cell capillary of 48.5 cm (40 cm effective length) x 75 µm I.D. with an extended path length of 200 µm I.D. *BGE:* 40 mM sodium tetraborate buffer (pH 9.5) *Capillary voltage:* -25 kV (sample matrix removal), +15 kV (electrophoretic separation)	UV: 200 nm	10 min	10-20 µg/L	[58]

Compounds	Samples	LVSS conditions	CZE conditions	Detection	Analysis time	LODs	Ref.
7 β-lactam antibiotics	Milk	Sample matrix: Milli-Q water Hydrodynamic injection: 1 min, 7 bar	Uncoated fused silica bubble-cell capillary of 64.5 cm (56 cm effective length) x 75 µm I.D. with an extended path length of 200 µm I.D. *BGE:* 175 mM Tris (pH 8 with HCl) and 20% (*v/v*) ethanol *Capillary voltage:* -20 kV (sample matrix removal), +25 kV (electrophoretic separation)	UV: 220 nm	24 min	2-10 µg/L	[59]
Albumin	Urine	Sample matrix: Milli-Q water Hydrodynamic injection: 300 s, 50 mbar	Uncoated fused silica capillary of 45 cm (37.5 cm effective length) x 30 µm I.D. *BGE:* 150 mM borate buffer (pH 10.2) *Capillary voltage:* -20 kV (sample matrix removal), +20 kV (electrophoretic separation)	UV: 214 nm	10 min	15 µg/L	[60]
5 Sulfonylurea herbicides	Groundwater and grape samples	Sample matrix: Methanol:water (1:9 *v/v*) Hydrodynamic injection: 1 min, 7 bar	Uncoated fused silica bubble-cell capillary of 48.5 cm (40 cm effective length) x 50 µm I.D. with an extended path length of 200 µm I.D. *BGE:* 90 mM ammonium acetate buffer (pH 4.8 with acetic acid) *Capillary voltage:* -25 kV (sample matrix removal), +20 kV (electrophoretic separation)	UV: 226 and 240 nm	20 min	45-116 ng/L (water) 0.97-8.3 µg/kg (grape)	[61]
Haloacetic acids	Drinking water samples	Sample matrix: Milli-Q water Hydrodynamic injection: 15 s, 140 kPa	Uncoated fused silica capillary of 57 cm (50 cm effective length) x 50 µm I.D. *BGE:* 20 mM acetic acid-ammonium acetate (pH 5.5) with 20% acetonitrile *Capillary voltage:* -25 kV (sample matrix removal), −25 kV (electrophoretic separation)	UV: 200 nm	12.5 min	49-200 µg/L (standards)	[62]
Flavonoids	Broccoli	Sample matrix: methanol Hydrodynamic injection: 50 s, 50 mbar	Uncoated fused silica capillary of 85 cm (77 cm effective length) x 50 µm I.D. *BGE:* 10 mM sodium borate buffer (pH 8.4) *Capillary voltage:* -5 kV (sample matrix removal), +30 kV (electrophoretic separation)	UV: 320 and 360 nm	9 min	0.6-0.9 mg/kg	[63]
Barbiturates	Urine	Sample matrix: Milli-Q water Hydrodynamic injection: 80 s, 3 psi	Uncoated fused silica capillary of 60.5 cm (50 cm effective length) x 75 µm I.D. *BGE:* 40 mM borate solution with 20% methanol (pH 8.0 adjusted with 0.5 M boric acid) *Capillary voltage:* -20 kV (sample matrix removal), +20 kV (electrophoretic separation)	UV: 214 nm	9 min	15-57 µg/L (LOQs)	[64]

A critical point of this method is to know when to switch the capillary voltage. For that purpose the electrophoretic current is monitored until it reaches approximately 95-99% of its original value (the one observed when working under normal conditions with the BGE). Under these reverse polarity conditions, cations and neutral compounds should exit the capillary into the waste buffer reservoir before the polarity is returned to normal. Moreover, if the electrophoretic current is not carefully monitored, some anionic compounds may also be lost by the EOF, especially those with lower electrophoretic velocities than the EOF velocity. For this reason, LVSS is a selective method and only analytes with electrophoretic mobilities lower than the EOF mobility and in the opposite direction (that is anions when working with fused-silica capillaries) can be preconcentrated [51].

LVSS is quite a demanding on-line electrophoretic-based preconcentration method since the current must be closely monitored by the analyst in order to achieve reproducible results. Moreover, LVSS is a preconcentration technique that will not allow separation of anions and cations simultaneously.

The application of LVSS with polarity switching for the analysis of cations using uncoated fused-silica capillaries has also been described. For that purpose, EOF direction must be reversed, fact that can be achieved by using capillary wall surfactant modifiers such as cetyltrimethylammonium bromide (CTAB) [52]. CTAB is a cationic surfactant that coats the internal capillary wall changing total charge to positive and reversing EOF direction.

LVSS without polarity switching has also been reported for the analysis of high mobility anions [53]. In this case, also an EOF modifier such as CTAB is present in the BGE. When the capillary is filled with the sample matrix and a reversed polarity is applied to the inlet end, the EOF pushes the sample matrix out of the capillary. Meanwhile, BGE from the outlet end (detector position) of the capillary is pulled into the capillary. The CTAB present in the BGE coats the capillary and reversed the direction of the EOF eliminating the need for polarity switching.

The number of LVSS applications in CZE is huge. Table 1 is summarizing a selection of LVSS-CZE methods employed in environmental, food and bio-analytical applications [52, 54-64].

Núñez et al. [52] reported the application of LVSS for the determination of several quaternary ammonium herbicides (paraquat, diquat and difenzoquat) in water by CZE. Due to the cationic nature of these quaternary ammonium salts, the authors employed CTAB to coat the capillary and reversed the EOF direction. For that purpose, a 50 mM acetic acid-ammonium acetate buffer solution (pH 4.0) with 0.8 mM CTAB and 5% methanol was used as BGE. Figure 8a shows the electropherograms obtained for a standard solution of these compounds at 0.2 mg/L in Milli-Q water. In this work the authors applied the proposed LVSS-CZE method for the analysis of these herbicides in drinking waters. However, although initially the results were quite good when dissolving the analytes in Milli-Q water, with LODs in the range 10-15 μg/L, thus achieving a 100-fold sensitive enhancement in comparison to the conventional CZE separation without preconcentration, the stacking process was not so effective when dealing with real drinking water samples because of their higher matrix salinity. For instance, the analyzed still mineral water and tap drinking water had a conductivity of 474 and 883 $\mu\Omega^{-1}cm^{-1}$, respectively, becoming similar to that of the BGE used for the separation. This effect can be observed in Figures 8b and 8c, where the electropherograms and the monitoring of the electrophoretic current is shown when LVSS-CZE was applied to the analysis of a drinking tap water from Barcelona (Spain) spiked with the studied quaternary ammonium herbicides at 1 mg/L. When directly analyzing the tap water by LVSS-CZE (signals 1 and 2 in the figures), only a difference of approximately 3 μA was observed between the current provided by the capillary after the hydrodynamic injection of the sample with respect to that of the capillary completely filled with BGE solution. In this situation, the stacking process is not very effective because electrophoretic velocities of target analytes in the sample region and in the BGE region become very similar.

Under these conditions, only PQ and DQ were detected but no signal for DF was observed (Figure 8b, electropherogram 2), which was probably due to the fact that DF is a mono-charged compound in comparison to PQ and DQ that have two positive charges.

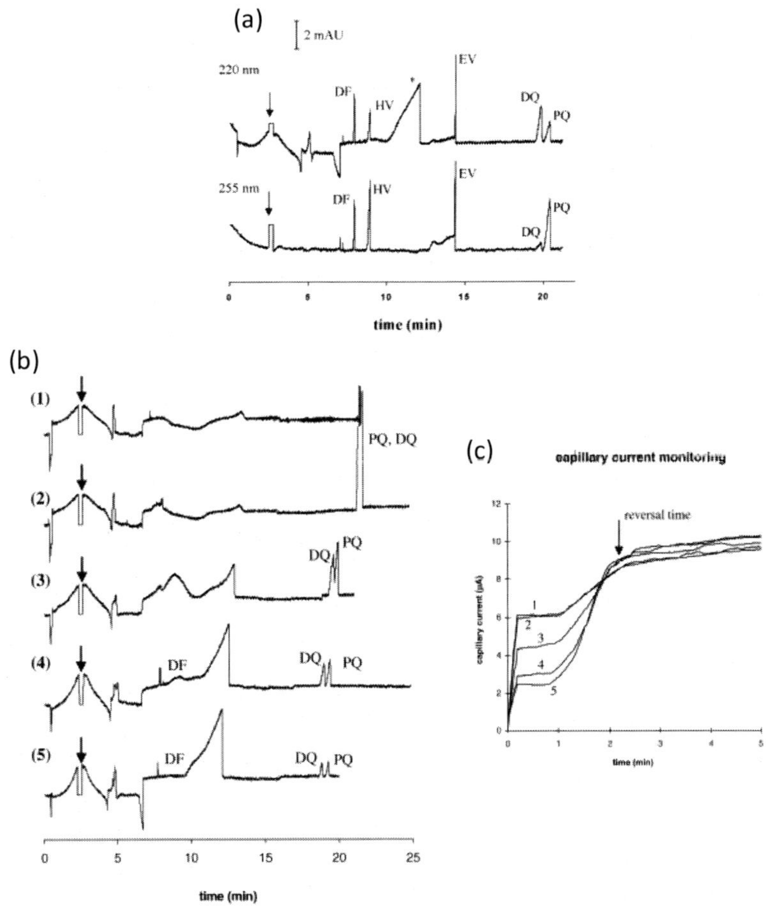

Figure 8. (a) Electropherograms of a standard solution of PQ, DQ and DF (0.2 mg/L) and the internal standards EV and HV (0.8 mg/L) in Milli-Q water. Hydrodynamic injection: 0.25 min (13.7.9 kPa). Capillary voltage: +20 kV (sample matrix removal), -20 kV (electrophoretic separation). BEG: 50 mM acetic acid-ammonium acetate (pH 4.0) buffer solution with 0.8 mM CTAB and 5% methanol. (*) system peak. The arrow shows the time at which capillary polarity was reversed. (b) Electropherograms of tap water and (c) capillary current monitoring. (1) Non-spiked water; (2) spiked water at 1000 μg/L; (3) spiked water diluted 1:1; (4) spiked water diluted 1:4; (5) spiked water diluted 1:9. Peak identification: PQ, paraquat; DQ, diquat; DF, difenzoquat; EV, ethyl viologen (internal standard); HV, hepthyl viologen (internal standard). Reprinted with permission from reference [52]. Copyright (2001) Elsevier.

In order to solve this problem and make the stacking process more effective the authors decided to analyze the drinking tap water samples after dilution with Milli-Q water as a way of reducing sample salinity (see Figures 8b and 8c, electropherograms 3, 4 and 5). From a 1:4 (*v/v*) dilution of drinking water with Milli-Q water, good electrophoretic separation was observed after LVSS-CZE analysis (electropherogram 4) and detection of DF was also possible. Although obviously dilution of the sample will decrease analyte concentration, good sensitivity in drinking water samples was achieved with the proposed on-line preconcentration method, with LODs in the range 18-62 μg/L and 48-154 μg/L for mineral water and tap water, respectively. It should be commented that although good sensitive enhancement was observed, the attainable LOD values were not enough to analyze this family of compounds at the levels required by EU legislation (0.1-0.5 μg/L). So, in a later work, the authors combined the application of an off-line SPE method using porous graphite carbon cartridges with LVSS-CZE for the analysis of these compounds achieving LOD values down to 0.2 μg/L [65].

In another interesting contribution, Quesada-Molina et al. [58] proposed the use of LVSS-CZE for the monitoring of the degradation products of metribuzin in environmental samples (waters and soils). Metribuzin is a selective systemic herbicide used for pre- and post-emergence control of many grasses and broad-leaved weeds in soy beans, potatoes, tomatoes, sugar cane, alfalfa, asparagus, maize and cereals. The decomposition of metribuzin in the environment is due to microbiological and chemical processes, and the primary products of its transformation are deaminometribuzin (DA), dietometribuzin (DK) and deaminodiketometribuzin (DADK). As an example, Figure 9 shows the electropherograms of a blank soil sample and a soil sample, spiked with 200 μg/kg with DK, DA and DADK, obtained by the proposed LVSS-CZE method and performing a hydrodynamic sample injection of 200 s at 50 mbar. LODs in the range 10-20 μg/L in water were reported with the proposed LVSS-CZE method. In the case of environmental soil samples, target compounds were extracted by pressurized liquid extraction with methanol.

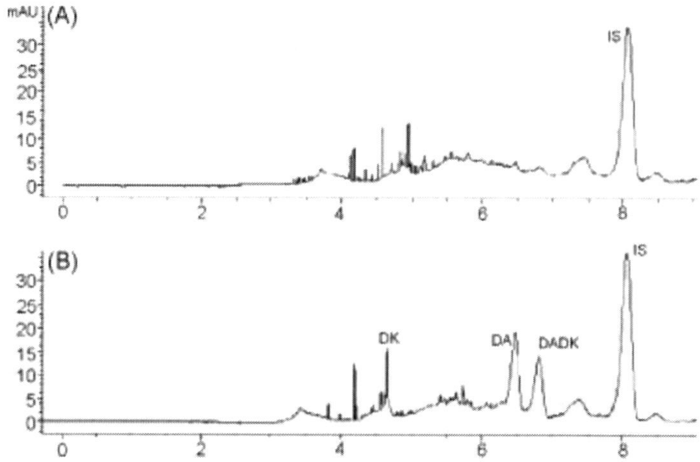

Figure 9. Electropherogram of (A) blank soil sample and (B) soil sample spiked with 200 μg/kg of diketometribuzin (DK), deaminometribuzin (DA) and deaminodiketometribuzin (DADK). LVSS-CZE experimental conditions as indicated in Table 1. Reprinted with permission from reference [58]. Copyright (2007) Elsevier.

However, in order to be able to analyze these compounds at the expected concentration levels in environmental samples, the authors also required the combination of LVSS-CZE with an off-line SPE method of the groundwater samples and the soil extracts obtained after PLE extraction. Taking into account that the applied SPE procedure involved a 500-fold preconcentration for the case of water samples and a 2.5-fold preconcentration in the case of soil samples, the proposed SPE-LVSS-CZE method allowed the detection of this family of compounds in the lower ng/L range in the case of water samples and at very low μg/L in the case of soils.

Regarding the application of LVSS methods in food analysis, Bermudo et al. [56] evaluated the application of on-line preconcentration methods for the analysis of acrylamide in food products. Acrylamide is a genotoxic and carcinogenic residue generated in many carbohydrate-rich foods when they are subjected to heating, such as fried and baked products, via the Maillard reaction between amino acids (mainly asparagines) and reducing sugars such as glucose or fructose. Acrylamide is a neutral compound so a derivatization step with 2-mercaptobenzoic acid was required to provide a negatively charged compound. In order to remove the derivatizing reagent, which increased sample matrix salinity preventing the application of LVSS, a LLE procedure using dichloromethane was proposed. After evaporation and reconstitution in water (at pH 10), the extract was analyzed by LVSS-CZE using 40 mM monohydrogen phosphate-dihydrogen phosphate buffer (pH 8.5) as BGE, and hydrodynamically injecting the sample for 20 s at 140 kPa. As depicted in Figure 10, the authors showed that increasing buffer concentration in the BGE improved acrylamide signal due to a more effective stacking process, although 40 mM was selected as optimum value due to the raise in capillary current and analysis time. Under these conditions, LODs of 7 μg/L and 20 μg/kg for acrylamide in a standard solution and a crisp bread sample, respectively, were obtained.

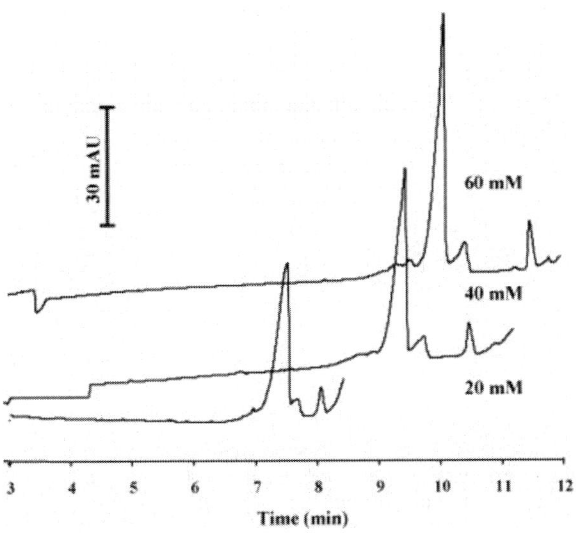

Figure 10. Effect of BGE buffer concentration on the separation of acrylamide by LVSS-CZE. BGE: monohydrogen phosphate-dihydrogen phosphate buffer. Acquisition wavelength: 210 nm. Sample: acrylamide standard of 1 mg/L derivatized with 2-mercaptobenzoic acid. Hydrodynamic injection: 20 s (140 kPa). Applied potential: -25 kV (sample matrix removal), +25 kV (electrophoretic separation). Reprinted with permission from reference [56]. Copyright (2006) Elsevier.

Bailón-Pérez et al. [59] reported de application of LVSS-CZE for the determination of seven β-lactam antibiotics in milk samples by using a 175 mM Tris buffer with 20% ethanol at pH 8.0 as BGE and a 64.5 cm x 75 μm I.D. bubble cell capillary. In order to be able to quantify these compounds in milk samples at the levels established by European Union legislation, the authors also proposed the application of a solvent extraction/SPE method as off-line preconcentration and sample clean-up. Under working conditions (see Table 1) LODs between 2 and 10 μg/L were reported. Satisfactory recoveries ranging from 86 to 93% were obtained in milk samples of different origins.

LVSS-CZE procedures have also been applied to the analysis of biological samples. For instance, Bessonova et al. [60] compared the application of several on-line preconcentration methods for the determination of albumin in urine samples, being LVSS-CZE the most sensitive one with a LOD of 15 μg/L. As an example, Figure 11a shows the improvement observed on albumin electrophoretic signal when injecting the sample by LVSS with reversed polarity procedure (300 s, 50 mbar) in comparison with a normal hydrodynamic injection (2 s, 50 mbar).

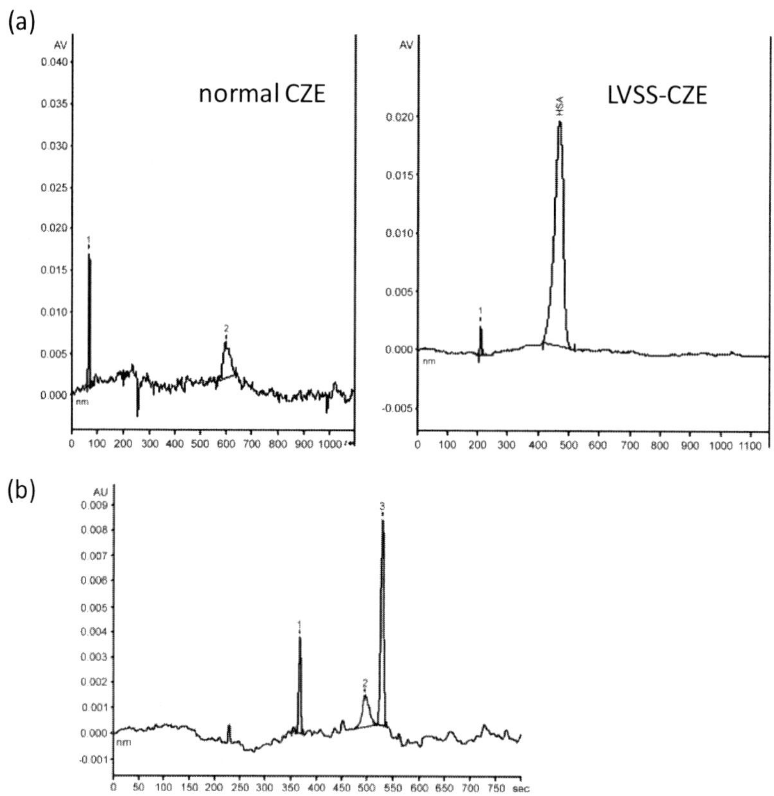

Figure 11. (a) Electrophorograms showing focusing of albumin (2 mg/L). (1) Neutral marker DMFA; (2) HAS-albumin. Capillary: 45 cm (37.5 cm effective length) x 30 μm I.D. BGE: 150 mM borate buffer (pH 10.2). Separation voltage: +20 kV. Left: Normal hydrodynamic injection of the sample (2 s, 50 mbar); Right: Injection of the sample using LVSS with reversed polarity procedure (300 s, 50 mbar). (b) LVSS-CZE electropherogram for urine sample from a normal male after desalting step. Sample matrix: water; Other conditions as in (a). Reprinted with permission from reference [60]. Copyright (2007) Elsevier.

One of the problems encountered when dealing with the application of on-line electrophoretic-based preconcentration methods such as LVSS in the analysis of biological samples such as urine is sample matrix salinity. In order to achieve an effective stacking procedure desalting of these samples is mandatory. Bessonova et al. [60] carried out desalting and depigmentation of urine samples by gel-filtration on Sephadex G25 Medium with a cut mass of 10 kDa. As an example, Figure 11b shows the electropherogram obtained by LVSS-CZE of a urine sample from a normal male after this desalting step.

3.3. Field-Amplified Sample Injection

Among the on-line electrophoretic-based preconcentration procedures, field-amplified sample injection (FASI), also known as field-amplified sample stacking (FASS), is very popular since it is quite simple only requiring the electrokinetic injection of the sample after the introduction of a short plug of a high-resistivity solvent such as methanol or water. This method is also based on the fact that ions electrophoretically migrating through a low-conductivity solution into a high-conductivity solution slow down dramatically at the boundary of the two solutions, as previously described. But in contrast to LVSS where sample is hydrodynamically injected, FASI is taking advantage of the higher amount of analytes introduced into the capillary when electrokinetic injections are used. A schematic representation of the application of FASI for the in-line enrichment of positively charged analytes is shown in Figure 12.

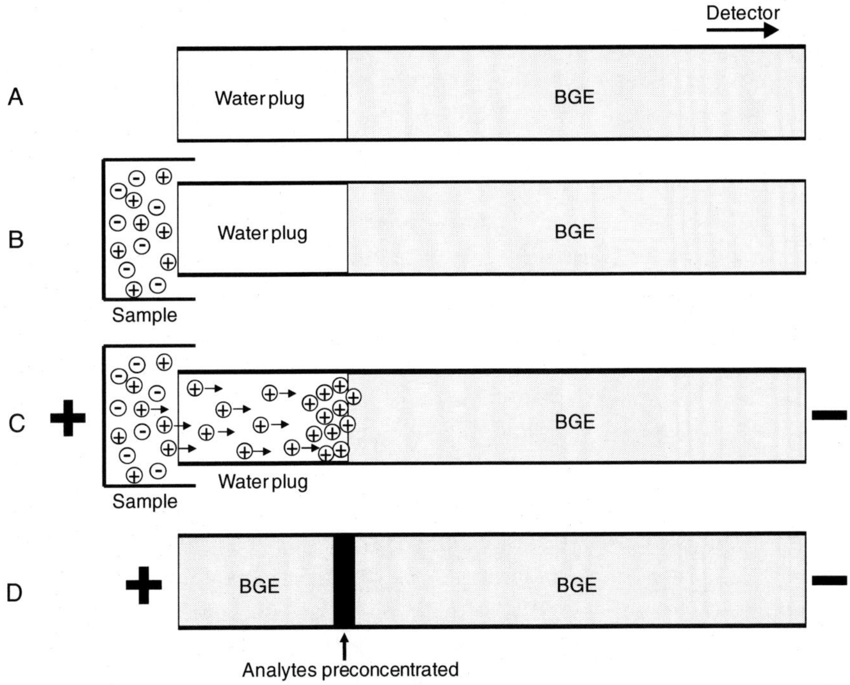

Figure 12. Scheme of field-amplified sample injection in CZE for the preconcentration of cations.

After filling the capillary with the BGE solution, a pre-injection of a short plug of a high-resistivity solvent such as water is hydrodynamically introduced into the capillary (Figure 12A). Then, a sample vial is set in the capillary inlet position (Figure 12B) and electrokinetic injection is carried out by applying a normal polarity (cathode in the outlet position). The short plug of water allows the enhancement of the sample electrokinetic injection because of the conductivity differences between sample and the water plug (Figure 12C). Moreover, long electrokinetic injection times can be employed while the analytes stack-up at the boundary between the high-resistivity solvent and the BGE solution because they slow down due to the important decrease on their migration velocity in the BGE region. Finally, a BGE vial is set in the inlet position and electrophoretic separation takes place with the cationic analytes being concentrated in a narrow zone (Figure 12D).

Electroosmotic flow must also be taken into account when negatively charged analytes are being analyzed by FASI, in order to prevent removal of low electrophoretic mobility compounds from the capillary when the enhanced sample electrokinetic injection is performed. Moreover, as FASI is based on electrokinetic injection mode discrimination occurs for ionic species since compounds that migrate more rapidly in the electrical field will be over-represented in the sample introduced into the capillary compared to slower moving components.

Because of its simplicity of application in comparison to LVSS where the current must be closely monitored by the analyst in order to achieve reproducible results, FASI is one of the most employed on-line electrophoretic-based preconcentration methods in CZE in multiple application fields. Table 2 summarizes a selection of FASI-CZE methods in environmental, food and bio-analytical applications [56, 62, 66-75].

As previously commented, analysts can take advantage of EOF to remove the water plug from the capillary when negatively charged analytes are being injected through FASI. Although part of the analytes will be also removed from the capillary, the long electrokinetic injection times normally used and the FASI enhancement produced due to the differences in migration mobilities in the sample and the water plug will allow to stack enough anions in the boundary region to improve sensitivity. For instance, Zhu and Lee [68] described the application of FASI combined with water removal by EOF pump in acidic buffer for the analysis of phenoxy acid herbicides by CZE. To achieve this, a non-ionic hydroxylic polymer (HEC) was used to modify the inner wall of the capillary and to suppress the EOF at low pH. The application of this FASI with water removal by EOF pump and then suppression of EOF to perform the separation is quite interesting, and the schematic of the process in shown in Figure 13a. During FASI injection employing a negative polarity (-10 kV) and because a long water plug was previously introduced into the capillary, EOF is pumping the water out of the capillary while the anions are being stacked in the boundary region with the BGE. When all the water was removed from the capillary and it is again filled with the BGE, EOF is suppressed by the presence of HEC allowing performing the electrophoretic separation of these phenoxy acids in negative polarity (-30 kV). As an example, Figure 13b shows the electropherogram obtained by FASI-CZE for a river water sample spiked with the studied phenoxy acid herbicides. This method afforded a sensitivity enhancement greater than 3,000 times. With the combination of this method with an off-line SPE procedure LODs for the phenoxy acid herbicides as low as 10 ng/L were obtained.

Table 2. Selection of FASI-CZE methods in environmental, food and bio-analytical applications

Compounds	Samples	FASI conditions	CZE conditions	Detection	Analysis time	LODs	Ref.
Abused drugs	Hair	*High resistivity solvent*: water *Hydrodynamic injection*: 5 s (external rinse step) *Sample electrokinetic injection*: 10 s, +10 kV	Uncoated fused silica capillary of 57 cm (50 cm effective length) x 75 µm I.D. *BGE:* 100 mM potassium phosphate (pH 2.5, adjusted with phosphoric acid) *Capillary voltage:* +10 kV	UV: 200 nm	20 min	2-8 µg/L	[66]
Opiate drugs	Hair	*High resistivity solvent*: water *Hydrodynamic injection*: 0.5 mm plug *Sample electrokinetic injection*: 99 s, +10 kV	Uncoated fused silica capillary of 47 cm (40 cm effective length) x 75 µm I.D. *BGE:* 0.1 M sodium phosphate, pH 2.5, with 40% ethylene glycol *Capillary voltage:* +20 kV	UV: 214 nm	30 min	100 ng/L	[67]
8 phenoxy acid herbicides	River water	*High resistivity solvent*: water *Hydrodynamic injection*: 1 min, 400 mbar *Sample electrokinetic injection*: 12 min, -10 kV	Uncoated fused silica capillary of 76 cm (63 cm effective length) x 50 µm I.D. *BGE:* 48 mM borate-phosphoric acid buffer (pH 3.2) with 0.1% of a nonionic hydroxylic polymer *Capillary voltage:* -30 kV	UV: 240 nm	20 min	1-5 µg/L	[68]
Acrylamide	Food products (biscuits, crisp breads, cereals, potato crisps, snacks, coffee)	*High resistivity solvent*: water *Hydrodynamic injection*: 35 s, 0.5 psi *Sample electrokinetic injection*: 35 s, -10 kV *Post-injection of water plug*: 6 s, 0.5 psi	Uncoated fused silica capillary of 60 cm (50 cm effective length) x 50 µm I.D. *BGE:* 40 mM phosphate buffer (pH 8.5) *Capillary voltage:* +25 kV	UV: 210 nm	6 min	3 µg/kg	[56]
Acrylamide	Foodstuffs (potato crisps, biscuits, crisp bread, breakfast cereals, coffee)	*High resistivity solvent*: water *Hydrodynamic injection*: 35 s, 0.5 psi *Sample electrokinetic injection*: 35 s, -10 kV *Post-injection of water plug*: 6 s, 0.5 psi	Uncoated fused silica capillary of 80 cm x 50 µm I.D. *BGE:* 35 mM formic acid-ammonium formate (pH 10) *Capillary voltage:* +25 kV	MS ion-trap mass analyzer (-) electrospray Product ion scan	8 min	8 µg/kg	[69]

Compounds	Samples	FASI conditions	CZE conditions	Detection	Analysis time	LODs	Ref.
Metal ions	Wine	Pre-injection of derivatizing reagent: Nitro-PAPS, 7 s, 0.5 psi *High resistivity solvent*: water *Hydrodynamic injection*: 1 s, 0.1 psi *Sample electrokinetic injection*: 5 s, +10 kV	Uncoated fused silica capillary of 60 cm (50 cm effective length) x 50 µm I.D. *BGE:* 45 mM borate pH 9.7 and 0.01 mM Nitro-PAPS with 20% acetonitrile *Capillary voltage:* +20 kV	UV: 241, 325, 232, 248 nm	22 min	25-100 µg/kg	[70]
Haloacetic acids	Drinking water	*High resistivity solvent*: water *Hydrodynamic injection*: 20 s, 3.5 kPa *Sample electrokinetic injection*: 20 s, -10 kV	Uncoated fused silica capillary of 60 cm (50 cm effective length) x 50 µm I.D. *BGE:* 200 mM formic acid-ammonium formate buffer (pH 3.0) *Capillary voltage:* -25 kV	UV: 200 nm	12.5 min	6-52 µg/L µg/L	[62]
Abuse drugs	Human urine	*High resistivity solvent*: water *Hydrodynamic injection*: 30 s (external rinse step) *Sample electrokinetic injection*: 20 s, +5 kV	Uncoated fused silica capillary of 80 cm (72 cm effective length) x 50 µm I.D. *BGE:* 100 mM phosphate buffer (pH 6.0) *Capillary voltage:* +25 kV	UV: 205 and 310 nm	8 min	0.4-7.2 µg/kg	[71]
Ephedrines	Human urine	*High resistivity solvent*: water *Hydrodynamic injection*: 5 s, 50 mbar *Sample electrokinetic injection*: 20 s, +15 kV	Uncoated fused silica capillary of 64.5 cm (46 cm effective length) x 75 µm I.D. *BGE:* 25 mM borate buffer with 1.0 mM sodium dodecyl sulphate at pH 9.3 *Capillary voltage:* +15 kV	UV: 194 nm	20 min	5-100 µg/L	[72]
Acetylcholinesterase inhibitors with antipsychotic drugs for Alzheimer's disease	Plasma	*High resistivity solvent*: methanol *Hydrodynamic injection*: 6 s, 0.3 psi *Sample electrokinetic injection*: 50 s, +10 kV	Uncoated fused silica capillary of 60.2 cm (50 cm effective length) x 50 µm I.D. *BGE:* 120 mM phosphate buffer (pH 4.0) with 0.1% γ-cyclodextrin, 40% methanol, 0.02% polyvinyl alcohol *Capillary voltage:* +27 kV	UV: 214 nm	20 min	1-4 µg/L (low level in linear range)	[73]

Table 2. (Continued)

Compounds	Samples	FASI conditions	CZE conditions	Detection	Analysis time	LODs	Ref.
Amprolium	Eggs	*High resistivity solvent*: water *Hydrodynamic injection*: 40 s, 3.5 kPa *Sample electrokinetic injection*: 50 s, +10 kV	Uncoated fused silica capillary of 57 cm (50 cm effective length) x 50 µm I.D. *BGE:* 150 mM acetic acid-ammonium acetate buffer (pH 4.5):methanol (60:40 *v/v*) *Capillary voltage:* +30 kV	UV: 235 nm	6 min	0.25 µg/L	[74]
8 benzophenone UV-filters	Environmental waters	*High resistivity solvent*: water *Hydrodynamic injection*: 20 s, 3.5 kPa Sample electrokinetic injection: 25 s, -10 kV *Sample matrix*: 2.5 mM sodium tetraborate buffer (pH 9.2) solution	Uncoated fused silica capillary of 50 cm (40 cm effective length) x 75 µm I.D. *BGE:* 35 mM sodium tetraborate buffer (pH 9.2) *Capillary voltage:* +30 kV	UV: 240, 285 and 345 nm	8 min	21-136 µg/L	[75]

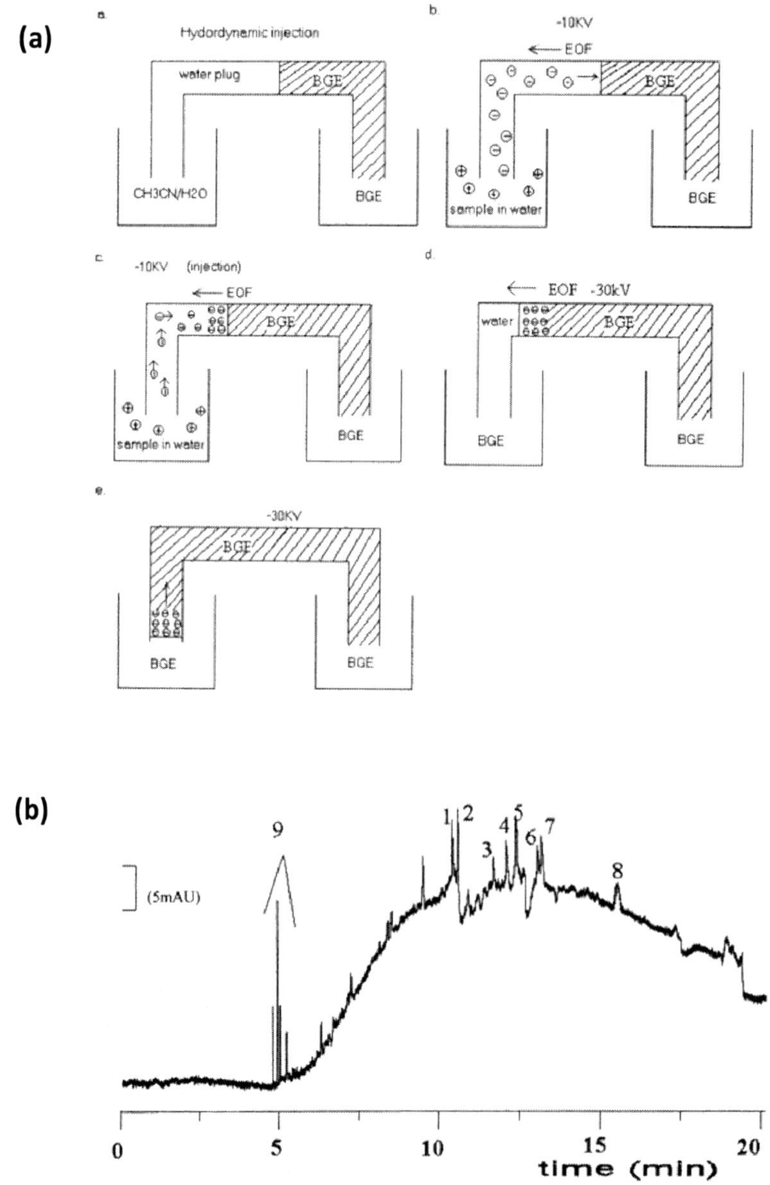

Figure 13. (a) Schematic illustration of the field-amplified sample injection with water removal: a, initial condition, injection of long plug of water into the capillary by pressure; b, field-amplified injection of anions into the capillary under negative voltage; c, water comes out from the inlet slowly during the injection process; d, process of anion stacking and removal of aqueous sample plug; and e, complete removal of aqueous sample plug and start of separation under negative voltage. (b) Electropherogram of an extract of river water sample spiked with 0.05 ng/mL of phenoxy herbicides. Water plug hydrodynamic injection: 1 min, 400 mbar. Sample injection: 16 min at -10 kV. BGE: 48 mM borate-phosphoric acid (pH 3.2) with 0.1% HEC. Capillary voltage: -30 kV. Peak identification: (1) picloram; (2) 2,4-dichlorobenzoic acid; (3) 4-chlorophenoxy acetic acid; (4) 2,4-dichlorophenoxy acetic acid; (5) 2,4,5-trichlorophenoxy acetic acid; (6) Dichlorprop; (7) Fenoprop; (8) Mecroprop), and (9) humic acids. Reprinted with permission from reference [68]. Copyright (2001) American Chemical Society.

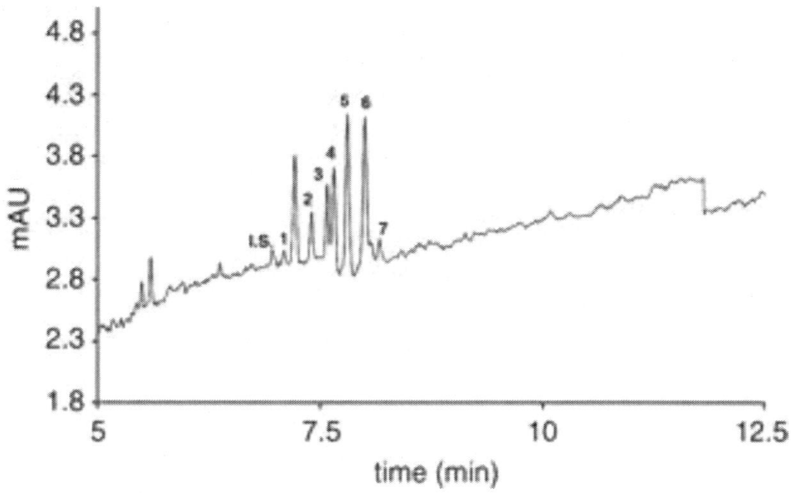

Figure 14. Electropherogram of the analysis of Barcelona (Spain) tap water by SPE-FASI-CZE. FASI-CZE acquisition conditions as described in Table 2. Peak identification: 1, dichloroacetic acid; 2, bromochloroacetic acid; 3, trichloroacetic acid; 4, dibromoacetic acid; 5; bromodichloroacetic acid; 6, chlorodibromoacetic acid; and 7, tribromoacetic acid. Reprinted with permission from reference [62]. Copyright (2011) John Wiley and Sons.

A similar approach was described by Bernad et al. [62] for the analysis of haloacetic acids in drinking water samples by FASI-CZE. The authors prevented the removal of haloacetic acids by working at low pH values. Additionally, injection times for both the plug of water (hydrodynamic mode) and sample (electrokinetic mode) were simultaneously optimized. Under optimal conditions (see Table 2) sensitivity enhancements up to 300-fold for some haloacetic acids such as dibromoacetic acid were obtained. These enhancements were 10-fold higher than the ones described by the same authors by LVSS for the same family of compounds. Although the important decrease in LODs (4-52 μg/L), the sensitivity was not enough for the analysis of this family of compounds in real water samples, so the combination of the proposed method with an off-line SPE step was necessary.

By using Oasis WAX (anion-exchange) SPE cartridges, specifically proposed for the preconcentration of acidic compounds, sample salinity was considerably removed, and with the combination of both SPE and FASI, sensitivity enhancements between 6,250 and 26,000 were obtained. The method was applied to the analysis of Barcelona (Spain) tap water being able to quantify seven haloacetic acids at concentration bellow 13 μg/L (Figure 14).

FASI-CZE has also been recently described for the analysis of benzophenone UV-filters in environmental water samples [75]. Benzophenone UV-filters (BPs) are frequently used in personal care products such as sunscreens because of their excellent absorbing abilities of UVA radiation. However, these chemicals than can cause hormonal disruption to the reproduction of fish and possess endocrine activity can easily reach the aquatic environment by direct sources (e.g., swimming) and/or indirect sources (wastewater treatment plants, showering or domestic washing). In order to analyze these compounds with phenolic groups by FASI-CZE, a sodium tetraborate buffer (pH 9.8) solution was necessary (with a pH value higher than BP pk_a values) in order to obtain anions. However, at the proposed pH value BP electrophoretic mobilities were lower than the EOF velocity. For this reason, although FASI injection was carried out in negative polarity, electrophoretic separation once the anions are preconcentrated was performed with a positive polarity, working in counter-EOF conditions. The proposed FASI-CZE method allowed the analysis of these compounds with LODs in the range 21-136 μg/L, providing a 9- to 25-fold enhancement in comparison to conventional CZE without preconcentration. Nevertheless, and it is usually common when dealing with environmental analysis, the sensitivity achieved with the proposed FASI-CZE method was not enough for the determination of BPs at the expected concentrations in environmental water samples. For this reason, the authors applied an off-line SPE preconcentration step using polymeric reversed phase (Strata X) cartridges prior to FASI-CZE analysis. Sample extracts after SPE preconcentration were reconstituted with a 2.5 mM sodium tetraborate buffer (pH 9.2) solution. As an example, Figure 15 shows the electropherograms obtained by the developed SPE-FASI-CZE method in the analysis of several water samples.

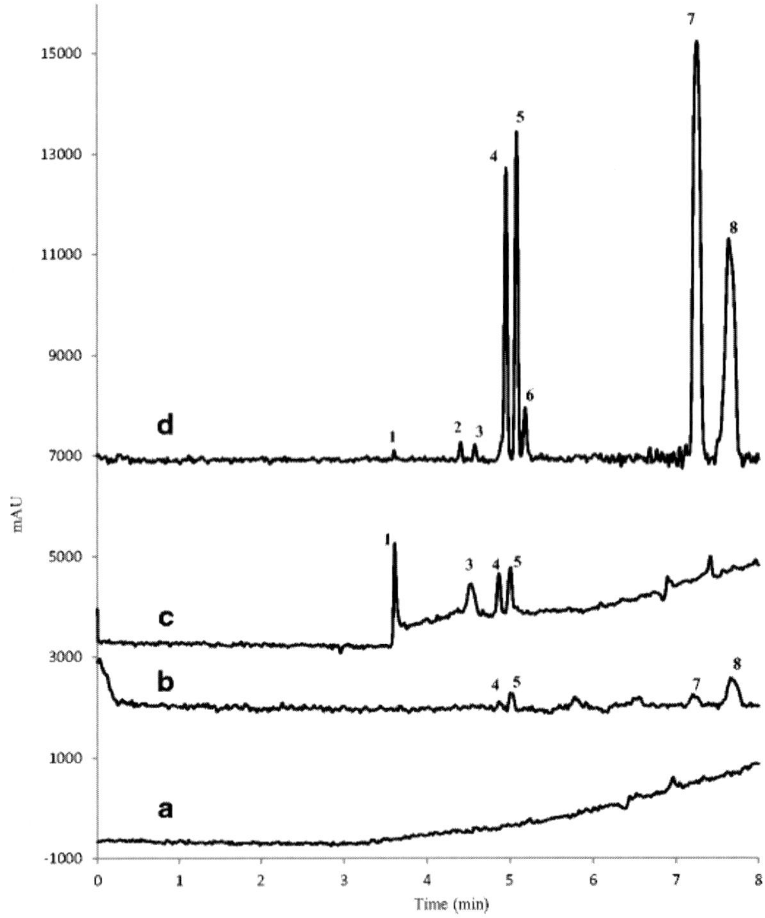

Figure 15. Off-line SPE-FASI-CZE electropherograms of (a) blank river water sample, (b) Barcelona (Spain) tap water, (c) Segre River (Spain) tap water, and (d) SPE extract of a blank river water sample spiked with BPs at ca. 1 mg/L. Peak identification: (1) 2-hydroxy-4-methoxybenzophenone; (2) 2,2'-dihydroxy-4-methoxybenzophenone; (3) 2,2'-dihydroxy-4,4'-dimethoxybenzophenone; (4) 4-hydroxybenzophenone; (5) 2,4-dihydroxybenzophenone; (6) 2,3,4-trihydroxybenzophenone; (7) 4,4'-dihydroxybenzophenone; and (8) 2,2',4,4'-tetrahydroxybenzophenone. Reprinted with permission from reference [75]. Copyright (2014) Springer.

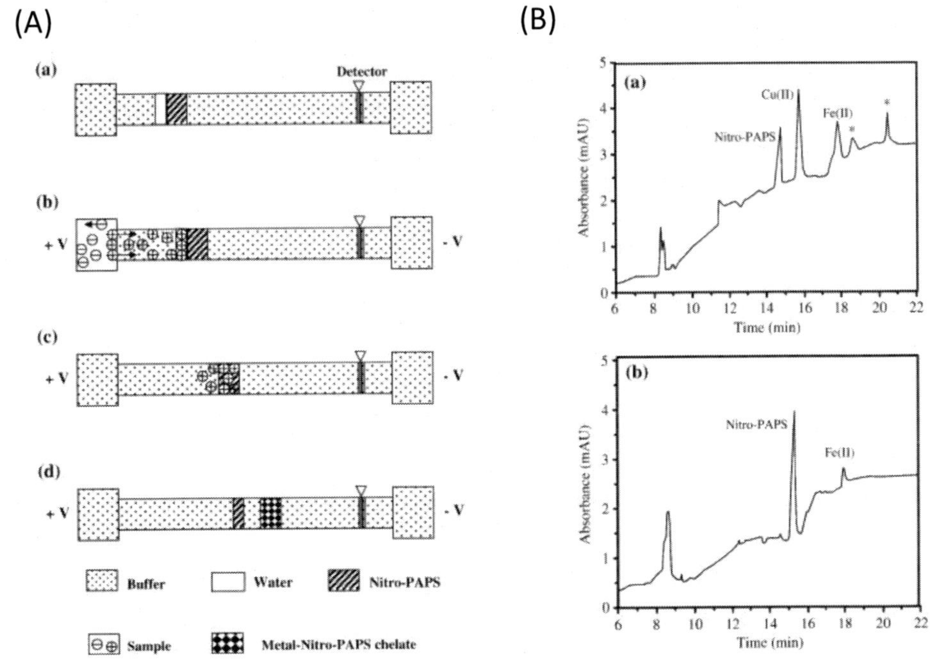

Figure 16. (A) Schematic reaction of metal ions and Nitro-PAPS base on in-capillary derivatization with FASI-CZE. (a) Hydrodynamic introduction of BGE, Nitro-PAPS and water plug, (b) electrokinetic injection of the sample, (c) mixing and reaction of metals with Nitro-PAPS, and (d) separation of the metal-Nitro-PAPS chelates. (B) Electropherograms of wine sample under optimum conditions. (a) Emblic fruit wine and (b) white grape wine. (*) are unknown peaks. Reprinted with permission from reference [70]. Copyright (2007) Elsevier.

None of the analyzed BPs was detected in the mineral water sample, as expected. However, Barcelona tap water showed the presence of several benzophenones although all of them at the LOD of the proposed SPE-FASI-CZE method or below the LOQ. The authors indicated that the presence of some BPs in Barcelona tap water was detected only occasionally, and in most cases negative results were achieve after analyzing this kind of sample. River water samples were also analyzed and sampling was carried out in two locations: (i) at the beginning of the river course upstream of industrialized and urban areas and (ii) at the middle of the river course downstream of some industrialized and urban areas. BPs were not detected in those river water samples collected upstream of industrialized and urban areas, as expected, whereas some BPS were found at quantified levels after these areas, showing the necessity of controlling environmental water samples for the presence of this kind of compounds.

Regarding the application of FASI-CZE methods in food analysis, Bermudo et al. [56] also proposed this on-line preconcentration method for the analysis of acrylamide in foodstuffs. In this case, after FASI injection of acrylamide in negative polarity (after derivatization with 2-mercaptobenzoic acid to provide a negatively charged compound) the authors proposed a post-injection of an additional small water plug (6 s, 0.5 psi) to prevent the complete removal of acrylamide from the capillary by the EOF. Then, separation was performed in positive polarity in counter-EOF conditions. A limit of detection of 3 μg/L in food product samples was achieved. In a later work, the authors developed a FASI-capillary electrophoresis-tandem mass spectrometry method (FASI-CE-MS/MS) for the analysis of the same compound in foodstuffs [69]. BGE solution was modified (formic acid/ammonium formate buffer) in order to use a compatible solution with mass spectrometry.

An interesting work is the one described by Santalad et al. [70] for the analysis of metal ions in wine samples by FASI-CZE and in-capillary derivatization. In this work, several metal ions (Co(II), Cu(II), Ni(II) and Fe(II)) were determined by using 2-(5-Nitro-2-Pyridylazo)-5-(*N*-Propyl-*N*-Sulfopropylamino)Phenol (Nitro-PAPA) as the derivatizing reagent, but derivatization was carried out into the capillary in combination with FASI. Figure 16A shows the schematic process used by the authors in this work. After conditioning the capillary with the BGE, a plug of derivatizing agent was hydrodynamically introduced into the capillary (a), and then FASI was performed, first with the hydrodynamic injection of a water plug and the electrokinetic injection of the sample (b). Once FASI injection is finished, mixing and reaction of the metal ions with the derivatizing agent is taking place inside the capillary (c), and finally the metal-Nitro-PAPS chelates generated are electrophoretically separated (d). Figure 16B shows the application of the proposed method to the analysis of wine samples. LOD values 3 to 28 times better than those from pre-capillary derivatization were obtained with the proposed FASI-CZE method.

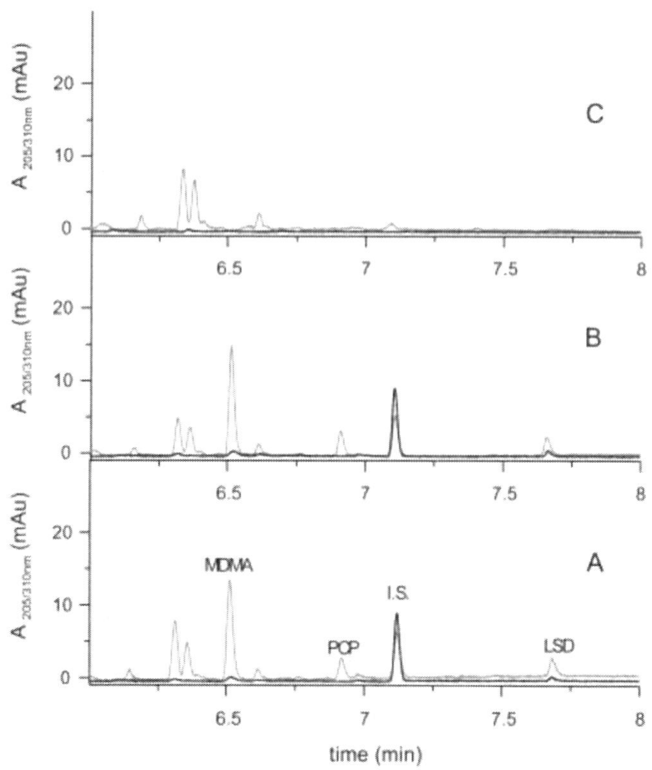

Figure 17. FASI-CZE electropherograms at 205 nm (grey) and 310 nm (black) for urine samples submitted to the DLLME treatment and spiked with 47.5 ng/mL of MDMA, PCP and LSD (A) directly in the sample solution and before applying the treatment, (B) just before injection and (C) blank sample. Reprinted with permission from reference [71]. Copyright (2012) Elsevier.

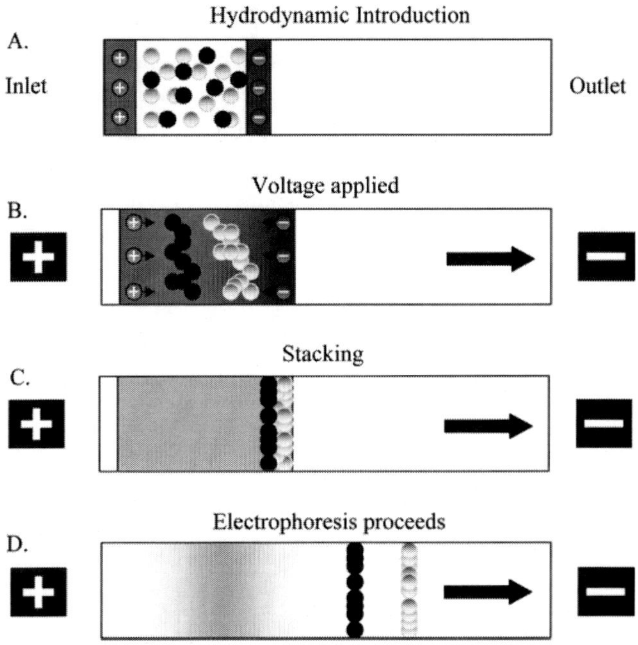

Figure 18. Schematic illustration of the pH-mediated sample stacking procedure for cationic analytes. Reprinted with permission from reference [79]. Copyright (2008) American Chemical Society.

FASI-CZE has also been employed for the analysis of biological matrices. For instance, Airado-Rodríguez et al. [71] applied a dispersive liquid-liquid microextraction (DLLME) prior to FASI-CZE for the sensitive analysis of 3,4-methylenedioxymethamphetamine (MDMA), phencyclidine (PCP) and lysergic acid diethylamide (LSD) in human urine by CZE. As an example, Figure 17 shows the FASI-CZE electropherograms obtained in the analysis of these abuse drugs in human urine with the proposed DLLME treatment.

High resemblance was observed between electropherograms A and B, which indicates the high efficiency of the proposed DLLME procedure and the absence of significant losses of analytes. The absence of interfering compounds at the migration times of MDMA, PCP and LSD can also be observed in the figure. However, urine presented an interfering compound which migration time is almost the same as that of the I.S. (tetracaine). The authors solved this problem by acquiring also at 310 nm where I.S. presented an absorption maximum and the interfering compound from urine did not absorb.

3.4. pH-Mediated Sample Stacking

The manipulation of electrophoretic mobilities by changing pH values between the sample region and the BGE region can also be employed as a way of preconcentrating weakly acidic and basic analytes. This on-line electrophoretic-based preconcentration method is referred as pH-mediated sample stacking.

This method is very useful in order to achieve field-amplified stacking of analytes in high-ionic strength samples without the need for a dilution or any other extraction step. pH-mediated sample stacking allows the on-line titration of a high-ionic-strength sample matrix to a low ionic strength one. The method can be performed to either cationic or anionic analytes by "acid stacking" [76, 77] or "base stacking" [78], respectively. Figure 18 shows the schematic illustration of the pH-mediated sample stacking procedure for cationic analytes. First, a NH_4OH plug, the sample, and a 4 M formic acid plug are sequentially introduced, under pressure, to a fused-silica capillary that has been previously filled with the BGE (A). Then, upon the application of a voltage in positive polarity (cathode in the outlet position), H^+ ions from the 4 M formic acid plug enters the sample zone and, together with H^+ already present in the sample zone, are titrated against OH^- ions from the NH_4OH plug (B). At the point of neutrality, solute ions are stacked into narrow bands at the boundary of the titrated region and BGE (C). Finally, electrophoretic separation proceeds (D).

With this kind of on-line preconcentration technique, mass-loading capacity can be increased without degradation in peak shape, and resolution is dramatically improved.

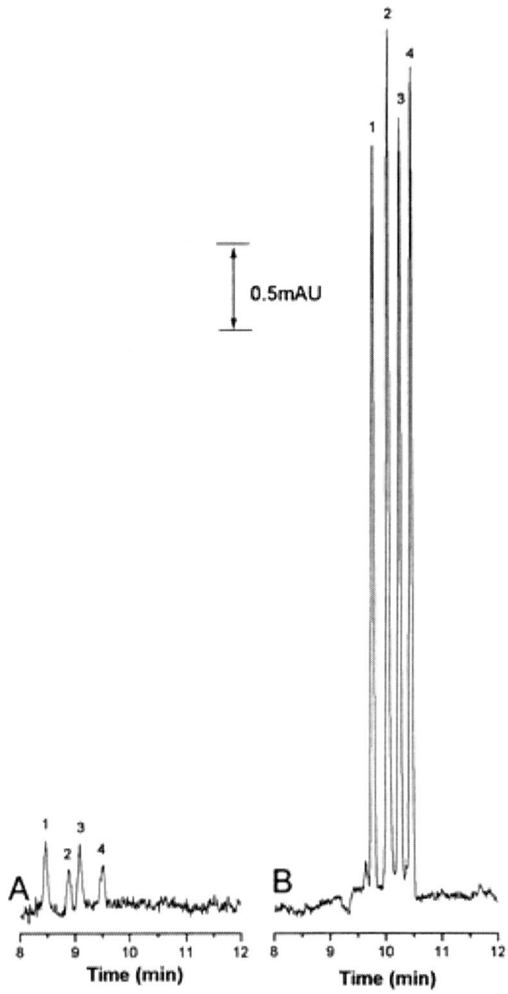

Figure 19. Comparison of normal electrokinetic injection (A) and pH-mediated field amplification stacking (B). (A) Without stacking, sample was injected for 2 s at 0.5 kV with 15 kV for separation. (B) With stacking, sample was injected for 30 s and hydroxide was injected for 65 s, both at 15 kV. Sample: 10 μM analytes in 90% Ringer's solution. Peak identification: (1), p-hydroxybenzoic acid; (2), vanillic acid; (3), p-coumaric acid; and (4) syringic acid. Reprinted with permission from reference [78]. Copyright (1999) American Chemical Society.

Table 3. Selection of methods using pH-mediated sample stacking in CZE

Compounds	Samples	pH-mediated sample stacking conditions	CZE conditions	Detection	Analysis time	LODs	Ref.
Phenolic acids (*p*-hydroxybenzoic, vanillic, *p*-coumaric and syringic acids)	Physiological samples	*Sample matrix*: 90% Ringer's solution *Sample electrokinetic injection*: 30 s, -15 kV *Hydroxide electrokinetic injection*: 65 s, -15 kV (Ringer's solution: 155 mM NaCl, 5.5 mM KCl, 2.3 mM $CaCl_2$ at pH 7.4)	Uncoated fused silica capillaries of 75 µm i.d. Double-capillary: 50 cm x 70 cm one (effective length 50 cm on the second capillary) *BGE:* 100 mM ammonium hydroxide (pH 9.3) with 0.5 mM TTAB *Capillary voltage:* -15 kV	UV: 275 nm	11 min	0.3 µM	[78]
7 noncatecholamine cations	Physiological samples	*Sample matrix*: BGE *Sample electrokinetic injection*: 10 s, +5 kV *Acid injection*: 16 s, +5 kV	Uncoated fused silica capillary of 70 cm (50 cm effective length) x 75 µm I.D. *BGE:* 100 mM sodium acetate buffer pH 4.75 *Capillary voltage:* +20 kV	UV: 220 nm	7 min	-	[77]
Coumarin metabolites	Microsomal incubations	*Sample matrix*: BGE *Sample electrokinetic injection*: 45 s, -10 kV *Hydroxide injection*: 90 s, -10 kV	Uncoated fused silica capillary of 61.2 cm (50 cm effective length) x 50 µm I.D. *BGE:* 25 mM phosphate buffer (pH 7.5) *Capillary voltage:* -20 kV	UV: 214 nm	8 min	0.1-0.5 µM	[80]
Glutathione (GSH) and glutathione disulfide (GSSG)	Rat liver microdialysate samples	*Sample matrix*: Ringer's solution *Sample electrokinetic injection*: 30 s, -10 kV *Hydroxide injection*: 60 s, -10 kV	Uncoated fused silica capillary of 60 cm (45 cm effective length) x 50 µm I.D. *BGE:* 100 mM ammonium chloride with 0.5 mM TTAB pH 8.4 (adjusted with 0.1 M sodium hydroxide) *Capillary voltage:* -10 kV	UV: 214 nm	10 min	0.25-0.75 µM	[81]
Amino acids and organic acids	Urine	*Sample matrix*: BGE *14% stronger ammonia water solution (v/v) hydrodynamic injection*: 3 s, 50 mbar *Sample hydrodynamic injection*: 40 s, 50 mbar	Uncoated fused silica capillary of 60 cm x 50 µm I.D. *BGE:* 4.0% aqueous formic acid solution containing 2.5% methanol *Capillary voltage:* +20 kV	MS single quadrupole analyzer (-) electro-spray	14 min	0.004-1.27 µM	[82]
Homocysteine and cysteine	Human plasma	*Sample matrix*: BGE *Sample electrokinetic injection*: 60 s, +20 kV *Hydrochloric acid injection*: 96 s, +20 kV	Uncoated fused silica capillary of 100 cm (50 cm effective length) x 75 µm I.D. *BGE:* 0.1 M lithium acetate buffer (pH 4.75) *Capillary voltage:* +30 kV	UV: 355 nm	16 min	2 µM (LOQ)	[83]

Table 3 summarizes a selection of CZE methods employing pH mediated sample stacking [77, 78, 80-83]. Many of the applications of this on-line electrophoretic-based preconcentration method focus on the analysis of samples from biological origin because of their high-ionic strength matrices. For instance, Zhao et al. [78] described the application of pH-mediated field amplification on-column preconcentration of anions in physiological samples by CZE. This requires reversal of the EOF direction using TTAB and the separation in negative polarity in order that anions electromigrate in the same direction as the EOF and toward the detector end of the capillary. The BGE was made from the salt of a weak base, such as ammonium, and neutralization is achieve by electrokinetic injection of hydroxide ions. A limitation of pH-mediated field amplification stacking is that a significant portion of the separation capillary is used for the stacking process of the analytes, leaving little capacity for the separation. This limits the amount of sample that can be injected while maintaining a sufficient separation. The authors overcome this limitation by using a double-capillary system in which one capillary was used for stacking and the other was used for the separation. As an example, Figure 19 shows the comparison of electropherograms using normal electrokinetic injection in CZE relative to pH-mediated field-amplification stacking.

A LOD value of 0.3 μM was achieved for the phenolic acids in Ringer's solution using simple UV-absorbance detection by pH-mediated stacking. This represented a 66-fold sensitivity enhancement relative to normal electrokinetic injection, and a 100-fold improvement relative to hydrodynamic injection.

Weiss et al. [77] investigated the pH-mediated field-amplified sample stacking of seven pharmaceutical noncatecholamine cations such as eletripan, dofetilide, doxazosin or sildenafil in high-ionic strength samples such as those of physiological origin. These compounds were chosen because they were cationic at the working BGE pH. In this work, the authors compared the capillary electrophoretic behavior of samples in BGE with those of samples in Ringer's solution (155 mM NaCl, 5.5 mM KCl, 2.3 mM $CaCl_2$ at pH 7.4) with and without pH-mediated acid stacking. Results indicated that the peak heights and efficiencies for acid-stacked samples increased compared to the unstacked samples in Ringer's solution or BGE. For example, the peak efficiencies for 5 s injections of eletriptan in BGE and Ringer's solution were 138,000 and 72,000 plates, respectively. In contrast, a 10 s injection of the same compound followed by acid injection for 16 s (pH-mediated sample stacking) produces a peak with 246,000 plates. Using the proposed pH-mediated acid stacking method a 10- to 27-fold sensitivity enhancement for the seven studied cations was achieved.

Hoque et al. [81] used pH-mediated base stacking for the determination of glutathione (GSH) and glutathione disulfide (GSSG) in the analysis of rat liver microdialysis samples by CZE. A 26-fold increase in sensitivity was achieved for both GSH and GSSG using this on-line preconcentration method in comparison with normal injection without stacking. LODs down to 0.75 μM and 0.25 μM for GSH and GSSG, respectively, were obtained. As an example, Figure 20 shows the electropherograms obtained for unspiked liver microdialysate (A) and a spiked liver microdialysate (B) samples with the proposed pH-mediated base stacking procedure.

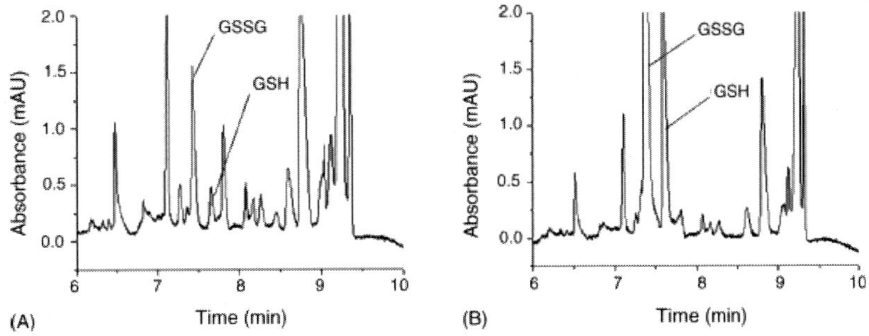

Figure 20. Electropherograms of (A) unspiked liver microdialysate and (B) spiked liver microdialysate (5 μL microdialysate spiked with 1 μL of 100 μM GSH and 1 μL of 100 μM GSSG standard solutions). Conditions as indicated in Table 3. Reprinted with permission from reference [81]. Copyright (2005) Elsevier.

The method was successfully used to determine GSH and GSSG in liver microdialysates of Sprague-Dawley male rats and could be employed in the future to monitor the GSH and GSSG concentration change during oxidative stress (e.g., ischemia and reperfusion) for the better understanding of antioxidant activity of GSH.

3.5. Electrokinetic Supercharging

A relatively recent on-line electrophoretic-based preconcentration method for CZE that has great potential is electrokinetic supercharging (EKS). This method is the combination of electrokinetic injection under field-amplified stacking conditions (FASI) and transient isotachophoresis (tITP) and was first described for the analysis of rare-earth ions by the group of Professor Hirokawa [84, 85]. EKS was developed to extend the range of FASI and is performed by hydrodynamic injection of a leading electrolyte (L), followed by electrokinetic injection of the analytes, and finally hydrodynamic injection of a terminating electrolyte (T). Figure 21 shows a schematic representation of the steps used in EKS.

Upon applying the separation voltage the diffuse band of analytes introduced during electrokinetic injection is stacked between the leading and the terminating electrolytes by tITP until the ITP stage destacks and the analytes are allowed to separate by conventional CZE. EKS is an exceptionally simple but powerful approach to on-line sample preconcentration and has been shown to improve the sensitivity of analytical response by several orders of magnitude.

Table 4 summarizes a selection of publications using electrokinetic supercharging in CZE [86-96]. As an example, Busnel et al. [88] applied EKS for the highly efficient preconcentration of β-lactoglobulin tryptic digest peptides in CZE. Sensitivity enhancement factors between 1,000 and 10,000 whilst maintaining a satisfactory resolution were achieved.

EKS has been employed for the analysis of non-steroidal anti-inflammatory drugs (nSAIDs) or hypolipidaemic drugs in water samples. For instance, Professor Haddad's group proposed the use of EKS on-line preconcentration for the analysis of seven NSAIDs in wastewater samples [86].

Table 4. Selection of publications using electrokinetic supercharging in CZE

Compounds	Samples	EKS conditions	CZE conditions	Detection	Analysis time	LODs	Ref.
Fe(II), Co(II) and Ni(II)	Water	*L*: BGE containing 0.1 mM *o*-Phe (1 min, 1 bar) + 0.1 mM *o*-Phe in 50 mM imidazole and 53 mM lacit acid, pH 4.9 (5 s, 50 mM) + 1 M HCl (1.5 s, 50 mM) + BGE (50 s, 50 mbar) + 200 mM ammonia-215 mM lactic acid, pH 4.9 (7 s, 50 mbar) *Sample EK injection*: 60 s, +30 kV *T*: 1 M HCl (3 s, 50 mbar)	Uncoated fused silica capillary of 68.5 cm (60 cm effective length) x 75 µm I.D. *BGE:* 30 mM creatinine and 18 mM lactic acid (pH 4.9) *Capillary voltage:* +30 kV	indirect-UV: 214 nm	8 min	1-30 ng/L	[87]
Peptides	β-lactoglobulin tryptic digest	*L*: 935 mM ammonium acetate pH 9.3 (90 s, 83 mbar) + BGE (20 s, 30 mbar) *Sample EK injection*: 20 min, +30 kV *T*: BGE	Uncoated fused silica capillary of 60 cm (50 cm effective length) x 50 µm I.D. *BGE:* 115 mM ammonium acetate buffer (pH 4.0) *Capillary voltage:* +30 kV	UV: 200 nm	20 min	-	[88]
NSAIDs	Wastewater	*L*: 100 mM sodium chloride(30 s, 50 mbar) *Sample EK injection*: 200 s, -10 kV *T*: 100 mM 2-(cyclohexylamino)ethanosulphonic acid (40 s, 50 mbar)	Uncoated fused silica capillary of 85 cm (76.6 cm effective length) x 50 µm I.D. *BGE:* 15 mM sodium tetraborate (pH 9.2) with 0.1% (*w/v*) hexadimethrine (HDMB) and 10% (*v/v*) methanol *Capillary voltage:* -28 kV	UV: 214 nm	10 min	50-180 ng/L	[86]
NSAIDs	Wastewater	*Counter-flow EKS* *L*: BGE + water *Sample EK injection*: 220 s, -16 kV combined with negative hydrodynamic pressure of 50 mbar to counter-balance EOF *T*: 100 mM 2-(cyclohexylamino)ethanosulphonic acid (48 s, 50 mbar)	Uncoated fused silica capillary of 85 cm (76.6 cm effective length) x 50 µm I.D. *BGE:* 15 mM sodium tetraborate (pH 9.2) with 0.1% (*w/v*) hexadimethrine (HDMB) and 10% (*v/v*) methanol *Capillary voltage:* -28 kV	UV: 214 nm	10 min	10-47 ng/L	[89]

Compounds	Samples	EKS conditions	CZE conditions	Detection	Analysis time	LODs	Ref.
NSAIDs	River water and human plasma	*L*: BGE + methanol (3 s, 50 mbar) *Sample EK injection*: 700 s, -2 kV *T*: 50 mM 2-(cyclohexylamino)ethanosulphonic acid (12 s, 50 mbar)	Uncoated fused silica capillary of 88.5 cm (80 cm effective length) x 50 µm I.D. *BGE:* 10 mM sodium tetraborate (pH 8 adjusted with NaOH) + 50 mM sodium chloride with 10% (*v/v*) methanol *Capillary voltage:* -30 kV	UV: 214 nm	9.5 min	0.9-2 µg/L	[90]
Hypolipidaemic drugs	Water	*L*: BGE + water (5 s, 40 mbar) *Sample EK injection*: 170 s, -10 kV *T*: 1 mM 3-(cyclohexylamino)-1-propanesulphonic acid (10 s, 50 mbar)	Uncoated fused silica capillary of 88 cm x 75 µm I.D. *BGE:* 60 mM ammonium hydrogen carbonate (ph 9.0) with 60% methanol *Capillary voltage:* +25 kV	MS ion trap mass analyzer (-) electrospray	24 min	180 ng/L	[91]
7 rare-earth metal ions	Water	*Sample EK injection*: 17 mL inlet vials with stirring at 10 kV for 250 s	Uncoated fused silica capillary of 50 cm (37.7 cm effective length) x 75 µm I.D. *BGE:* 10 mM 4-methylbenzylamine, 4 mM 2-hydroxyisobutyric acid, 0.4 mM malonic acid and 0.1% hydroxypropyl cellulose with pH 4.8 adjusted by adding 2-ethylbutyric acid *Capillary voltage:* +30 kV	UV: 214 nm	15 min	1 ng/L	[92]
Flavonoids	Aqueous extract of *Clematis hexapetala* pall	*L*: BGE *Sample EK injection*: 130 s, -10 kV *T*: 100 mM 2-(cyclohexylamino)ethanesulfonic acid (17 s, 0.5 psi)	Uncoated fused silica capillary of 60.2 cm (50 cm effective length) x 50 µm I.D. *BGE:* 30 mM sodium tetraborate (pH 9.5) containing 5% (*v/v*) of methanol *Capillary voltage:* -20 kV	UV: 254 nm	12 min	2.0-6.8 µg/L	[93]

Table 4. (Continued)

Compounds	Samples	EKS conditions	CZE conditions	Detection	Analysis time	LODs	Ref.
NSAIDs	Water	*Pressure-assisted EKS* *L*: BGE + water (3 s, 50 mbar) *Sample EK injection*: 20 min, -14 kV combined with positive hydrodynamic pressure of 50 mbar *T*: 8 mM 3-(cyclohexylamino)-1-propanesulphonic acid (20 s, 50 mbar)	Uncoated fused silica capillary of 80 cm (71.5 cm effective length) x 50 μm I.D. *BGE:* 50 mM ammonium hydrogen carbonate (pH 9.2) with 10% methanol *Capillary voltage:* -28 kV	UV: 214 nm	10 min	6.7-18.7 ng/L	[94]
Catecholamines	Standards	*Counter-flow EKS* *L:* BGE (3 min, 40 psi) *Sample EK injection:* 90 min, +30 kV combined with a counter pressure of 1.2 psi *T:* 75 mM β-alanine tritrated to pH 4.0 with 130 mM acetic acid, with 0.05% (wt) of FC-430	μ-Sil-FC coated capillary of 50 cm x 50 μm I.D. *BGE:* 100 mM triethylamine titrated to pH 5.0 with 150 mM acetic acid, with 0.05% (wt) of a fluorocarbon surfactant (FC-430) *Capillary voltage:* +30 kV	UV: 200 nm	12 min	1.2-1.4 nM	[95]
Barbiturate drugs	Urine	*L*: 50 mM sodium chloride (120 s, 50 mbar) *Sample EK injection*: 300 s, -8.5 kV *T*: 100 mM 2-(cyclohexylamino)ethanesulfonic acid (140 s, 50 mM)	Uncoated fused silica capillary of 100 cm (91.5 cm effective length) x 50 μm I.D. *BGE:* 20 mM sodium tetraborate (pH 9.15) buffer *Capillary voltage:* -30 kV	UV: 214 nm	10.5 min	1.5-2.1 μg/L	[96]

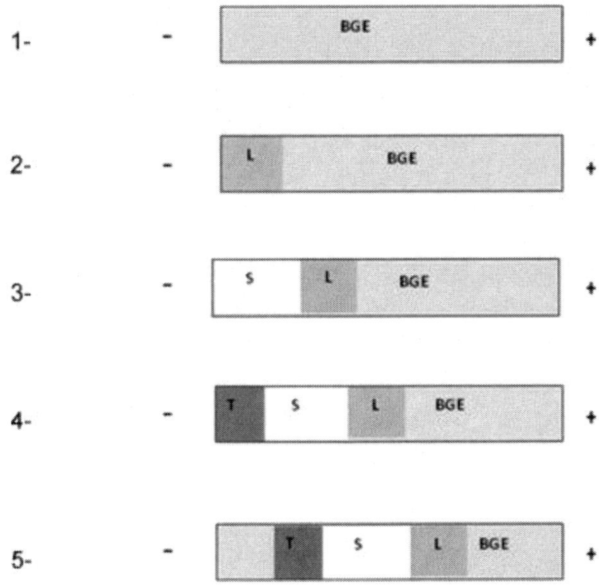

Figure 21. Schematic representation of the steps used in EKS: (1) filling the capillary with BEG, (2) hydrodynamic injection of leading electrolyte (L), (3) electrokinetic injection of sample (S), (4) hydrodynamic injection of terminating electrolyte (T), and (5) starting tITP-CZE. Reprinted with permission from reference [86]. Copyright (2008) Elsevier.

They examined the application of FASI and found an improvement in detection limits by 200-fold providing LODs down to 0.6-2.0 μg/L, which were insufficient for the determination of NSAIDs as environmental pollutants in water samples. Sensitivity was then improved by EKS. The optimum EKS method involved the hydrodynamic injection of a leading electrolyte (L: 100 mM of sodium chloride, 30 s, 50 mbar), the electrokinetic injection of the sample for a long time (200 s, -10 kV), and finally the hydrodynamic injection of a terminating electrolyte (T: 100 mM of 2-(cyclohexylamino) ethanesulphonic acid, 40 s, 50 mbar). A 2,400-fold sensitivity enhancement was achieved with this method, with LODs ranging from 50 to 180 ng/L. The proposed method was validated and applied to the analysis of wastewater samples. As an example, Figure 22 shows the electropherogram obtained with EKS of a wastewater sample spiked with 20 μg/L of targeted NSAIDs, as well as the one from a wastewater blank sample.

In a later work, some modification of the method by combining the application of an additional pressure during the electrokinetic injection of the sample was proposed [89]. For instance, counter-flow electrokinetic supercharging (CF-EKS) performed by applying a negative hydrodynamic pressure of 50 mbar to counter-balance the EOF was also evaluated for the analysis of NSAIDs in wastewater, achieving a 11,800-fold sensitivity enhancement and LODs ranging from 10.7 to 47 ng/L [89]. Pressure-assisted electrokinetic supercharging (PA-EKS) was also evaluated for the analysis of this family of compounds [94]. In this case, a positive hydrodynamic pressure of 50 mbar during sample injection to improve stacking of NSAIDs was used. Sensitivity enhancements up to 50,000 fold were observed with LODs down to 6.7 ng/L. As an example, Figure 23 shows an example of PA-EKS application for the analysis of 1 μg/L of ibuprofen, ketoprofen and diflunisal.

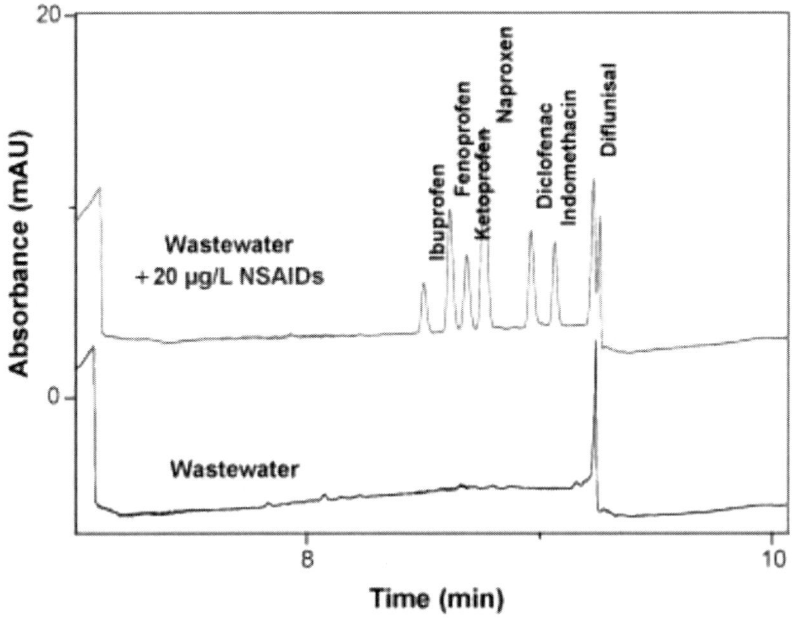

Figure 22. Electropherogram obtained from electrokinetic supercharging of (A) wastewater sample spiked with 20 μg/L of the NSAIDs and (B) blank wastewater sample. CZE conditions as described in Table 4. Reprinted with permission from reference [86]. Copyright (2008) Elsevier.

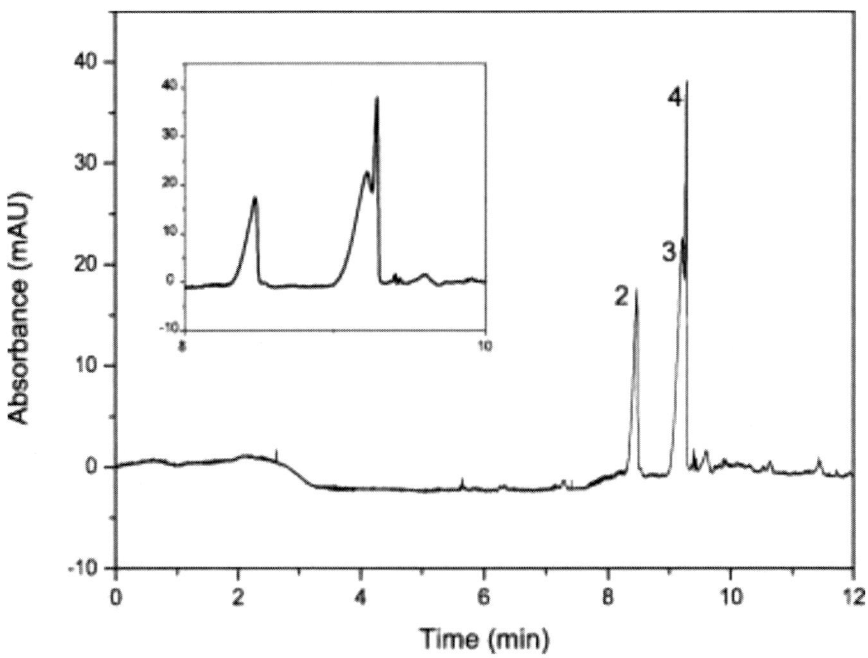

Figure 23. PA-EKS analysis of 1 μg/L of ibuprofen (2), ketoprofen (3) and diflunisal (4). Inset: Enlarged image of the peaks. Conditions as described in Table 4. Reprinted with permission from reference [94]. Copyright (2011) Elsevier.

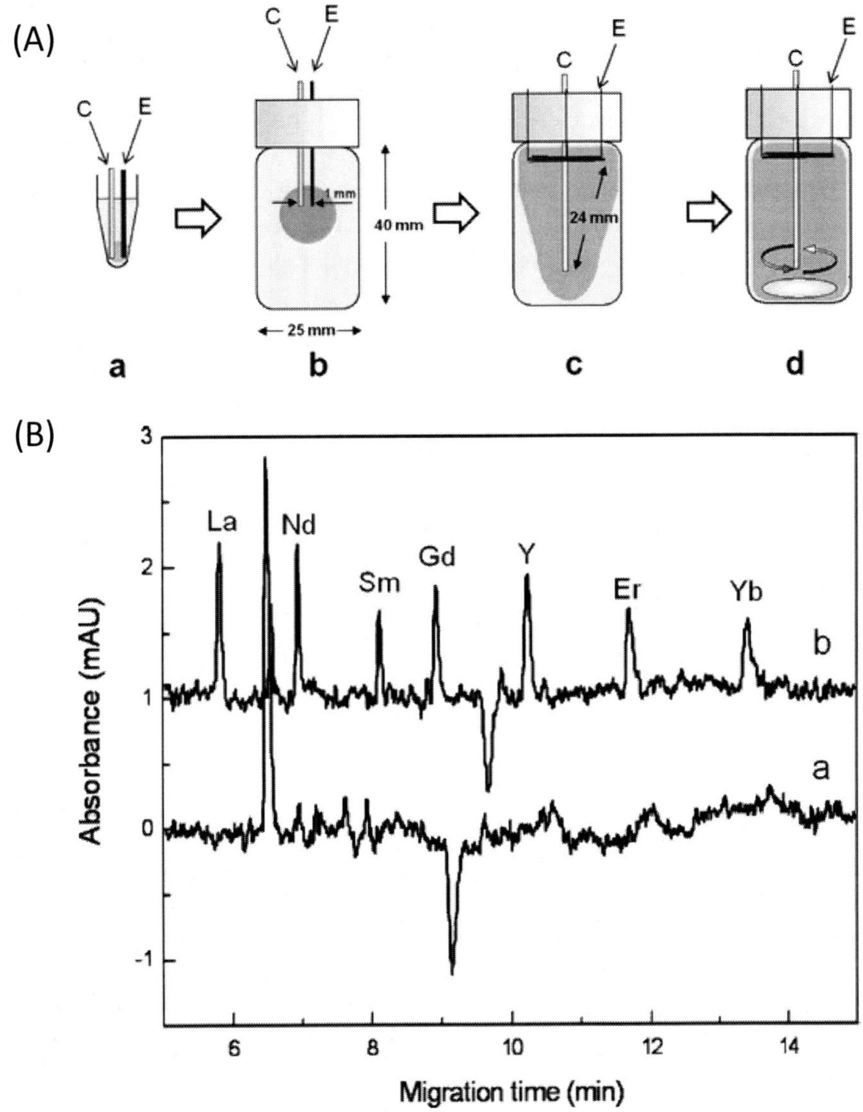

Figure 24. (A) Schematic of sequential alterations of the injection setup implemented to improve the sensitive of EKS-CZE: a wire electrode in (a) a standard and (b) a large-volume sample vial; a ring electrode in a large-volume sample vial (c) without and (d) with stirring. The darker area depicts the actual part of the sample solution subjected to injection (not to scale). C and E stand for capillary and electrode, respectively. (B) Electropherograms of (a) blank and (b) 25 pM sample analyzed at the optimized EKS conditions (see Table 4). Sample: 100,000-fold diluted at 25 pM. Reprinted with permission from reference [92]. Copyright (2011) American Chemical Society.

A similar methodology was proposed by Professor Haddad's group for the analysis of hypolipidaemic drugs in water samples using CE-MS with electrospray as ionization source and an ion trap as mass analyzer [91]. The electrophoretic separation was carried out by counter-EOF conditions by reversing EOF with hexadimethrine bromide. Using EKS, the sensitivity of the method was improved 1,000-fold in comparison to injection under FASI conditions, obtaining LODs down to 180 ng/L.

Xu et al. [92] proposed an interesting approach to achieve over 100,000-fold sensitivity increase in CZE by electrokinetic supercharging with an optimized sample injection. Figure 24A shows a schematic of the sequential alterations of the injection setup evaluated to improve EKS-CZE. The authors increased the volume of the sample vial from typical 500 μL to 17 mL, replaced the common wire electrode by a ring electrode, and stirred the sample solution during the electrokinetic injection. This increases the area of the sample solution subjected to injection and, consequently, more analyte ions are accumulated within the effective electric field and then maintained as focused zones due to the transient isotachophoresis. The versatility of this customized EKS-CZE approach for sample concentration was demonstrated for a mixture of seven rare-earth metal ions. Figure 24B shows the electropherograms obtained with the EKS-CZE analysis of these metals in (a) a blank sample and (b) a 25 pM sample. An enrichment factor of 500,000 was achieved with LODs at or even below 1 ng/L. These LOD values are over 100,000 times better than those that can be achieved by normal hydrodynamic injection, 1000 times better than the sensitivity thresholds of inductively coupled plasma atomic emission spectrometry (ICP-AES) (0.1-2 μg/L), and even close to those of inductively coupled plasma mass spectrometry (ICP-MS) (0.1-0.9 ng/L).

Zhong et al. [93] proposed EKS-CZE for the analysis of four flavonoids (Naringenin, Hesperetin, Naringin, and Herperidin) in a Chinese herbal medicine (*Clematis hexapetala* pall). Under EKS-CZE conditions (see Table 4) the four flavonoids could be separated with a sample-to-sample time of 15 minutes and LODs from 2.0 to 6.8 ng/L. When compared to a conventional hydrodynamic injection the sensitivity was between 824 and 1,515 times which is 7.6-16 times higher than other CZE methods used for the on-line concentration of flavonoids.

Regarding the use of EKS-CZE in the analysis of biological fluids, Botello et al. [96] recently reported an EKS-CZE method for the separation and preconcentration of barbiturate drugs in urine samples. The obtained results showed that the EKS strategy enhanced detection sensitivity around 1,050-fold compared with normal hydrodynamic injection, providing LODs ranging from 1.5 and 2.1 ng/mL for standard samples with good repeatability in terms of peak area (RSD values lower than 3%). The applicability of the optimized method was demonstrated by the analysis of human urine samples spiked with the studied compounds. LODs obtained in urine samples, after a liquid-liquid extraction step used as clean-up procedure, ranged between 8 and 15 ng/mL.

CONCLUSION AND FUTURE TRENDS

Fundamentals aspects of capillary zone electrophoresis regarding theoretical principles (electrophoretic mobility and electroosmotic flow) and sample introduction (hydrodynamic *vs* electrokinetic injection) have been addressed. CZE is becoming a popular technique because of the simplicity of the instrumentation required and its versatility of applications. A good selection of the voltage configuration (cathodic or anodic separation) and the magnitude of the EOF velocity (by changing BGE composition –buffer type, concentration, pH– and separation temperature) will allow analysts to achieve good electrophoretic separations of complex matrices under CZE.

CZE is becoming very popular in multiple application fields such as bio-analysis, food control and safety and environmental applications, and many publications can be found in the literature. However, when low concentration samples are expected to be found, CZE requires the application of off-line and/or on-line preconcentration methods to improve sensitivity. However, today the low sensitivity characteristic of conventional CZE techniques where UV-detection is performed by using the inner-diameter of the capillary as optical path length cannot be considered at all a real handicap of CZE. Many electrophoretic-based on-line preconcentration methods are available, and the fundamentals of some of them based on stacking phenomena, such as normal sample stacking, large-volume sample stacking, field-amplified sample injection, pH-mediated sample stacking, and electrokinetic supercharging have been presented and discussed. Examples of relevant applications in bio-analytical, food and environmental analysis of these on-line preconcentration methods have also been addressed. Huge sensitivity enhancements (for instance up to 100,000-fold by using EKS) can be achieved with CZE for environmental applications without any special instrumental requirement.

The time has arrived for CZE techniques for the practical and routine ultra-trace analysis by using on-line electrophoretic-based preconcentration techniques. Today analysts are playing an important role in exploring multiple possibilities of on-line preconcentration methods in CZE by combining existing procedures with new ones, which is making CZE a very promising technique for future applications in many disciplines, and sure the number of publications will be increasing in the future.

ACKNOWLEDGMENTS

This work has been funded by the Spanish Ministry of Economy and Competitiveness under the project CTQ2012-30836, and from the Agency for Administration of University and Research Grants (Generalitat de Catalunya, Spain) under the project 2014 SGR-539.

REFERENCES

[1] Reuss, F. F. (1809). Memoires de la Societe Imperiale des Naturalistes de Moscou, vol *2*, p. 327.

[2] Landers, J. P. (1994) Handbook of Capillary Electrophoresis. CRC Press, Boca Raton, USA.

[3] Kuhn, R. & Hoffstetter-Kuhn, S. (1993) Capillary Electrophoresis: Principles and Practice. Springer-Verlag, Berlin, Germany.

[4] Hjerten, S. (1967). Free zone electrophoresis. *Chromatogr Rev*, *9*, 122-219.

[5] Hjerten, S. (1970). Free zone electrophoresis. Theory, equipment, and applications. *Methods Biochem Anal*, *18*, 55-79.

[6] Jorgenson, J. W. & Lukacs, K. D. (1981). Zone electrophoresis in open-tubular glass capillaries: preliminary data on performance. *HRC CC, J. High Resolut. Chromatogr. Chromatogr. Commun.*, *4*, 230-231.

[7] Jorgenson, J. W. & Lukacs, K. D. (1981). Zone electrophoresis in open-tubular glass capillaries. *Anal. Chem.*, *53*, 1298-1302.
[8] Holcomb, R. E. (2010). Bioanalytical applications of capillary electrophoresis and microfluidics: From metabolomics to biofuels. Thesis Dissertation. Ann Arbor, USA.
[9] Kostal, V., Katzenmeyer, J. & Arriaga, E. A. (2008). Capillary electrophoresis in bioanalysis. *Anal Chem*, *80*, 4533-4550.
[10] Kohler, I., Schappler, J. & Rudaz, S. (2013). Microextraction techniques combined with capillary electrophoresis in bioanalysis. *Anal Bioanal Chem*, *405*, 125-141.
[11] Locatelli, M. & Carlucci, G. (2010). Advanced capillary electrophoresis techniques in the analytical quantification of drugs, metabolites and biomarkers in biological samples. *Global J. Anal. Chem.*, *1*, 244-261.
[12] Tseng, H. M., Li, Y. & Barrett, D. A. (2010). Bioanalytical applications of capillary electrophoresis with laser-induced native fluorescence detection. *Bioanalysis*, *2*, 1641-1653.
[13] Astefanei, A., Núñez, O. & Galceran, M. T. (2012). Non-aqueous capillary electrophoresis separation of fullerenes and C60 fullerene derivatives. *Anal. Bioanal. Chem.*, *404*, 307-313.
[14] Xie, N., Ding, X., Wang, X., Wang, P., Zhao, S. & Wang, Z. (2014). Determination of thioglycolic acid in cosmetics by capillary electrophoresis. *J. Pharm. Biomed. Anal.*, *88*, 509-512.
[15] Dolzan, M. D., Spudeit, D. A., Azevedo, M. S., Costa, A. C., Leal de Oliveira, M. A. & Micke, G. A. (2013). A fast method for simultaneous analysis of methyl, ethyl, propyl and butylparaben in cosmetics and pharmaceutical formulations using capillary zone electrophoresis with UV detection. *Anal. Methods*, *5*, 6023-6029.
[16] Tsai, I. C., Su, C. Y., Hu, C. C. & Chiu, T. C. (2014). Simultaneous determination of whitening agents and parabens in cosmetic products by capillary electrophoresis with on-line sweeping enhancement. *Anal. Methods*, *6*, 7615-7620.
[17] Ma, H., Wang, L., Liu, H., Luan, F. & Gao, Y. (2014). Application of a non-aqueous capillary electrophoresis method to the analysis of triclosan in personal care products. *Anal. Methods*, *6*, 4723-4728.
[18] Simo, C., Barbas, C. & Cifuentes, A. (2005). Capillary electrophoresis-mass spectrometry in food analysis. *Electrophoresis*, *26*, 1306-1318.
[19] Cifuentes, A. (2006). Recent advances in the application of capillary electromigration methods for food analysis. *Electrophoresis*, *27*, 283-303.
[20] Herrero, M., Simo, C., Garcia-Canas, V., Fanali, S. & Cifuentes, A. (2010). Chiral capillary electrophoresis in food analysis. *Electrophoresis*, *31*, 2106-2114.
[21] Herrero, M., Garcia-Canas, V., Simo, C. & Cifuentes, A. (2010). Recent advances in the application of capillary electromigration methods for food analysis and Foodomics. *Electrophoresis*, *31*, 205-228.
[22] Ravelo-Perez, L. M., Asensio-Ramos, M., Hernandez-Borges, J. & Rodriguez-Delgado, M. A. (2009). Recent food safety and food quality applications of CE-MS. *Electrophoresis*, *30*, 1624-1646.
[23] Pinero, M. Y., Bauza, R. & Arce, L. (2011). Thirty years of capillary electrophoresis in food analysis laboratories: Potential applications. *Electrophoresis*, *32*, 1379-1393.

[24] Bald, E., Kubalczyk, P., Studzinska, S., Dziubakiewicz, E. & Buszewski, B. (2013). Application of electromigration techniques in environmental analysis. *Springer Ser. Chem. Phys.*, *105*, 335-353.

[25] Ali, I. & Aboul-Enein, H. Y. (2010). Environmental pollutants: CE analysis. *Encycl. Chromatogr. (3rd Ed.)*, 1, 792-801.

[26] Garcia-Campana, A. M., Gamiz-Gracia, L., Lara, F. J., Iruela, M. d. O. & Cruces-Blanco, C. (2009). Applications of capillary electrophoresis to the determination of antibiotics in food and environmental samples. *Anal. Bioanal. Chem.*, *395*, 967-986.

[27] Smoluchowski, M. V. (1905). Elektrosche Kataphorese. *Physik. Z.*, 6, 529-

[28] Zhu, M., Hansen, D. L., Burd, S. & Gannon, F. (1989). Factors affecting free zone electrophoresis and isoelectric focusing in capillary electrophoresis. *J. Chromatogr.*, *480*, 311-319.

[29] CE Expert computer program, Beckman Coulter Inc. Available at: https://www.beckmancoulter.com/wsrportal/wsr/industrial/products/capillary-electrophoresis/ce-expert/index.htm (Assessed January 2015)

[30] Jorgenson, J. W. & Lukacs, K. D. (1983). Capillary zone electrophoresis. *Science*, *222*, 266-272.

[31] Huang, X., Gordon, M. J. & Zare, R. N. (1988). Bias in quantitative capillary zone electrophoresis caused by electrokinetic sample injection. *Anal. Chem.*, *60*, 375-377.

[32] Whatley, H. Basic principles and modes of capillary electrophoresis. In Clinical and Forensic Applications of Capillary Electrophoresis; Petersen, J.R. and Mohammad, A.A. Eds; Humana Press Inc.: Totowa, NJ 2001; p 21.

[33] Nickerson, B. & Jorgenson, J. W. (1988). High sensitivity laser-induced fluorescence detection in capillary zone electrophoresis. *HRC CC, J. High Resolut. Chromatogr. Chromatogr. Commun.*, *11*, 878-881.

[34] Nickerson, B. & Jorgenson, J. W. (1988). High speed capillary zone electrophoresis with laser induced fluorescence detection. *HRC CC, J. High Resolut. Chromatogr. Chromatogr. Commun.*, *11*, 533-534.

[35] Brumley, W. C., Grange, A. H., Kelliher, V., Patterson, D. B., Montcalm, A., Glassman, J. & Farley, J. W. (2000). Environmental screening of acidic compounds based on capillary zone electrophoresis/laser-induced fluorescence detection with identification by gas chromatography/mass spectrometry and gas chromatography/high-resolution mass spectrometry. *J. AOAC Int.*, 83, 1059-1067.

[36] Deng, Y. H., Zhang, H. S. & Wang, H. (2008). Rapid and sensitive determination of phosphoamino acids in phosvitin by N-hydroxysuccinimidyl fluorescein-O-acetate derivatization and capillary zone electrophoresis with laser-induced fluorescence detection. *Anal. Bioanal. Chem.*, *392*, 231-238.

[37] Cai, J. & Henion, J. (1995). Capillary electrophoresis-mass spectrometry. *J. Chromatogr. A*, 703, 667-692.

[38] Schmitt-kopplin, P. & Frommberger, M. (2003). Capillary electrophoresis - mass spectrometry: 15 years of developments and applications. *Electrophoresis*, *24*, 3837-3867.

[39] Robledo, V. R. & Smyth, W. F. (2009). The application of CE-MS in the trace analysis of environmental pollutants and food contaminants. *Electrophoresis*, *30*, 1647-1660.

[40] Robledo, V. R. & Smyth, W. F. (2014). Review of the CE-MS platform as a powerful alternative to conventional couplings in bio-omics and target-based applications. *Electrophoresis*, *35*, 2292-2308.

[41] Hempel, G. (2000). Strategies to improve the sensitivity in capillary electrophoresis for the analysis of drugs in biological fluids. *Electrophoresis*, *21*, 691-698.

[42] Pedersen-Bjergaard, S., Rasmussen, K. E. & Gronhaug Halvorsen, T. (2000). Liquid-liquid extraction procedures for sample enrichment in capillary zone electrophoresis. *J. Chromatogr. A*, *902*, 91-105.

[43] Lucci, P., Pacetti, D., Núñez, O. & Frega, N.G. (2012). Current Trends in Sample Treatment Techniques for Environmental and Food Analysis. In *Chromatography: the Most Versatil Method of Chemical Analysis*, ed. L.A. Calderon, InTech Publisher, Rijeka (CR), pp. 127-164, ISBN: 978-953-51-0813-9

[44] Qin, W., Wei, H. & Li, S. F. Y. (2002). Determination of acidic herbicides in surface water by solid-phase extraction followed by capillary zone electrophoresis. *J. Chromatogr. Sci.*, *40*, 387-391.

[45] Veronica, C., Maria de los Angeles, F., Claudio, R. & Fernanda, S. M. (2014). Analysis of phenolic compounds in onion nectar by miniaturized off-line solid phase extraction-capillary zone electrophoresis. *Anal. Methods*, *6*, 4878-4884.

[46] Kitagawa, F. & Otsuka, K. (2014). Recent applications of on-line sample preconcentration techniques in capillary electrophoresis. *J. Chromatogr. A*, *1335*, 43-60.

[47] Breadmore, M. C. & Haddad, P. R. (2001). Approaches to enhancing the sensitivity of capillary electrophoresis methods for the determination of inorganic and small organic anions. *Electrophoresis*, *22*, 2464-2489.

[48] Shihabi, Z. K. (1998). Serum procainamide analysis based on acetonitrile stacking by capillary electrophoresis. *Electrophoresis*, *19*, 3008-3011.

[49] Heller, M., Vitali, L., Oliveira, M. A. L., Costa, A. C. & Micke, G. A. (2011). A Rapid Sample Screening Method for Authenticity Control of Whiskey Using Capillary Electrophoresis with Online Preconcentration. *J. Agric. Food Chem.*, *59*, 6882-6888.

[50] Chien, R. L. & Burgi, D. S. (1992). On-column sample concentration using field amplification in CZE. *Anal. Chem.*, *64*, 489A-496A.

[51] Albert, M., Debusschere, L., Demesmay, C. & Rocca, J. L. (1997). Large-volume stacking for quantitative analysis of anions in capillary electrophoresis. I. Large-volume stacking with polarity switching. *J. Chromatogr. A*, *757*, 281-289.

[52] Núñez, O., Moyano, E., Puignou, L. & Galceran, M. T. (2001). Sample stacking with matrix removal for the determination of paraquat, diquat and difenzoquat in water by capillary electrophoresis. *J. Chromatogr. A*, *912*, 353-361.

[53] Albert, M., Debusschere, L., Demesmay, C. & Rocca, J. L. (1997). Large-volume stacking for quantitative analysis of anions in capillary electrophoresis. II. Large-volume stacking without polarity switching. *J. Chromatogr. A*, *757*, 291-296.

[54] Cugat, M. J., Borrull, F. & Calull, M. (2001). Large-volume sample stacking for on-capillary sample enrichment in the determination of naphthalene- and benzenesulfonates in real water samples by capillary zone electrophoresis. *Analyst (Cambridge, U. K.)*, *126*, 1312-1317.

[55] Ding, W. H. & Liu, C. H. (2001). Analysis of linear alkylbenzenesulfonates by capillary zone electrophoresis with large-volume sample stacking. *J. Chromatogr. A*, *929*, 143-150.

[56] Bermudo, E., Nunez, O., Puignou, L. & Galceran, M. T. (2006). Analysis of acrylamide in food products by in-line preconcentration capillary zone electrophoresis. *J. Chromatogr. A*, *1129*, 129-134.

[57] Soto-Chinchilla, J. J., Garcia-Campana, A. M., Gamiz-Gracia, L. & Cruces-Blanco, C. (2006). Application of capillary zone electrophoresis with large-volume sample stacking to the sensitive determination of sulfonamides in meat and ground water. *Electrophoresis*, *27*, 4060-4068.

[58] Quesada-Molina, C., Garcia-Campana, A. M., Olmo-Iruela, L. & del Olmo, M. (2007). Large volume sample stacking in capillary zone electrophoresis for the monitoring of the degradation products of metribuzin in environmental samples. *J. Chromatogr. A*, *1164*, 320-328.

[59] Bailon-Perez, M. I., Garcia-Campana, A. M., Cruces-Blanco, C. & Olmo Iruela, M. (2007). Large-volume sample stacking for the analysis of seven Î^2-lactam antibiotics in milk samples of different origins by CZE. *Electrophoresis*, *28*, 4082-4090.

[60] Bessonova, E. A., Kartsova, L. A. & Shmukov, A. U. (2007). Electrophoretic determination of albumin in urine using on-line concentration techniques. *J. Chromatogr. A*, *1150*, 332-338.

[61] Quesada-Molina, C., Olmo-Iruela, M. & Garcia-Campana, A. M. (2010). Trace determination of sulfonylurea herbicides in water and grape samples by capillary zone electrophoresis using large volume sample stacking. *Anal. Bioanal. Chem.*, *397*, 2593-2601.

[62] Bernad, J. O., Damascelli, A., Nunez, O. & Galceran, M. T. (2011). In-line preconcentration capillary zone electrophoresis for the analysis of haloacetic acids in water. *Electrophoresis*, *32*, 2123-2130.

[63] Lee, I. S. L., Boyce, M. C. & Breadmore, M. C. (2012). Extraction and on-line concentration of flavonoids in Brassica oleracea by capillary electrophoresis using large volume sample stacking. *Food Chem.*, *133*, 205-211.

[64] Fan, L. Y., He, T., Tang, Y. Y., Zhang, W., Song, C. J., Zhao, X., Zhao, X. Y. & Cao, C. X. (2012). Sensitive determination of barbiturates in biological matrix by capillary electrophoresis using online large-volume sample stacking. *J. Forensic Sci.*, *57*, 813-819.

[65] Núñez, O., Moyano, E. & Galceran, M. T. (2002). Solid-phase extraction and sample stacking-capillary electrophoresis for the determination of quaternary ammonium herbicides in drinking water. *J. Chromatogr. A*, *946*, 275-282.

[66] Tagliaro, F., Manetto, G., Crivellente, F., Scarcella, D. & Marigo, M. (1998). Hair analysis for abused drugs by capillary zone electrophoresis with field-amplified sample stacking. *Forensic Sci. Int.*, *92*, 201-211.

[67] Manetto, G., Tagliaro, F., Crivellente, F., Pascali, V. L. & Marigo, M. (2000). Field-amplified sample stacking capillary zone electrophoresis applied to the analysis of opiate drugs in hair. *Electrophoresis*, *21*, 2891-2898.

[68] Zhu, L. & Lee, H. K. (2001). Field-amplified sample injection combined with water removal by electroosmotic flow pump in acidic buffer for analysis of phenoxy acid herbicides by capillary electrophoresis. *Anal. Chem.*, *73*, 3065-3072.

[69] Bermudo, E., Núñez, O., Moyano, E., Puignou, L. & Galceran, M. T. (2007). Field amplified sample injection-capillary electrophoresis-tandem mass spectrometry for the analysis of acrylamide in foodstuffs. *J. Chromatogr. A*, *1159*, 225-232.

[70] Santalad, A., Burakham, R., Srijaranai, S. & Grudpan, K. (2007). Field-amplified sample injection and in-capillary derivatization for capillary electrophoretic analysis of metal ions in local wines. *Microchem. J.*, *86*, 209-215.

[71] Airado-Rodriguez, D., Cruces-Blanco, C. & Garcia-Campana, A. M. (2012). Dispersive liquid-liquid microextraction prior to field-amplified sample injection for the sensitive analysis of 3,4-methylenedioxymethamphetamine, phencyclidine and lysergic acid diethylamide by capillary electrophoresis in human urine. *J. Chromatogr. A*, *1267*, 189-197.

[72] Alshana, U., Goeger, N. G. & Ertas, N. (2012). Ultrasound-assisted emulsification microextraction for the determination of ephedrines in human urine by capillary electrophoresis with direct injection. Comparison with dispersive liquid-liquid microextraction. *J. Sep. Sci.*, *35*, 2114-2121.

[73] Wang, Y. R., Yang, Y. H., Lu, C. Y., Lin, S. J. & Chen, S. H. (2013). Trace analysis of acetylcholinesterase inhibitors with antipsychotic drugs for Alzheimer's disease by capillary electrophoresis with on column field-amplified sample injection. *Anal. Bioanal. Chem.*, 405, 3233-3242.

[74] Martínez-Villalba, A., Nuñez, O., Moyano, E. & Galceran, M. T. (2013). Field amplified sample injection-capillary zone electrophoresis for the analysis of amprolium in eggs. *Electrophoresis*, *34*, 870-876.

[75] Purrà, M., Cinca, R., Legaz, J. & Núñez, O. (2014). Solid-phase extraction and field-amplified sample injection-capillary zone electrophoresis for the analysis of benzophenone UV filters in environmental water samples. *Anal. Bioanal. Chem.*, *406*, 6189-6202.

[76] Hadwiger, M. E., Torchia, S. R., Park, S., Biggin, M. E. & Lunte, C. E. (1996). Optimization of the separation and detection of the enantiomers of isoproterenol in microdialysis samples by cyclodextrin-modified capillary electrophoresis using electrochemical detection. *J. Chromatogr. B: Biomed. Sci. Appl.*, *681*, 241-249.

[77] Weiss, D. J., Saunders, K. & Lunte, C. E. (2001). pH-Mediated field-amplified sample stacking of pharmaceutical cations in high-ionic strength samples. *Electrophoresis*, *22*, 59-65.

[78] Zhao, Y. & Lunte, C. E. (1999). pH-Mediated Field Amplification On-Column Preconcentration of Anions in Physiological Samples for Capillary Electrophoresis. *Anal. Chem.*, *71*, 3985-3991.

[79] Baidoo, E. E. K., Benke, P. I., Neusuess, C., Pelzing, M., Kruppa, G., Leary, J. A. & Keasling, J. D. (2008). Capillary Electrophoresis-Fourier Transform Ion Cyclotron Resonance Mass Spectrometry for the Identification of Cationic Metabolites via a pH-Mediated Stacking-Transient Isotachophoretic Method. *Anal. Chem. (Washington, DC, U. S.)*, *80*, 3112-3122.

[80] Ward, E. M., Smyth, M. R., O'Kennedy, R. & Lunte, C. E. (2003). Application of capillary electrophoresis with pH-mediated sample stacking to analysis of coumarin metabolites in microsomal incubation. *J. Pharm. Biomed. Anal.*, *32*, 813-822.

[81] Hoque, M. E., Arnett, S. D. & Lunte, C. E. (2005). On-column preconcentration of glutathione and glutathione disulfide using pH-mediated base stacking for the analysis

of microdialysis samples by capillary electrophoresis. *J. Chromatogr. B: Anal. Technol. Biomed. Life Sci.*, *827*, 51-57.

[82] Yu, L., Jiang, C., Huang, S., Gong, X., Wang, S. & Shen, P. (2013). Analysis of urinary metabolites for breast cancer patients receiving chemotherapy by CE-MS coupled with on-line concentration. *Clin. Biochem.*, *46*, 1065-1073.

[83] Kubalczyk, P., Bald, E., Furmaniak, P. & Glowacki, R. (2014). Simultaneous determination of total homocysteine and cysteine in human plasma by capillary zone electrophoresis with pH-mediated sample stacking. *Anal. Methods*, *6*, 4138-4143.

[84] Okamoto, H. & Hirokawa, T. (2003). Application of electrokinetic supercharging capillary zone electrophoresis to rare-earth ore samples. *J. Chromatogr., A*, *990*, 335-341.

[85] Hirokawa, T., Okamoto, H. & Gas, B. (2003). High-sensitive capillary zone electrophoresis analysis by electrokinetic injection with transient isotachophoretic preconcentration: Electrokinetic supercharging. *Electrophoresis*, *24*, 498-504.

[86] Dawod, M., Breadmore, M. C., Guijt, R. M. & Haddad, P. R. (2008). Electrokinetic supercharging for on-line preconcentration of seven non-steroidal anti-inflammatory drugs in water samples. *J Chromatogr A*, *1189*, 278-284.

[87] Urbanek, M., Delaunay, N., Michel, R., Varenne, A. & Gareil, P. (2007). Analysis of sub-ppb levels of Fe(II), Co(II), and Ni(II) by electrokinetic supercharging preconcentration, CZE separation, and in-capillary derivatization. *Electrophoresis*, *28*, 3767-3776.

[88] Busnel, J. M., Lion, N. & Girault, H. H. (2008). Electrokinetic supercharging for highly efficient peptide preconcentration in capillary zone electrophoresis. *Electrophoresis*, *29*, 1565-1572.

[89] Dawod, M., Breadmore, M. C., Guijt, R. M. & Haddad, P. R. (2009). Counter-flow electrokinetic supercharging for the determination of non-steroidal anti-inflammatory drugs in water samples. *J. Chromatogr. A*, *1216*, 3380-3386.

[90] Botello, I., Borrull, F., Aguilar, C. & Calull, M. (2010). Electrokinetic supercharging focusing in capillary zone electrophoresis of weakly ionizable analytes in environmental and biological samples. *Electrophoresis*, *31*, 2964-2973.

[91] Dawod, M., Breadmore, M. C., Guijt, R. M. & Haddad, P. R. (2010). Electrokinetic supercharging-electrospray ionization-mass spectrometry for separation and on-line preconcentration of hypolipidaemic drugs in water samples. *Electrophoresis*, *31*, 1184-1193.

[92] Xu, Z., Nakamura, K., Timerbaev, A. R. & Hirokawa, T. (2011). Another Approach Toward over 100 000-Fold Sensitivity Increase in Capillary Electrophoresis: Electrokinetic Supercharging with Optimized Sample Injection. *Anal. Chem. (Washington, DC, U. S.)*, *83*, 398-401.

[93] Zhong, H., Yao, Q., Breadmore, M. C., Li, Y. & Lu, Y. (2011). Analysis of flavonoids by capillary zone electrophoresis with electrokinetic supercharging. *Analyst (Cambridge, U. K.)*, *136*, 4486-4491.

[94] Meighan, M. M., Dawod, M., Guijt, R. M., Hayes, M. A. & Breadmore, M. C. (2011). Pressure-assisted electrokinetic supercharging for the enhancement of non-steroidal anti-inflammatory drugs. *J. Chromatogr. A*, *1218*, 6750-6755.

[95] Kwon, J. Y., Chang, S. B., Jang, Y. O., Dawod, M. & Chung, D. S. (2013). Highly sensitive analysis of catecholamines by counter-flow electrokinetic supercharging in the constant voltage mode. *J. Sep. Sci.*, *36*, 1973-1979.
[96] Botello, I., Borrull, F., Calull, M. & Aguilar, C. (2013). Electrokinetic supercharging in CE for the separation and preconcentration of barbiturate drugs in urine samples. *J. Sep. Sci.*, *36*, 524-531.

In: Capillary Electrophoresis (CE)
Editor: Christian Reed

ISBN: 978-1-63483-122-2

Chapter 5

ON-LINE ELECTROPHORETIC-BASED PRECONCENTRATION METHODS IN MICELLAR ELECTROKINETIC CAPILLARY CHROMATOGRAPHY: PRINCIPLES AND RELEVANT APPLICATIONS

Oscar Núñez*

Department of Analytical Chemistry, University of Barcelona, Martí i Franquès, Barcelona, Spain

Serra Húnter Fellow, Generalitat de Catalunya, Spain

ABSTRACT

Micellar electrokinetic capillary chromatography (MECC or MEKC) is maybe the most intriguing mode of capillary electrophoresis (CE) techniques for the determination of small molecules, and it is considered a hybrid of electrophoresis and chromatography. The use of micelle-forming surfactant solutions can give rise to separations that resemble reversed-phase liquid chromatography (LC) with the benefits of CE techniques. Introduced by Professor Shigeru Terabe in 1984, MECC is today, together with capillary zone electrophoresis (CZE), one of the most widely used CE modes, and its main strength is that it is the only electrophoretic technique that can be used for the separation of neutral analytes as well as charged ones.

In MECC, a suitable charged or neutral surfactant, such as sodium dodecyl sulfate (SDS), is added to the separation buffer in a concentration sufficiently high to allow the formation of micelles. Surfactants are long chain molecules (10-50 carbon units) and are characterized as possessing a long hydrophobic tail and a hydrophilic head group. When surfactant concentration in the buffer solution reach a certain level (known as critical micelle concentration), they aggregate into micelles which are, in the case of normal micelles, arrangements that will have a hydrophobic inner core and a hydrophilic outer surface. Micelles are dynamic and constantly form and break apart, constituting a pseudo-stationary phase in solution within the capillary. It is the interaction between the micelles and the solutes (neutral or charged ones) that causes their separation.

* Corresponding author: oscar.nunez@ub.edu.

However, as in the case of other CE techniques, one of MECC handicaps is sensitivity due to the short path length (capillary inner diameter) when on-capillary detection is performed, and the low volume of samples frequently used. In order to improve MECC sensitivity, off-line and/or on-line preconcentration methods can be employed. Among them, on-line electrophoretic-based preconcentration techniques are also becoming very popular in MECC because no special requirement but a CE instrument is necessary. These on-line preconcentration methods are designed to compress analyte bands within the capillary, thereby increasing the volume of sample that can be injected without an important loss in electrophoretic efficiency. In MECC, these on-line preconcentration methods are based on either the manipulation of differences in the electrophoretic mobility of analytes at the boundary of two buffers with differing resistivities and the partitioning of analytes into a micellar pseudostationary phase.

This chapter will address the principles of on-line electrophoretic-based preconcentration methods in micellar electrokinetic capillary chromatography. Coverage of all kind of on-line electrophoretic-based preconcentration methods is beyond the scope of the present contribution, so only the most frequently used in MECC such as sweeping, field-amplified sample injection (FASI), ion-exhaustive sample injection-sweeping (IESI-sweeping) and dynamic pH junction-sweeping will be discussed. Relevant applications of these preconcentration methods in several fields (bio-analysis, food safety, environmental analysis) will also be presented.

1. Introduction

1.1. Micellar Electrokinetic Capillary Chromatography

Electrophoresis is a means of separating charged analytes under the influence of an electric field. The transformation of conventional electrophoresis to modern capillary electrophoresis (CE) took place by the production of narrow-bore capillaries for gas chromatography (GC) and the development of highly sensitive on-line detection systems for high performance liquid chromatography (HPLC). So today, electrophoresis is mainly performed within the confines of narrow-bore capillaries from 20 to 200 μm inner diameter (i.d.) that are usually filled only with a solution containing electrolytes.

Among the different CE modes available, micellar electrokinetic capillary chromatography (MECC), also known as micellar electrokinetic chromatography (MEKC), is considered a hybrid of electrophoresis and chromatography. The use of micelle-forming surfactant solutions can give rise to separations that resemble reversed-phase LC with the benefits of CE techniques. Introduced by Professor Shigeru Terabe in 1984 [1], MECC is today, together with capillary zone electrophoresis (CZE), one of the most widely used CE modes. The same instrumentation that is used for CZE is used for MECC, which demonstrates the versatility and adaptability of the method. MECC differs from CZE because it uses an ionic micellar solution (in general) instead of the simpler buffer salt solution used in CZE. One of the main strength of MECC is that it is the only electrophoretic technique that can be used for the separation of both ionic and neutral substances while CZE typically separates only ionic substances. Thereby MECC has a great advantage over CZE in the separation of mixtures containing both ionic and neutral analytes [2]. So, the separation principle of MECC is based on the differential partition of the analytes between micelles and water while CZE is based on the differences between the own electrophoretic mobility of the analytes.

A suitable charged or neutral surfactant is added to the background electrolyte (BGE) in a concentration sufficiently high to allow the formation of micelles. The surfactants are long chain molecules (10-50 carbon units) and are characterized as possessing a long hydrophobic tail and a hydrophilic head group. When the concentration of surfactants in the BGE reach a certain level, known as critical micelle concentration (8 to 9 mM for SDS, for example), they aggregate into micelles which are, in the case of normal micelles, arrangements that will have a hydrophobic inner core and a hydrophilic outer surface. These micelle aggregates are formed as a consequence of the hydrophobic effect, that is, they rearranged to reduce the free energy of the system. For this reason micelles are essentially spherical with the hydrophobic tails of the surfactant oriented towards the center to avoid interaction with the hydrophilic BGE, and the charged heads oriented toward the buffer. A representation of a normal micelle is shown in Figure 1.

Micelles are dynamic and constantly form and break apart, constituting a pseudo-stationary phase in solution within the capillary. It is the interaction between the micelles and the solutes (neutral or charged ones) that causes their separation. For any given analyte, there is a probability that the molecules of that analyte will associate within the micelle at any given time. This probability is the same as the partition coefficient in classical chromatography. For neutral compounds, it will only be partitioning in and out of the micelle that affects the separation. When associated with the micelle, the analyte will migrate at the velocity of the micelle. When not in the micelle, the analyte will migrate with the electroosmotic flow (EOF) (if present). Differences in the time that the analytes spend in the micellar phase will determine the separation. For charged compounds, variations in micelle electrophoretic mobility when the analyte is associated with the micelle and the analyte electrophoretic mobility when not associated with the micelle will play an important role in the separation, together with their partitioning in and out of the micelle. The overall MECC separation process is depicted schematically in Figure 2.

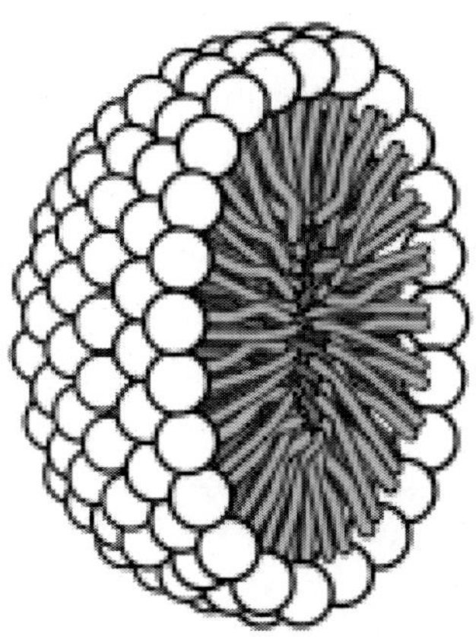

Figure 1. Representation of a normal micelle containing a hydrophobic core and a hydrophilic outer surface.

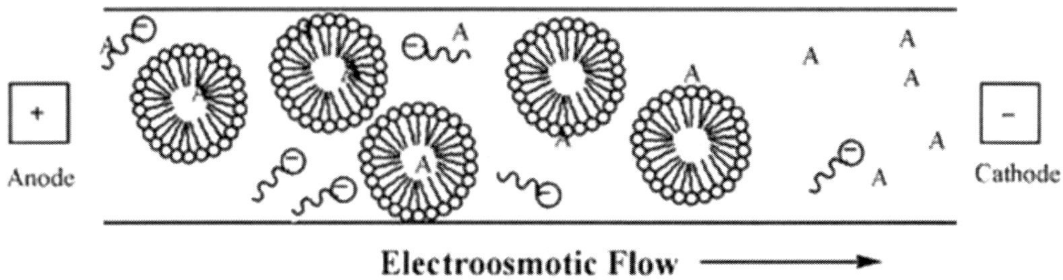

Figure 2. Schematic of a MECC separation process with partitioning between a solute A and a negatively charged micelle in the presence of EOF.

The most commonly used surfactant in MECC is sodium dodecyl sulfate (SDS), an anionic surfactant such as the one depicted in Figure 2. The anionic SDS micelles are electrostatically attracted towards the anode. The EOF transports the bulk solution towards the negative electrode due to the negative charge on the internal surface of fused-silica capillaries. But the EOF is usually stronger than the electrophoretic migration of the micelles and therefore the micelles will also migrate toward the negative electrode with a retarded velocity (Figure 2).

1.2. Separation Principles of MECC

As previously commented, MECC behaves as a hybrid between capillary electrophoresis and chromatography. In MECC, the ionic micelle functions as the stationary phase in chromatography, and the surrounding BGE solution acts as the mobile phase. The micellar solubilization is the partition mechanism. When a neutral analyte is injected into the micellar solution, a fraction of it is incorporated into the micelle and it migrates at the velocity of the micelle. The remaining fraction of the analyte remains free from the micelle and migrates at the electroosmotic velocity. Thus, the migration velocity of the analyte depends on the distribution coefficient between the micellar and the non-micellar (aqueous) phase. The greater the percentage of analyte that is distributed into the micelle, the slower it migrates. The analyte must migrate at a velocity between the EOF velocity and the velocity of the micelle (see scheme in Figure 3A), if a non-charged analyte is being separated [3, 4].

Because MECC has many similarities to chromatographic techniques, nomenclature of several chromatographic parameters is still employed. So, the migration time (t_m) or the retention time (t_R) of the analyte is limited between the migration time of the bulk solution (t_o) and that of the micelle (t_{mc}) as shown in Figure 3B. This is often referred to in the literature as the *migration time window* in MECC.

In MECC, the retention factor of an analyte (k) can be defined, similarly to that of chromatography, as:

$$k = \frac{n_{mc}}{n_{aq}}$$

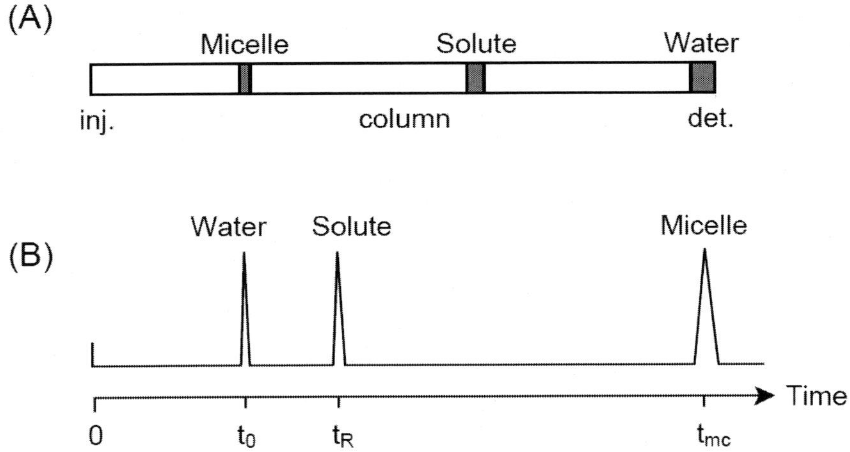

Figure 3. Schematic of the zone separation in MECC (A) and electropherogram (B). Reprinted with permission from reference [3]. Copyright (1985) American Chemical Society.

where n_{mc} is the amount of the analyte incorporated into the micelle and n_{aq} is the amount free of the micelle, or the amount of the analyte in the aqueous phase of the BGE. Considering a neutral analyte, the migration time can then be given by:

$$t_R = \frac{1 + k}{1 + (\frac{t_o}{t_{mc}})k} t_o$$

where t_o and t_{mc} are the migration times of the EOF marker and the micelle marker, respectively (Figure 3B). Some generally employed markers are methanol and Sudan III or IV for the EOF and micelle, respectively. Considering this equation, the range of the migration time for a neutral analyte is limited to between t_o ($k = 0$) and t_{mc} ($k = \infty$).

When EOF is completely suppressed, the migration time of a neutral analyte can be calculated as:

$$t_R = \left(1 + \frac{1}{k}\right) t_{mc}$$

So, in the absence of EOF, the micelle migrates through the surrounding aqueous phase, although it corresponds to the stationary phase in conventional chromatography. In this case, it can be assumed that the micelle is the mobile phase and that the aqueous phase is the stationary phase.

The retention factor is a fundamental term in chromatography and the previously commented equations are derived from a chromatographic perspective. From an electrophoretic point of view, the electrophoretic velocity in MECC is modified by the micellar additive in the BGE [5]. Under electrophoretic conditions, the ionic micelle migrates via both electrophoresis and EOF. The migration velocity of the micelle (v_{mc}) differs from that of the EOF velocity (v_{eo}) by the electrophoretic velocity of the micelle ($v_{ep}(mc)$), as indicated in the next equation:

$$v_{mc} = v_{eo} + v_{ep}(mc)$$

The velocity is a vector quantity and is positive when directed toward the cathode and v_{eo} and $v_{ep}(mc)$ have generally different signs. So, the migration velocity of a neutral analyte (v_a) can be expressed as:

$$v_a = \frac{(v_{eo} + v_{mc}k)}{(1+k)}$$

where $1/(1 + k)$ and $k/(1 + k)$ are the fraction of the analyte free from the micelle and the fraction of the analyte incorporated into the micelle, respectively. So, the velocity of the neutral analyte is limited between v_{eo} ($k = 0$) and v_{mc} ($k = \infty$).

The resolution (R_S) between two neutral analytes in MECC can be calculated with the next equation:

$$R_S = \frac{\sqrt{N}}{4}\left(\frac{\alpha - 1}{\alpha}\right)\left(\frac{k_2}{1+k_2}\right)\left(\frac{1 - t_o/t_{mc}}{1 + (t_o/t_{mc})k_1}\right)$$

where N is the theoretical plate number and α is the selectivity factor defined by k_2/k_1 ($k_2 \geq k_1$), where subscripts 2 and 1 refer to the analyte number, respectively. Retention factors can be directly calculated from experimental conditions by determining the migration times of the analyte (t_R), the micelle (t_{mc}) and the EOF (t_o) using adequate markers as previously commented, and following the next equation:

$$k = \frac{t_R - t_o}{t_o(1 - t_R/t_{mc})}$$

In general, resolution is influenced by four parameters: the plate number, the selectivity factor, the retention factor, and the migration time window factor. Although the resolution equation in MECC is similar to that derived for chromatographic separations, the last parameter in the right-hand side of the equation is superfluous; it arises from the variable length of the micellar zone, where the analyte can interact with the micelle.

1.3. Composition of the Micellar Solution

Ionic surfactants are an essential component for micellar electrokinetic capillary chromatography. Although a large number of surfactants are commercially available, a limited number are widely used in MECC separations. This is because surfactants suitable for MECC should meet some properties such as having enough solubility in the BGE buffer solution used to form micelles, the micellar solution they form must be homogeneous and UV transparent (if UV detection is employed), and they must have a low viscosity.

Table 1. Critical micelle concentration, surfactant classes, aggregation number (n) and Krafft point (K_p) of some selected surfactants

Surfactant	Type	CMC (mM)[a]	n	K_p (°C)
Sodium dodecyl sulphate (SDS)	Anionic	8.1	62	16
Sodium tetradecyl sulphate (STS)	Anionic	2.1 (50°C)	138[b]	32
Sodium decanesulfonate	Anionic	40	40	-
Sodium dodecanesulphate	Anionic	7.2	54	37.5
Sodium cholate	Anionic	13-15	2-4	-
Sodium deoxycholate	Anionic	4-6	4-10	-
Sodium taurocholate	Anionic	10-15	5	-
Dodecyltrimethylammonium chloride (DTAC)	Cationic	16 (30°C)	-	-
Dodecyltrimethylammonium bromide (DTAB)	Cationic	15	-	-
Tetradecyltrimethylammonium bromide (TTAB)	Cationic	3.5	75	-
Cetyltrimethylammonium bromide (CTAB)	Cationic	0.92	91	-
Brij-35	Non-ionic	0.1	40	-
Sulfobetaine	Zwitterionic	3.3	55	-

[a] In pure water and at 25 °C.
[b] In 0.10 M NaCl.

There are mainly four major classes of surfactants: anionic, cationic, zwitterionic and non-ionic surfactants [6]. Table 1 shows a list of some selected ionic surfactants available for MECC together with several properties such as the critical micelle concentration (CMC, lowest surfactant concentration required to form micelles), the aggregation number (number of surfactant units in a micelle), and the Krafft point (temperature above which the solubility of the surfactant increases steeply due to the formation of micelles).

In order to obtain a micellar solution, the concentration of the surfactant must be higher than its CMC. The surfactant has enough solubility to form micelles only at temperatures above the Krafft point. The micelles used in MECC are generally charged on the surface, so an analyte with the opposite charge will strongly interact with the micelle through electrostatic forces while an analyte with the same charge will interact weakly due to the electrostatic repulsion. Thus, ionic surfactants are generally used in MECC.

SDS is the most widely employed surfactant used to generate the micelle in MECC because it has several advantages over other surfactants, including its well-characterized properties, high solubilization capacity, easy availability, low ultraviolet absorbance, and high solubility to aqueous solutions. Minor disadvantages of SDS are its relatively low CMC (8 mM in pure water, although it is a little lower in buffer solutions) and its relatively high Krafft point (16 °C), which causes precipitation of SDS at low temperatures. The counter ion of the ionic surfactant does affect the Krafft point. For example, the Krafft point for potassium dodecyl sulfate is approximately 40 °C. So, if SDS (with a Krafft point of 16 °C) is dissolved in a BGE containing potassium ions, the solubility of SDS will be less than its CMC at ambient temperature because of the exchange reaction of counter ions. So the use of potassium ions as an electrolyte should be prevented when SDS is employed in MECC.

Cationic surfactants such as cetyl-, dodecyl-, and hexadecyltrimethylammonium salts can be used in MECC to reverse the charge on the capillary wall. These surfactants are absorbed on the capillary wall surface by a mechanism involving electrostatic attraction between the positively charged ammonium moieties and the negatively charged Si-O$^-$ groups on the fused-silica capillary wall. The non-polar chains of these surfactants (C10, C14, C16, etc) create a hydrophobic layer and, at a high enough surfactant concentration, the negative surface charge will be completely neutralized. If surfactant concentration increases a bi-layer can be formed through hydrophobic interaction between the non-polar chains as can be seen in Figure 4.

The cationic head groups are facing the BGE solution and the charge of the capillary wall is reversed from negative to positive. Consequently, a reversal of the EOF direction is achieved under the influence of an electric field. If surfactant concentration is even higher and reaches the CMC cationic micelles are then generated.

Non-ionic surfactants such as Brij-35 do not posses electrophoretic mobility and therefore cannot be used as pseudo-stationary phase in "conventional" MECC. However, they can be useful for the separation of charged analytes. This method using non-ionic micelles can be considered as an extension of MECC [6, 7].

Two different surfactants can also be combined to form a mixed micelle. Mixed micelles consisting of ionic and non-ionic surfactants are useful pseudo-stationary phases in MECC because they provide significantly different separation selectivity from that of standard micelles. The change in selectivity can be explained by the alteration of the surface structure of the mixed micelle. Since a mixed micelle of an ionic and a non-ionic surfactant has a lower surface charge and a larger size, its electrophoretic mobility will be lower than a single ionic micelle. The addition of a non-ionic surfactant to an ionic micellar solution causes a narrower migration time window in MECC.

The constituents of the aqueous phase of the micellar BGE have very little effect upon selectivity. Organic buffers usually have a relatively low conductivity and, therefore, are recommended to modify selectivity if they are stable and UV transparent. Care should be taken to prevent replacing the counter ion of the ionic surfactant with the buffer ion, which will modify the micelle Krafft point and could induce its precipitation into the capillary.

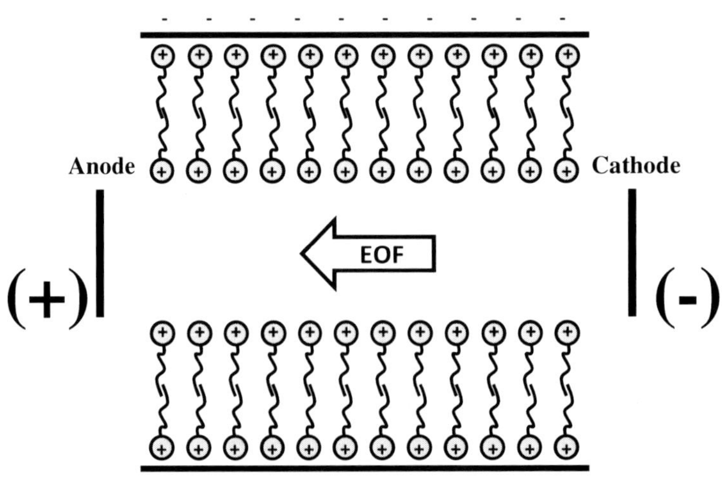

Figure 4. Schematic of the EOF reversal in a fused-silica capillary by using a cationic surfactant such as CTAB.

As it is well known, the pH of the buffer solution used in the BGE (if used) is a critical parameter for the separation of ionizable analytes, and can also be used as a tool to modify partition of the analyte with the micelle pseudo-stationary phase by changing the ionic state of the analyte.

A very effective way in manipulating selectivity in MECC is the modification of the BGE aqueous phase by adding additives such as cyclodextrins, ion-pair reagents, urea, organic solvents, etc. Cyclodextrins are oligosaccharides with truncated cylindrical molecular shapes. Their outside surfaces are hydrophilic, while their cavities are hydrophobic. In MECC, cyclodextrins are electrically neutral and have no electrophoretic mobility. They are assumed not to be incorporated into the micelle, because of the hydrophilic nature of their outside surface. However, a surfactant molecule may be included into the cyclodextrin cavity. The analyte molecule included into the cyclodextrin migrates at the same velocity as the EOF because, electrophoretically, cyclodextrins behave as the bulk aqueous phase. Therefore, the addition of cyclodextrins to the micellar BGE reduces the apparent distribution coefficient and enables the separation of highly hydrophobic analytes, which otherwise would be almost totally incorporated into the micelle in the absence of cyclodextrins.

Regarding organic solvents they could have a notable effect on selectivity in MECC due to the changes on BGE viscosity, but high concentrations should be prevented because organic solvents may break down the micellar structure. Generally, concentrations up to 20-35% (depending on the solvent) can be used without difficulty in MECC. In general, the addition of methanol, isopropanol or acetonitrile reduces the electroosmotic velocity and, hence, expands the migration time window.

2. On-Line Electrophoretic-Based Preconcentration Methods in Micellar Electrokinetic Capillary Chromatography

It is well known that detection sensitivity in CE techniques is low in terms of concentration sensitivity when photometric detectors are used. This is mainly due to the small amounts of sample injected into the capillary and the short path lengths for the photometric detection. With the small dimensions of CE capillaries (typically 20-200 μm I.D.) and capillary lengths (40-80 cm in most of the applications), only very small sample volumes may be loaded into the capillary. For instance, for conventional 50 μm I.D. x 50 cm total length capillary only 1.18 nL/s of sample are introduced into the capillary when hydrodynamic injection (0.5 psi) is performed. Regarding the path length when photometric detection is used, the effective length of the light path through the capillary is about 63.5% of the stated capillary internal diameter. Thus, a 50 μm I.D. capillary has an effective path length of only 32 μm.

To circumvent these disadvantages, several on-line preconcentration techniques have been developed [8-11]. The most convenient approach to improve sensitivity in MECC is to increase the amount of analyte injected into the capillary. This approach does not require any special instrument set-up configuration and, for this reason, is one of the most frequently proposed to improve sensitivity in MECC when photometric detection is used. For that purpose, a large volume of the sample solution is injected into the capillary before separation or it is selectively injected electrokinetically from the sample solution and concentrated at the injection end of the capillary before the separation. Most of these methods involve manipulating the migration velocity of the analyte during injection and separation on MECC and include techniques such as sweeping, field-amplified sample injection (FASI), ion-exhaustive sample injection-sweeping (IESI-sweeping) and dynamic pH junction-sweeping. The principles of each method and some relevant applications will be discussed in the next sections.

2.1. Sweeping

Sweeping is an on-line electrophoretic-based preconcentration method initially developed for the preconcentration of neutral analytes in MECC by Professor's Terabe research group [12]. Figure 5 shows a schematic of the principles of sweeping-MECC under suppressed electroosmotic conditions. In general, the principles of sweeping procedure differ from that of other on-line electrophoretic-based preconcentration techniques in that no field-enhancement effect occurs. This is because the sample solution is prepared in a matrix without micelles but its conductivity is adjusted to be nearly equal to that of the micellar background electrolyte (mBGE) by modifying the salt concentration.

For sweeping-MECC the capillary is first filled with the mBGE. Then the sample (with adjusted conductivity to equal that of the mBGE) is introduced hydrodynamically into the capillary (Figure 5a). Then, a mBGE vial is placed in the inlet position and a reverse voltage (anion in the outlet position) is applied (Figure 5b). Under the electric field strength, micelles are entering into the capillary by electrophoresis and are picking the analytes up and concentrating them in a narrow zone. So, as the micelles migrate towards the detector they "sweep" the neutral analytes along (Figure 5c). This effect is dependent on a uniform electric field and the absence of micelles in the sample solution. For this reason, sample solution conductivity must equal that of the mBGE. The effectiveness of this on-line sample preconcentration technique has been shown to be dependent on the analytes' affinity for the pseudostationary phase.

The concentration efficiency can be described as:

$$l_{inj}/l_{sweep} = (1 + k)$$

where l_{inj} and l_{sweep} are the injected sample zone length and the swept length, respectively, and k is the analyte retention factor.

This preconcentration procedure has been described in detail by Quirino and Terabe [13, 14] to illustrate its wide applicability and today it has been applied in multiple fields such as environmental analysis, food analysis and bio-analytical applications. Table 2 is summarizing a selection of sweeping-MECC methods in the mentioned application fields [15-27].

Regarding the analysis of environmental pollutants in water, Núñez et al. [16] developed a sweeping-MECC for the analysis of three quaternary ammonium herbicides (paraquat, diquat and difenzoquat). For that purpose, 80 mM SDS in 50 mM phosphate buffer (pH 2.5) with 20% acetonitrile was used as mBGE, and a sample matrix consisting on a phosphate buffer solution (pH 2.5) with a concentration to provide a conductivity similar to that of the mBGE (6.3 mS/cm) was used. As an example, Figure 6 shows the electropherograms obtained when analyzing these compounds with conventional MECC and the proposed sweeping-MECC method. The limits of detection, based on a signal-to-noise ratio of 3:1, were about 2.6-5.1 mg/L in purified water when MECC was applied for the standards. By using the on-line preconcentration method sweeping-MECC, up to a 500-fold increase in detection sensitivity was obtained, achieving LOD values around 10 μg/L. Good linearity (r^2 higher than 0.99) and good run-to-run (n = 6) and day-to-day (n = 6, two replicates in three different days) precisions were obtained, with RSD values lower than 7.9%.

Maijó et al. [25] proposed a sweeping-MECC method for the determination of five anti-inflammatory drugs (ibuprofen, fenoprofen, naproxen, diclofenac sodium, and ketoprofen) in river water samples. The authors proposed the use of a 75 mM SDS in 25 mM sodium dihydrogenphosphate solution (pH 2.5) with 40% (*v/v*) as micellar BGE, and the employment of 75 mM sodium dihydrogenphosphate solution (pH 2.5) as sample matrix devoid of micelles with conductivity similar to that of mBGE (6 mS/cm). With the developed sweeping method, about 143- and 401-fold improvements in peak height and peak area, respectively, were obtained. For the analysis of real water sample, river waters were diluted 1:5 with a solution of 95 mM of sodium dihydrogenphosphate (pH 2.5) in such a way that the conductivity of the sample was equal to the mBGE conductivity. Although a dilution was required, the proposed sweeping-MECC method showed to be capable enough for the analysis of environmental aquatic samples without any previous sample treatment, obtaining LODs ranging between 6.5 and 14.6 μg/L.

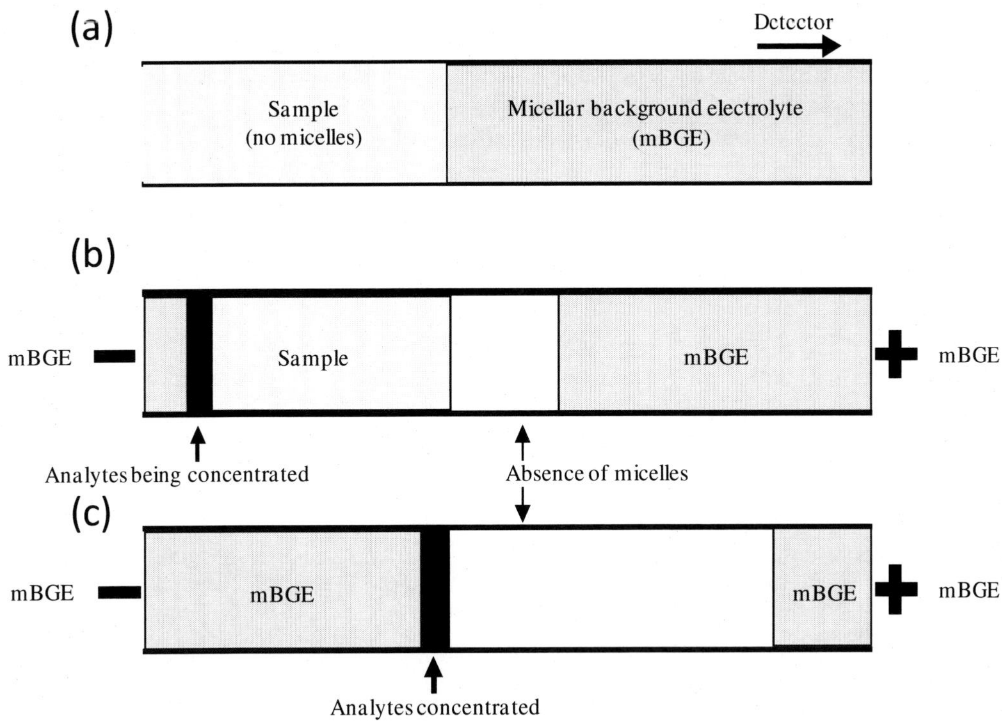

Figure 5. Schematic illustration of sweeping-MECC under suppressed electroosmotic flow conditions.

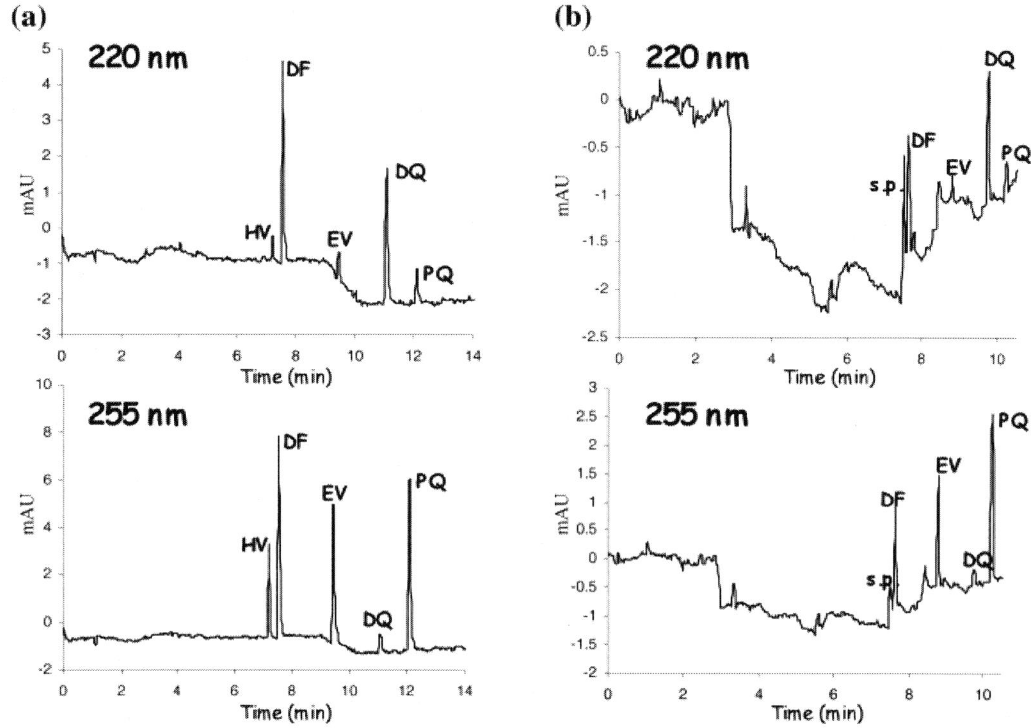

Figure 6. Conventional MECC and sweeping-MECC of quaternary ammonium herbicides. mBGE: 80 mM SDS in 50 mM phosphate buffer (pH 2.5) containing 20% acetonitrile. (a) MECC: sample prepared in mBGE; sample concentration, 100 mg/L; injection time, 1 s at 5 kPa. (b) Sweeping-MECC: sample prepared in a phosphate buffer (pH 2.5) with the same conductivity of mBGE (6.3 mS/cm); sample concentration, 100 μg/L; injection time, 500 s at 5 kPa. Separation conditions (a and b): separation voltage, -22 kV with the mBGE at both ends of the capillary. PQ, paraquat; DQ, diquat; DF, difenzoquat; EV, ethyl viologen (I).S.); HV, heptyl viologen (I.S.); s.p., system peak. Reprinted with permission from reference [16]. Copyright (2002) Elsevier.

Regarding food applications, several sweeping-MECC methods are described in the literature for the analysis of contaminants in food matrices. For instance, Tsai et al. [17] proposed the use of this on-line preconcentration method for the rapid analysis of melamine in infant formulas. Although melamine is not allowed as an additive in food or related ingredients, in 2008 more than 51,900 infants and young children in Chine suffered from urinary problem due to the consumption of melamine-contaminated infant formula [28].

The foul motivation of adding melamine to milk was to increase the amount of nitrogen which will result in higher measurement of protein. The outbreak of such calamity pushed the governments worldwide to set a limit of detection of melamine in infant formula. Since melamine is a raw material in manufacturing some plastic wares used for serving food, low-level migration of melamine into the food has been reported. Figure 7 shows the sweeping-MECC analysis of a melamine-contaminated milk samples. The authors also evaluated the application of field amplified sample injection (FASI) technique for the analysis of melamine standards. Although LOD of melamine standard was 0.5 μg/L with the FASI technique and 9.2 μg/L with sweeping-MECC, the authors observed that the matrix effect was higher with FASI. Thus, sweeping-MECC demonstrated to be most suitable for real sample analysis.

Table 2. Selection of sweeping-MECC methods in environmental, food and bio-analytical applications

Compounds	Samples	sweeping conditions	MECC conditions	Detection	Analysis time	LODs	Ref.
Environmental pollutants (organic amines and alkyl phthalates)	Standards	*Sample hydrodynamic injection:* 300 s	Uncoated fused silica capillary of 80 cm x 50 µm I.D. *mBGE:* 50 mM SDS in 30 mM phosphoric acid-5 mM phosphate (pH 2.2) with 20% (*v/v*) methanol (conductivity 2.9 mS/cm) *Capillary voltage:* -17 kV	Mass spectrometry Quadrupole analyzer (+)APCI	14-25 min	0.4-0.6 mg/L	[15]
Quaternary ammonium herbicides	Drinking water	*Sample matrix*: phosphate buffer at pH 2.5 with similar conductivity to mBGE (6.3 mS/cm) *Sample hydrodynamic injection:* 500 s (5 kPa)	Uncoated fused silica capillary of 60 cm (51.5 cm effective length) x 50 µm I.D. *mBGE:* 80 mM SDS in 50 mM phosphate buffer (pH 2.5) with 20% (*v/v*) acetonitrile *Capillary voltage:* -22 kV	UV: 220 and 255 nm	12 min	10.1-13.0 µg/L	[16]
Melamine	Infant formula	*Sample matrix*: 75 mM phosphoric acid *Sample hydrodynamic injection:* 1.7 min (50 mbar)	Uncoated fused silica capillary of 65 cm (50 cm effective length) x 50 µm I.D. *mBGE:* 175 mM SDS in 50 mM phosphoric acid *Capillary voltage:* -20 kV	UV: 218 nm	13 min	9.2 µg/L	[17]
Abused drugs and hypnotics	Human urine	*Sample matrix*: 15 mM phosphate buffer (pH 5) *Sample hydrodynamic injection:* 200 s (1 psi)	Uncoated fused silica capillary of 50.4 cm (40 cm effective length) x 50 µm I.D. *mBGE:* 65 mM SDS in 75 mM phosphate buffer (pH 2.5) with 10% (*v/v*) methanol *Capillary voltage:* -15 kV	UV: 200 nm	27 min	20-50 µg/L	[18]
Alkaloids	Human urine	*Sample matrix*: 50 mM phosphoric acid *Sample injection:* 300 s (15 cm height difference between sample vial and outlet vial)	Uncoated fused silica capillary of 70 cm (41 cm effective length) x 50 µm I.D. *mBGE:* 15 mM SDS in 100 mM phosphoric acid (pH 1.8) with 12% (*v/v*) tetrahydrofuran *Capillary voltage:* -28 kV	UV: 265 nm	11 min	0.2-1.5 µg/L	[19]
Alkaloids	Human urine	*Sample matrix*: 50 mM phosphoric acid *Sample injection:* 300 s (15 cm height difference between sample vial and outlet vial)	Uncoated fused silica capillary of 70 cm (41 cm effective length) x 50 µm I.D. *mBGE:* 15 mM SDS in 100 mM phosphoric acid (pH 1.8) with 12% (*v/v*) tetrahydrofuran *Capillary voltage:* -28 kV	UV: 265 nm	11 min	0.2-1.5 µg/L	[19]

Table 2. (Continued)

Compounds	Samples	sweeping conditions	MECC conditions	Detection	Analysis time	LODs	Ref.
Tricyclic antidepressant and β-blocker drugs	Wastewater	*Sample matrix*: 150 mM phosphoric acid *Sample injection:* 14.4 s (950 mbar)	Uncoated fused silica capillary of 50 cm (41.5 cm effective length) x 50 μm I.D. *mBGE:* 50 mM SDS in 50 mM phosphoric acid with 27.5% (*v/v*) acetonitrile *Capillary voltage:* -15 kV	UV: 214 nm	15-20 min	7-2 μg/L	[20]
Melamine	Food (milk, gluten, chicken feed and cookies)	*Sample matrix*: 65 mM phosphoric acid *Sample injection:* 1.2 s (50 mbar)	Uncoated fused silica capillary of 65 cm (50 cm effective length) x 50 μm I.D. *mBGE:* 175 mM SDS in 45 mM phosphoric acid with 15% (*v/v*) methanol *Capillary voltage:* -22 kV	UV: 218 nm	13 min	5 μg/ L	[21]
Steroid hormones	Urine	*Sample matrix*: 100 mM phosphoric acid (pH 2.5) with 30% (*v/v*) methanol *Sample hydrodynamic injection:* 90 s (31 kPa)	Uncoated fused silica capillary of 50.2 cm (40 cm effective length) x 50 μm I.D. *mBGE:* 50 mM SDS in 100 mM phosphoric acid (pH 2.5) with 30% (*v/v*) methanol *Capillary voltage:* -16.5 kV	UV: 240 nm	18 min	5-15 μg/L	[22]
Triazol antifungal drugs	Human plasma	*Sample matrix*: 167 mM phosphoric acid with 16.7% (*v/v*) methanol *Sample hydrodynamic injection:* 3 min (90 mbar)	Uncoated fused silica capillary of 71 cm (56 cm effective length) x 50 μm I.D. *mBGE:* 100 mM SDS in 25 mM phosphoric acid (pH 2.2) with 13% (*v/v*) acetonitrile and 13% (*v/v*) tetrahydrofuran *Capillary voltage:* -30 kV	UV: 254 nm	13 min	30-40 μg/L	[23]
Anti-histamines	Human urine	*Sample matrix*: 100 mM phosphoric acid *Sample hydrodynamic injection:* 300 s (15 cm height difference between sample vial and outlet vial)	Uncoated fused silica capillary of 70 cm (41 cm effective length) x 50 μm I.D. *mBGE:* 15 mM SDS in 75 mM phosphoric acid (pH 2.0) with 10% (*v/v*) tetrahydrofuran *Capillary voltage:* -20 kV	UV: 214 nm	16 min	0.12-0.95 μg/L	[24]
Anti-inflammatory drugs	River water	*Sample matrix*: 75 mM NaH_2PO_4 (pH 2.5) *Sample hydrodynamic injection:* 350 s (50 mbar)	Uncoated fused silica capillary of 60 cm (51.5 cm effective length) x 75 μm I.D. *mBGE:* 75 mM SDS in 25 mM NaH_2PO_4 (pH 2.5) with 40% (*v/v*) acetonitrile *Capillary voltage:* -26 kV	UV: 214 nm	22 min	6.5-14.6 μg/L	[25]

Compounds	Samples	sweeping conditions	MECC conditions	Detection	Analysis time	LODs	Ref.
Tea catechins	Human plasma	*Sample matrix*: phosphate buffer *Sample injection:* 150 s (50 mbar)	Uncoated fused silica capillary of 28 cm (19.5 cm effective length) x 50 µm I.D. *mBGE:* 5 mM 1-tetradecyl-3-methylimidazolium bromide in 15 mM phosphate buffer (pH 4.5) with 12% (*v/v*) THF *Capillary voltage:* 10 kV	UV: 200 nm	5 min	0.2-1.2 µg/L	[26]
Whitening agents and parabens	Cosmetic products	*Sample matrix*: dilution of cosmetic sample with deionized water *Sample injection:* 90 s (20 cm height difference between sample vial and outlet vial)	Uncoated fused silica capillary of 60 cm (50 cm effective length) x 75 µm I.D. *mBGE:* 40 mM SDS in 15 mM tetraborate buffer (pH 8.5) with 0.1% (*w/v*) poly(ethylene oxide) *Capillary voltage:* 15 kV	UV: 200 nm	10 min	1.1-21.0 µg/L	[27]

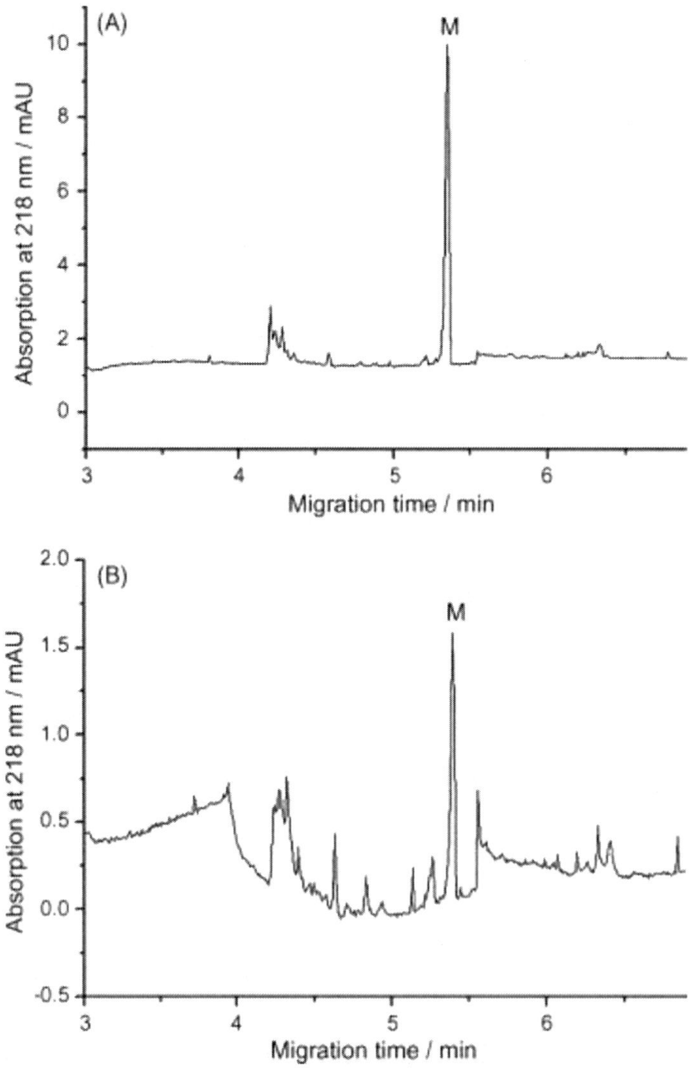

Figure 7. Sweeping-MECC electropherograms of melamine-contaminated milk provided by (A) the Joint research centre of the European Commission, and (B) the Bureau of food and drug analysis in Taiwan. Sweeping-MECC conditions are the same as those described in Table 2. Reprinted with permission from reference [17]. Copyright (2009) Elsevier.

A similar method was proposed some years later for the analysis of melamine in foodstuffs such as milk, gluten, chicken feed, and cookies [21]. In this case, sweeping-MECC separation was achieved by using a 175 mM SDS in a 45 mM phosphoric acid with 15% methanol solution as mBGE. Figure 8 shows the electropherograms obtained for several foodstuffs spiked with 1 μg/mL of melamine and their respective blank electropherograms. With the proposed method melamine content could be determined within 13 minutes with a LOD of 5 ng/mL.

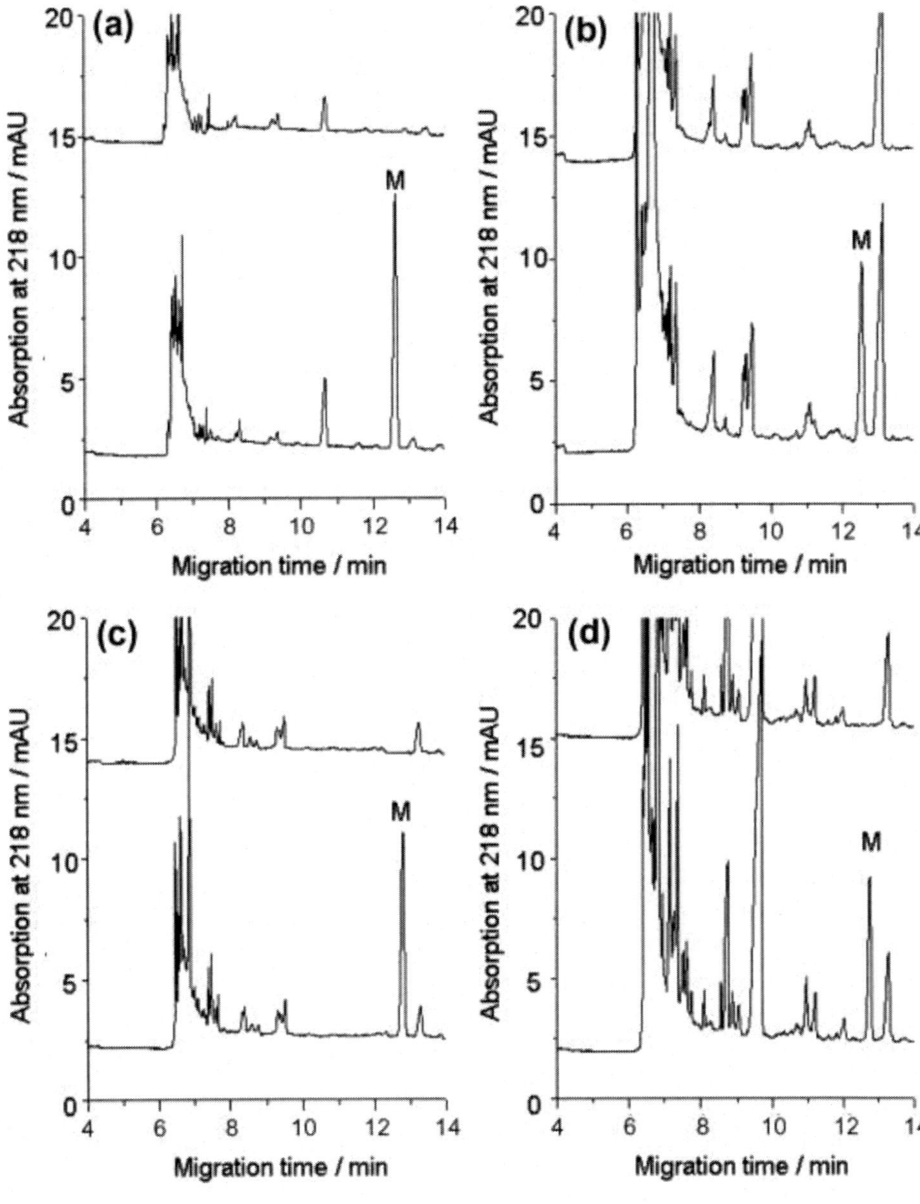

Figure 8. Sweeping-MECC electropherograms of (a) milk powder, (b) gluten, (c) cookie, and (d) chicken feed spiked with 1 µg/mL of melamine and their respective blank electropherograms. Other experimental conditions as described in Table 2. Reprinted with permission from reference [21]. Copyright (2011) Elsevier.

Sweeping-MECC preconcentration methods can also be very useful for the analysis of biological fluids in bio-analytical applications. For instance, Gao et al. [24] described the trace analysis of three antihistamines (mizolastine, chlorpheniramine and pheniramine) in human urine by on-line single drop liquid-liquid-liquid microextraction coupled to sweeping-MECC and its application to some pharmacokinetic studies. In this work, the unionized analytes were subsequently extracted into a drop of *n*-octanol layered over the urine sample, and then into a microdrop of an acceptor phase (100 mM phosphoric acid) suspended from a capillary inlet. This enriched acceptor phase was then on-line injected into the capillary with a height difference (see conditions in Table 2) and then analyzed directly by sweeping-MECC. Figure 9 shows the electropherograms of a blank urine sample and of a urine sample spiked at a concentration level of $5.0x10^{-5}$ g/L of each studied antihistamine. The proposed method allowed achieving limits of detection from 0.12 to 0.95 μg/L based on a signal-to-noise ratio of 3 (S/N 3), with involves a 751- to 1,372-fold increase in detection sensitivity for the analytes in comparison to conventional conditions. This method resulted to be a promising combination for the rapid trace analysis of antihistamines in human urine with the advantages of operation simplicity, high enrichment factors and little solvent consumption.

Recently, an interesting application of sweeping-MECC in the analysis of whitening agents and parabens in cosmetic products was reported by Tsai et al. [27]. The authors investigated in detail the optimum conditions of the on-line concentration and separation of arbutin, kojic acid, resorcinol, salicylic acid, and methyl-, ethyl-, propyl-, and butyl-parabens. Finally, sweeping-MECC was performed at 15 kV using a BGE containing 15 mM tetraborate buffer (pH 8.5), 40 mM SDS, and 0.1% (*v/v*) poly(ethylene oxide). LODs in the range 1.1 to 21.0 μg/L were obtained, corresponding to a 46- to 279-fold improvement in sensitivity in comparison to conventional sample injections. The authors validated the method and used it to determine whitening agents and parabens in five commercial cosmetic products, with average recoveries from 85.2 to 118.0%. This method showed to be a powerful alternative approach for identifying and determining whitening agents and parabens in commercial cosmetic samples.

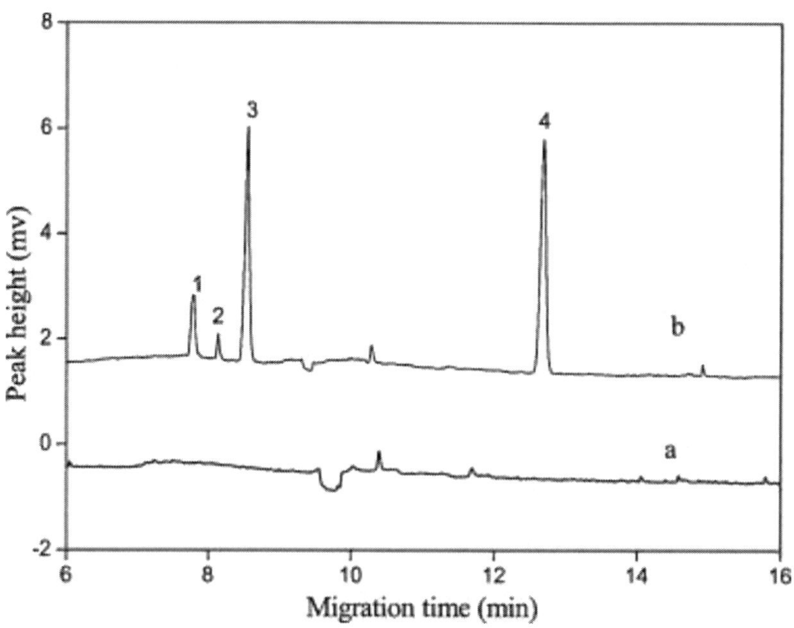

Figure 9. Electropherograms of urine from blank (a) and after spiking at a concentration level of 50 μg/L of each analyte and internal standard (I.S.) (b). Peak identification: 1, mizolastine; 2, chlorpheniramine; 3, pheniramine; and 4, strychnine (I.S.). 10% THF used as organic modifier and other sweeping-MECC conditions as those described in Table 2. Reprinted with permission from reference [24]. Copyright (2012) Elsevier.

2.2. Field-Amplified Sample Injection

Field-amplified sample injection (FASI), also known as field-amplified sample stacking or field-enhanced sample stacking, is one of the most popular on-line electrophoretic-based preconcentration methods in CE techniques because of its simplicity of application only requiring the electrokinetic injection of the sample after the introduction of a short plug of a high-resistivity solvent such as methanol or water. FASI was originally developed for the preconcentration of charged analytes. When a sample solution is prepared in a dilute electrolyte solution (or in a low-conductivity solution), and when the BGE is in a high-concentration electrolyte solution (or a high-conductivity solution), the analyte ions migrated rapidly in the sample solution and with a lower migration velocity in the BGE zone because the analyte electrophoretic velocity is proportional to the field strength (higher in the sample zone than in the BGE zone). Thus, the analytes will stack-up at the boundary region between the sample solution and the BGE solution.

There are several approaches for the application of FASI preconcentration techniques in CE [10] and, recently the progress on stacking techniques based on field amplification has been reviewed [29]. But the first application of this on-line preconcentration technique for the concentration of neutral analytes in MECC was described by Liu et al. [30]. Since then, Quirino and Terabe extensively studied and developed sample preconcentration techniques for neutral analytes using the field-enhanced technique [8, 31].

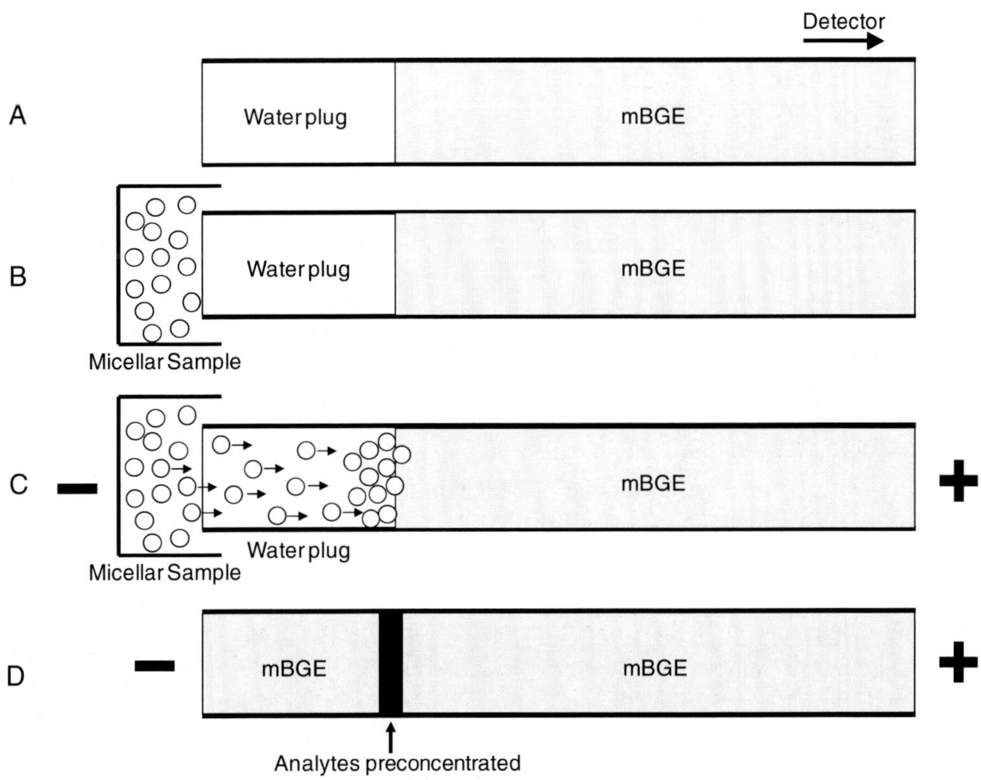

Figure 10. Scheme of field-amplified sample injection in MECC for the preconcentration of neutral compounds using SDS micelles.

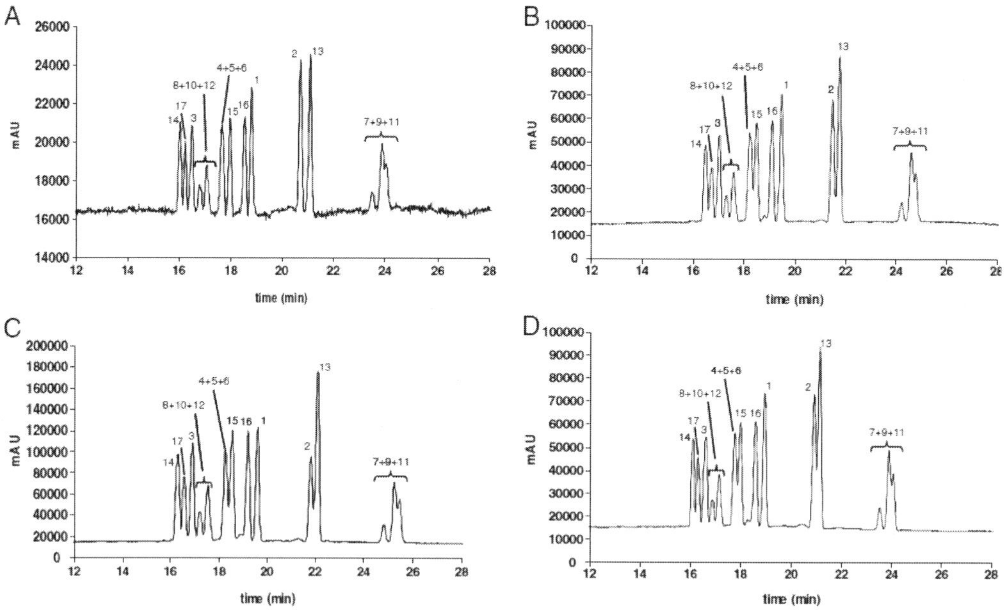

Figure 11. Effect of SDS concentration in the sample matrix injection on the electrophoretic separation when applying FASI-MECC. (A) 150 mM SDS, (B) 20 mM SDS, (C) 10 mM SDS, and (D) 5 mM SDS. All sample matrixes contain a 5% of ethanol. FASI injection: 2 s water plug (3.5 kPa) and 45 s electrokinetic injection (-10 kV). Other conditions are described in Table 3. Peak identification: 1, BPA; 2, BPF; 3, BADGE; 4, p,p-BFDEG; 5, o,o-BFDGE; 6, o,p-BFDGE; 7, o,o-BFDGE·2H2O; 8, o,o-BFDGE·2HCl; 9, o,p-BFDGE·2H2O; 10, o,p-BFDGE·2HCl; 11, p,p-BFDGE·2H2O; 12, p,p-BFDGE·2HCl; 13, BADGE·2H2O; 14, BADGE·2HCl; 15, BADGE·HCl·H2O; 16, BADGE·H2O; and 17, BADGE·HCl. Reprinted with permission from reference [32]. Copyright (2010) Wiley-VCH

As previously commented, FASI involves a field-enhanced electrokinetic sample injection of the analytes into the capillary. However, this method is not appropriate for neutral compounds. One approach to achieve the electrokinetic injection of neutral analytes when dealing with MECC methods is by adding micelles into the sample solution and performing the electrokinetic injection of this micellar sample solution. Figure 10 shows a schematic of this simple approach.

After filling the capillary with a micellar BGE solution (mBGE), a pre-injection of a short plug of a high-resistivity solvent such as water is hydrodynamically introduced into the capillary (Figure 10A). Then, a sample vial containing micelles (SDS) is set in the capillary inlet position (Figure 10B) and electrokinetic injection is carried out by applying a negative polarity (anode in the outlet position). Neutral analytes (as well as charged ones according to their charge and hydrophobicity) will interact with the SDS micelles. The short plug of water allows the enhancement of the sample electrokinetic injection because of the conductivity differences between sample and the water plug (Figure 10C). Hence, micelles containing the analytes will be introduced into the capillary. Moreover, long electrokinetic injection times can be employed while the SDS micelles with the analytes stack-up at the boundary between the high-resistivity solvent (water) and the mBGE solution because they slow down due to the important decrease on their migration velocity in the mBGE region. Finally, a mBGE vial is set in the inlet position and electrophoretic separation takes place with the analytes being concentrated in a narrow zone (Figure 10D).

Table 3. Selection of FASI-MECC methods in environmental, food and bio-analytical applications

Compounds	Samples	FASI conditions	MECC conditions	Detection	Analysis time	LODs	Ref.
Phenols	Water	*High resistivity solvent*: water *Hydrodynamic injection:* equivalent to 50 cm water plug *Sample electrokinetic injection:* 25 min (-10 kV) *Sample matrix:* 8 mM NaOH	Uncoated fused silica capillary of 70 cm (55 cm effective length) x 50 µm I.D. *mBGE:* 80 mM SDS in 50 mM phosphoric acid (pH 2.5) with 2 mM urea *Capillary voltage:* -20 kV	UV: 210 n m	15 min	2.5-8.0 µg/L	[33]
Fangchinoline and tetrandrine	Herbal medicine	*High resistivity solvent*: water *Hydrodynamic injection:* 20 µL *Sample electrokinetic injection:* 8 s (10 kV) *Sample matrix:* 50% (*v/v*) aqueous ethanol	Uncoated fused silica capillary of 29 cm (25.5 cm effective length) x 50 µm I.D. *mBGE:* 75 mM phosphoric acid-triethylamine, 2.5% (*v/v*) polyoxyethylene sorbitan monolaurate, 20% (*v/v*) methanol (pH 5.0) *Capillary voltage:* 10 kV	UV: 254 nm	20 min	61-98 µg/L	[34]
Steroids	Water	*High resistivity solvent*: water *Hydrodynamic injection:* equivalent to 4 cm water plug *Sample electrokinetic injection:* 450 s (-10 kV) pressure-assisted by 8966 Pa. *Sample matrix:* 15 mM phosphoric acid with 7.5 mM SDS	Uncoated fused silica capillary of 50.2 cm (40 cm effective length) x 50 µm I.D. *mBGE:* 50 mM SDS in 150 mM phosphate buffer (pH 2.4) with 30% (*v/v*) methanol *Capillary voltage:* -18 kV	UV: 220 and 240 nm	22 min	1-10 µg/L	[35]

Compounds	Samples	FASI conditions	MECC conditions	Detection	Analysis time	LODs	Ref.
Albumin and transferring	Human urine	*High resistivity solvent*: water *Hydrodynamic injection:* 20 s (0.5 psi) *Sample electrokinetic injection:* 90 s (10 kV) *Sample matrix:* 10 mM phosphate buffer pH 7.5	Uncoated fused silica capillary of 51 cm (43 cm effective length) x 75 μm I.D. *mBGE:* 50 mM Tris (pH 8.1) with 10 mM SDS *Capillary voltage:* 20 kV	UV: 214 nm	14 min	0.31-0.14 mg/L	[36]
Bisphenols, bisphenol-diglycidyl ethers and derivatives	Canned soft drinks	*High resistivity solvent*: water *Hydrodynamic injection:* 2 s (13.5 kPa) *Sample electrokinetic injection:* 45 s (-10 kV) *Sample matrix:* 10 mM SDS with 5% ethanol	Uncoated fused silica capillary of 57 cm (50 cm effective length) x 75 μm I.D. *mBGE:* 200 mM SDS in 25 mM phosphoric acid-monohydrogen phosphate buffer solution (pH 2.5) with 35% (*v*/v) 2-propanol *Capillary voltage:* -30 kV	UV: 214 nm	26 min	27-55 μg/L	[32]
Isonicotinamide and nicotinamide	Whitening cosmetics and supplemented foodstuffs	*High resistivity solvent*: water *Hydrodynamic injection:* 45 s (0.5 psi) *Sample electrokinetic injection:* 60 s (4 kV) *Sample matrix:* 1% (*v/v*) methanol	Uncoated fused silica capillary of 51 cm (43 cm effective length) x 50 μm I.D. *mBGE:* 150 mM SDS in 25 mM sodium borate (pH 8.3) *Capillary voltage:* 25 kV	UV: 214 nm	8 min	51-69 μg/L	[37]

This, for instance, was the approach used by Gallart-Ayala et al. [32] for the analysis of bisphenol A (BPA), bisphenol F (BPF), bisphenol A-diglycidyl ether (BADGE), bisphenol F-diglycidyl ether (BFDGE) and their derivatives in canned soft drinks by FASI-MECC. Other FASI-MECC environmental, food and bio-analytical application examples are summarized in Table 3 [32-37]. Gallart-Ayala et al. studied the effect of SDS concentration in the sample matrix injection on the electrophoretic separation when applying FASI-MECC for the analysis of bisphenol-kind compounds, and the results are shown in Figure 11.

When high SDS concentrations were added to the sample matrix, the application of FASI was not satisfactory enough probably due to the high conductivity of the standard matrix (Figure 11A). The authors evaluated then lower SDS concentrations in the sample matrix. FASI methodology improved considerably with the decrease in SDS concentration in the injection matrix, showing significant enhancement at a concentration of 10 mM SDS (Figure 11C). However, if lower SDS concentrations were used, a loss in sensitivity was again observed (Figure 11D), which was attributed to a significant decrease in SDS micelles to interact with the analytes. Thus, 10 mM SDS solution was selected as optimal concentration in the injection matrix for FASI-MECC analysis of this family of compounds. Figure 12 shows the electropherograms obtained by FASI-MECC for a non-spiked plastic bottle isotonic soft drink (A), a plastic bottle isotonic soft drink spiked at 100 µg/L (B), and a canned citrus soda soft drink sample (C).

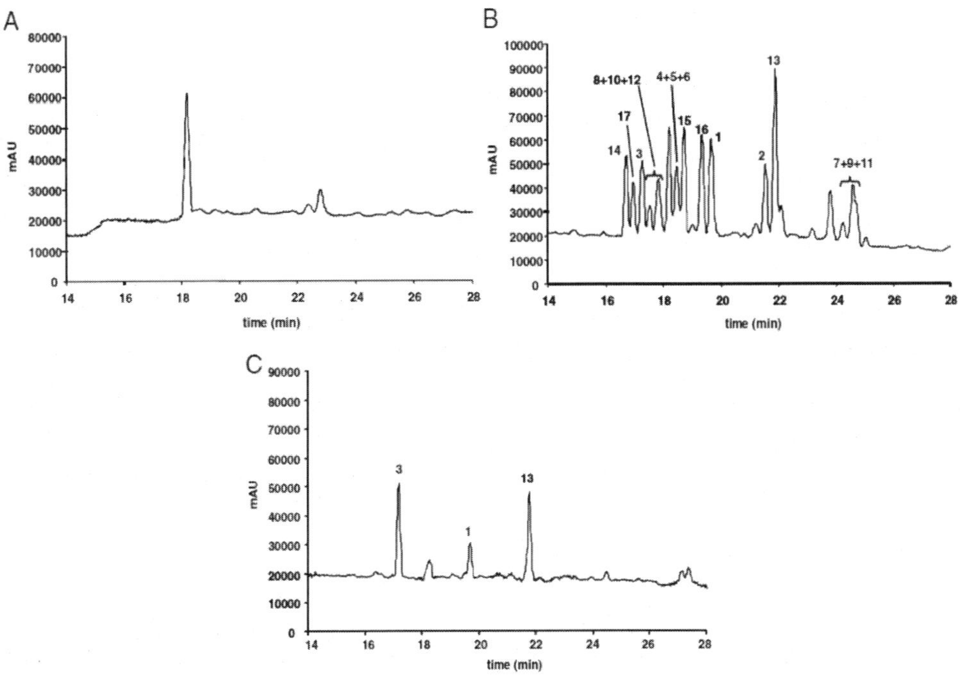

Figure 12. Electropherograms of a non-spiked plastic bottle isotonic soft drink (A), a plastic bottle isotonic soft drink spiked at 100 µg/L (B), and a canned citrus soda soft drink sample (C), obtained by FASI-MECC under optimal conditions (Table 3). Peak identification is the same as in Figure 11. Reprinted with permission from reference [32]. Copyright (2010) Wiley-VCH.

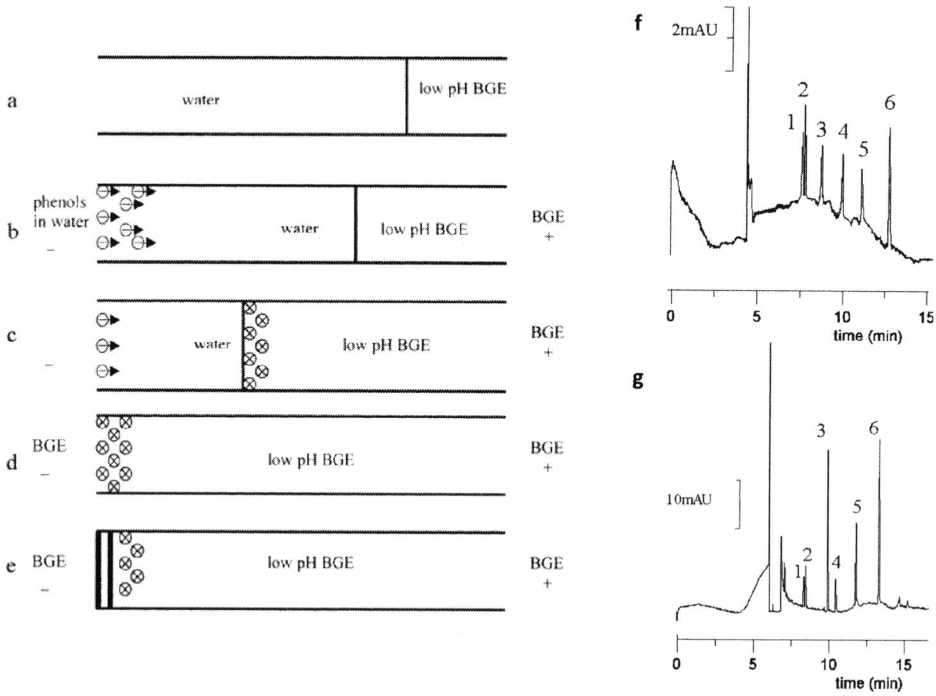

Figure 13. Schematic illustration of the sample stacking mechanism of FASI combined with the pH-mediated method in MEKC mode. (a) After filling the capillary with low-pH BGE containing SDS, a water plug was injected into the column to provide the high electric field at the injection point. (b) Electrokinetic sample injection into the capillary. Because of the high electric field, the anions move rapidly toward the outlet. At the same time, the water plug is moving out of the outlet of the capillary. (c) When the phenolic anions enter the boundary of water and low-pH BGE, the phenols are neutralized and cease moving. (d) Inlet is changed to low pH-BGE, and a negative potential of -20 kV is applied. The water plug continues to move out of the inlet. (e) SDS micelles enter the capillary, and the separation beings in MEKC mode. (f) Electropherograms of phenolic compounds in 8 mM NaOH. Original concentration of phenols, 25 ng/mL. Injection conditions: water plug length, 50 cm; sample electrokinetic injection, -10 kV x 25 min. (g) Electropherogram of a water sample extract after liquid-liquid-liquid microextraction. The extract was 8 mM NaOH. MECC condicions are as in (f). Peak identification: 1, 2,4-dimethylphenol; 2, 2,3,5-trimethylphenol; 3, 2,4-dichlorophenol; 4, 3-chlorophenol; 5, 2-chlorophenol; and 6, 2,4-dinitrophenol. Reprinted with permission from reference [33]. Copyright (2001) American Chemical Society.

A 50-fold sensitivity enhancement was achieved with FASI for most of the analyzed compounds, obtaining LODs in the range of 27-55 μg/L (for standards), and with good run-to-run and day-to-day precisions (RSD values lower than 12.5). The authors applied a simple solid-phase extraction (SPE) sample treatment and clean-up procedure using C18 cartridges for the analysis of these compounds in canned soft drinks by FASI-MECC, without affecting method performance, and achieving a 900-fold sensitivity enhancement for real sample compared with conventional MECC. LOD values in the range 3.0-5.4 μg/L for a plastic bottle isotonic drink were achieved, showing that the proposed FASI-MECC was a reliable and economic method for the analysis of this family of compounds in canned soft drinks at concentrations higher than 9-15 μg/L (limit of quantitation in real samples) and below the specific migration limits (SML) values established by the European Union.

Another interesting approach is the one described by Zhu et al. [33] for the analysis of phenolic compounds by FASI at low pH in MECC. Six phenolic compounds prepared in water or NaOH solution were used as test analytes. Sample was injected electrokinetically after the introduction of a plug of water (see scheme in Figure 13). During the injection, the water plug was pumped out of the capillary inlet by the EOF, and the phenolic anions migrated very quickly in the direction of the outlet. When the anions reached the boundary between the water plug and BGE, they were neutralized and ceased moving. Thereafter, MECC was initiated for the separation.

The electropherogram of phenolic compounds (25 ng/mL) in 8 mM NaOH obtained under optimal FASI-MECC conditions is shown in Figure 13f. The authors observed that as the concentration of NaOH in the sample matrix increased from 0 to 8 mM, the enrichment of the first five compounds increased gradually. This was induced by the increased dissociation of these compounds at higher pH. However, as the pH increased further the enrichment factor decreased slightly because of the higher ionic strength. 8 mM NaOH was then selected as optimum value. The method was validated by analyzing phenolic compounds in water samples. Analytes were extracted from the spiked water sample by a liquid-liquid-liquid microextraction (LLLME) method using a hollow fiber as a solvent support. This extraction procedure was used as a sample clean-up and preconcentration procedure. Figure 13g shows the electropherogram of a water sample extract after LLLME, where the extract was 8 mM NaOH. Method sensitivity was improved up to 2,600-fold compared with normal hydrodynamic injection.

2.3. Ion-Exhaustive Sample Injection-Sweeping (IESI-Sweeping)

Today, combining two on-line sample preconcentration techniques is a practice frequently used and that efficiently increases detection sensitivities, although sometimes the application conditions are rather limited. But, as well as other stacking techniques, the combination of sweeping and related techniques with the electrokinetic injection of a large volume of sample is significantly effective for obtaining higher enrichment factors. Ion-exhaustive sample injection-sweeping (IESI-sweeping) is a combination of FASI and sweeping, which can provide more than 100,000-fold increases in detection sensitivity [38, 39]. As FASI is involved, depending on the ions selectively introduced into the capillary, this method is called cation-exhaustive sample injection (CSEI) or anion-exhaustive sample injection (ASEI). For instance, Quirino and Terabe reported almost a million-fold sensitivity enhancement in capillary electrophoresis with direct ultraviolet detection by combining FASI and sweeping in MECC (CSEI-sweeping-MECC) [38]. The schematic illustration of this on-line preconcentration method is shown in Figure 14. First, a bare fused silica capillary is initially filled or conditioned with a low-pH buffer or non-micellar BGE (nmBGE). A zone of a high-conductivity buffer devoid or organic solvent (HCB) followed by a short zone of water is injected hydrodynamically (Figure 14A). The cationic sample prepared in a low-conductivity solution (or simply water) is injected using voltage at positive polarity (cathode in the outlet position, Figure 14B) for a period much longer than usual (e.g., 10 min). Here, the molecules enter the capillary through the water plug with high velocities. Once the molecules cross the stacking boundary or interface between the water and HCB zone, they will slow and focus at this interface. This procedure creates long zones of cationic analytes

(Figure 14C), which have concentrations greater than that in the original. The direction of EOF (considered small due to suppressed dissociation of the silanol groups at low-pH) and the cationic analytes is toward the cathode. At this moment, part of the sample matrix enters the capillary since the electroosmotic flow is directed toward the cathode. This step has involved FASI. The next step is to focus the injected zones by sweeping. This is accomplished by placing a low-pH buffer solution containing anionic micelles or micellar BGE (mBGE) in the inlet vial followed by application of voltage at negative polarity (anode in the outlet position, Figure 14D). Once voltage is applied at negative polarity with the mBGE in the inlet vial, anionic micelles will enter the capillary and sweep the analytes that were injected whether they are stacked or not. The micelles enter the low-conductivity zone consisting of the sample matrix introduced during FASI and the water plug, and then stack at the interface between the water plug and HCB. The stacked micelles then sweep the stacked cations. Once the stacked cations are completely swept, their separation is accomplished by MECC in reversed migration mode (Figure 14E). The mechanism of separation is due to partitioning of the analytes between the fast moving micellar phase and the very slow moving aqueous phase. The micellar phase carries the analytes toward the detector. The water zone and the sample matrix that was introduced are consequently removed from the capillary by the slow bulk EOF, which is now directed toward the inlet vial.

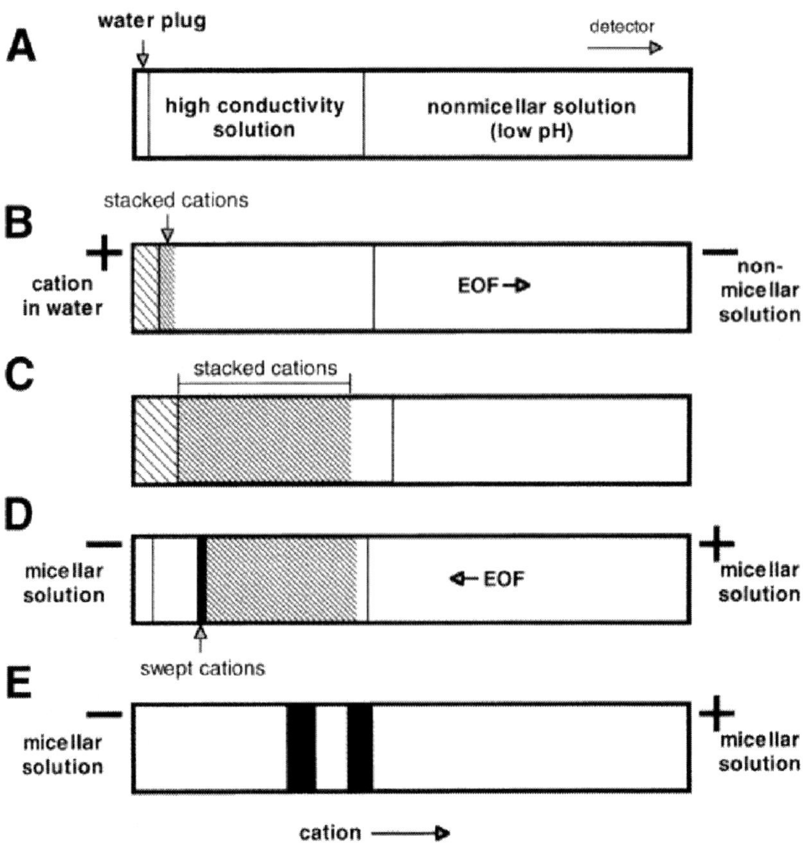

Figure 14. Schematic illustration of the evolution of analyte zones in CSEI-sweeping-MEKC. Reprinted with permission from reference [38]. Copyright (2000) American Chemical Society.

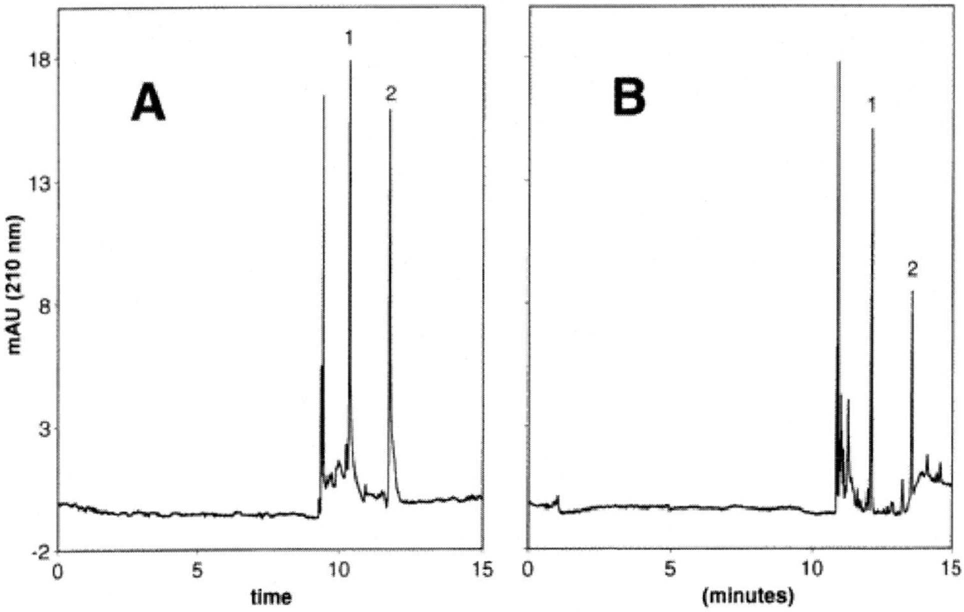

Figure 15. Almost a million-fold concentration of dilute cations by CSEI-sweeping-MECC. Conditions: nmBGE, 1 mM triethanolamine/15% acetonitrile/100 mM phosphoric acid; mBGE, 100 mM SDS/1 mM triethanolamine/15% acetonitrile/50 mM phosphoric acid; HCB, 100 mM phosphoric acid; sample solution, laudanosine (1) and 1-naphthylamine (2) in water; sample concentration, ~240 mg/L (A), ~240 ng/L (B); injection scheme, 0.6 mm of the sample solution (A), 30 cm of HCB and then 3 mm of water followed by 23 kV electrokinetic injection of the sample solution for 1,000 s (B); sweeping and MECC voltage, -23 kV with the mBGE at both ends of the capillary. Reprinted with permission from reference [38]. Copyright (2000) American Chemical Society.

Figure 15 shows the ~240 ng/L detection of two cationic analytes using the reported CSEI-sweeping-MECC method. Around 900,000- and 550,000-fold sensitivity enhancements for laudanosine and 1-naphthylamine, respectively, were obtained. An important precaution in performing CSEI-sweeping-MECC is that fresh samples should always be used for each injection. This is because of the decrease in the concentration of the sample after injection from a single sample vial.

One difficulty of this kind of methodology, which is also present in some other on-line electrophoretic-based preconcentration methods, will be the necessity of preparing the sample in a low-conductivity matrix. This is especially true in real world analysis, for instance when dealing with environmental water samples because of their high salt content on these samples.

Some ISEI-sweeping-MECC environmental, food and bio-analytical application examples are summarized in Table 4 [16, 40-48]. Núñez et al. [16] used CSEI-sweeping-MECC for the analysis of the herbicides paraquat, diquat and difenzoquat (quaternary ammonium salts) in drinking water samples. CSEI-sweeping was performed by employing a 100 mM phosphate buffer (pH 2.5) with 20% (*v/v*) acetonitrile as nmBGE and a 200 mM phosphate buffer (pH 2.5) as high conductivity buffer which was hydrodynamically introduced into the capillary at 5 kPa for 200 s, followed by a water plug. Sample was electrokinetically injected at +22 kV for 400 s. After the enhanced injection of cations into the capillary, a mBGE vial containing 80 mM SDS in 50 mM phosphate buffer (pH 2.5) with 20% (20% (*v/v*) acetonitrile solution was placed in the inlet position and separation was performed by sweeping-MECC. Figure 16 shows the comparison of conventional MECC and CSEI-sweeping-MECC for the analysis of these three herbicides. Between 3,000- and 51,000-fold sensitivity enhancement was achieved for difenzoquat and diquat, respectively. LOD values in the range 0.075-1 μg/L were obtained.

To demonstrate how the proposed CSEI-sweeping-MECC method can be applied for routine analysis of real samples, spiked tap water samples were analyzed. Figure 17 shows the electropherograms obtained when Japanese tap water (Harima Science Garden) spiked at 10 μg/L for PQ, DQ and EV and at 50 μg/L for DF was injected using the optimized method.

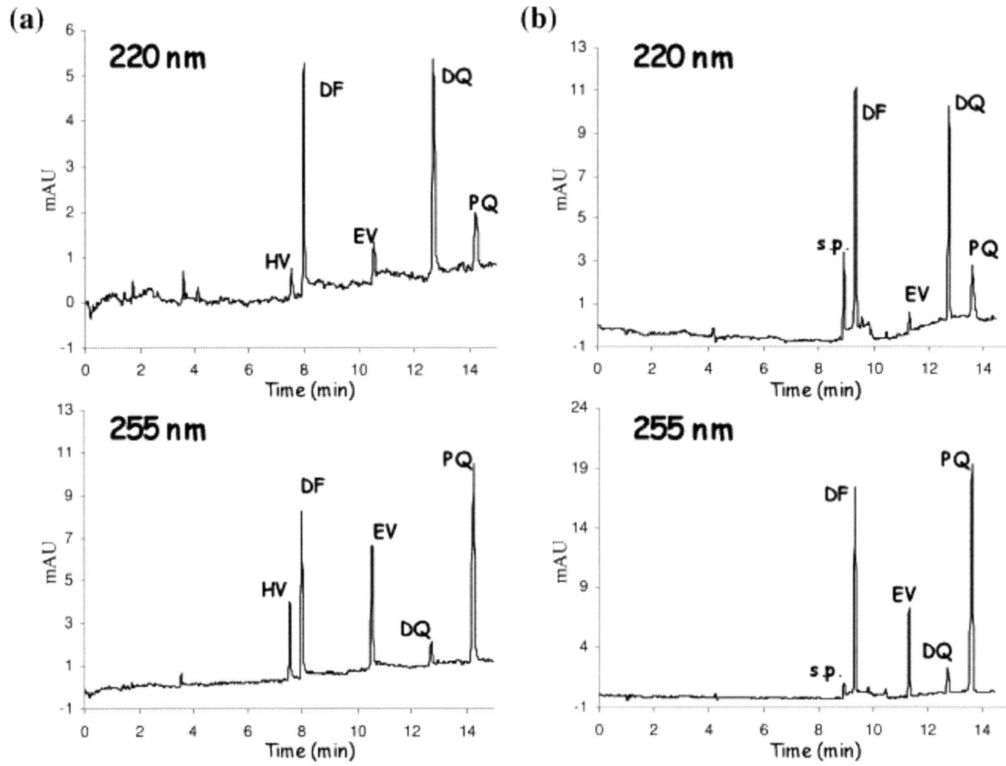

Figure 16. Conventional MECC and CSEI-sweeping-MECC of quaternary ammonium herbicides. (a) MECC: sample prepared in BGE; sample concentration, 100 mg/L; injection time, 1 s at 5 kPa. (b) CSEI-sweeping-MECC: sample prepared in water; sample concentration, 10 μg/L PQ, DQ and EV and 50 μg/L DF. Injection scheme: hydrodynamic injection of HCB for 200 s (5 kPa), hydrodynamic injection of water for 6 s (5 kPa), electrokinetic injection of sample for 400 s (+22 kV); Other experimental conditions as in Table 4. Peak identification as in Figure 6. Reprinted with permission from reference [16]. Copyright (2002) Elsevier.

Table 4. Selection of ISEI-sweeping-MECC methods in environmental, food and bio-analytical applications

Compounds	Samples	ISEI-sweeping conditions	MECC conditions	Detection	Analysis time	LODs	Ref.
Herbicides paraquat, diquat and difenzoquat	Water samples	*nmBGE*: 100 mM phosphate buffer (pH 2.5) with 20% (*v/v*) acetonitrile *HCB*: 200 mM phosphate buffer (pH 2.5) *HCB hydrodynamic injection:* 200 s (5 kPa) water plug hydrodynamic injection: 6 s (5 kPa) *Sample electrokinetic injection:* 400 s (+22 kV) *Sample matrix:* purified water or tap water	Uncoated fused silica capillary of 60 cm (51.5 cm effective length) x 50 µm I.D. *mBGE:* 80 mM SDS in 50 mM phosphoric acid (pH 2.5) with 20% (*v/v*) acetonitrile *Capillary voltage:* -22 kV	UV: 220 and 255 nm	15 min	0.075-1.0 µg/L	[16]
Phenoxy acid herbicides	Water samples	*nmBGE*: 100 mM phosphoric acid, 20% (*v/v*) acetonitrile, 1 M urea (pH 2.5) *High resistivity solvent*: water:acetonitrile (1:1), hydrodynamically injected for a length of 4.96 cm. *Sample electrokinetic injection:* 12 min (-20 kV) *Sample matrix:* purified water	Uncoated fused silica capillary of 70 cm (55 cm effective length) x 50 µm I.D. *mBGE:* 75 mM SDS in 25 mM phosphoric acid with 20% (*v/v*) acetonitrile and 1 M urea (pH 2.5) *Capillary voltage:* -20 kV	UV: 210 nm	30 min	0.1-0.5 µg/L	[40]
Ephedra-alkaloids	Chinese herbal drug Serum samples	*nmBGE*: 50 mM phosphoric acid with 20% (*v/v*) acetonitrile *HCB*: 100 mM phosphoric acid *HCB hydrodynamic injection:* 200 s (50 mbar) water plug hydrodynamic injection: 3 s (5 mbar) *Sample electrokinetic injection:* 600 s (+20 kV) *Sample matrix:* ethanol	Uncoated fused silica capillary of 70 cm (61.5 cm effective length) x 75 µm I.D. *mBGE:* 50 mM SDS in 25 mM phosphoric acid with 25% (*v/v*) acetonitrile *Capillary voltage:* -20 kV	UV: 200 nm	30 min	3.1-32.5 µg/L	[41]

Table 4. (Continued)

Compounds	Samples	ISEI-sweeping conditions	MECC conditions	Detection	Analysis time	LODs	Ref.
Methamphetamine, ketamine, morphine and codeine	Human urine	*nmBGE*: 50 mM phosphate buffer (pH 2.5) with 30% (*v/v*) methanol *HCB*: 100 mM phosphate buffer (pH 2.5) *HCB hydrodynamic injection:* 99.9 s (6.9 kPa) *Sample electrokinetic injection:* 500 s (+10 kV) *Sample matrix:* diluted urine samples	Uncoated fused silica capillary of 50.2 cm (40 cm effective length) x 50 µm I.D. *mBGE:* 100 mM SDS in 25 mM phosphate buffer (pH 2.5) with 20% (*v/v*) methanol *Capillary voltage:* -20 kV	UV: 200 nm	20 min	5-15 µg/L	[42]
Methamphetamine, ketamine, morphine and codeine	Hair	*nmBGE*: 50 mM phosphate buffer (pH 2.5) with 30% (*v/v*) methanol *HCB*: 100 mM phosphate buffer (pH 2.5) *HCB hydrodynamic injection:* 99.9 s (6.9 kPa) *Sample electrokinetic injection:* 600 s (+10 kV) *Sample matrix:* water	Uncoated fused silica capillary of 50.2 cm (40 cm effective length) x 50 µm I.D. *mBGE:* 100 mM SDS in 25 mM phosphate buffer (pH 2.5) with 20% (*v/v*) methanol *Capillary voltage:* -20 kV	UV: 200 nm	22 min	50-100 ng/kg	[43]
Morphine and four metabolits	Human urine	*nmBGE*: 75 mM phosphate buffer (pH 2.5) with 30% (*v/v*) methanol *HCB*: 120 mM phosphate buffer (pH 2.5) *HCB hydrodynamic injection:* 99.9 s (10.3 kPa) *Sample electrokinetic injection:* 600 s (+10 kV) *Sample matrix:* water	Uncoated fused silica capillary of 50.2 cm (40 cm effective length) x 50 µm I.D. *mBGE:* 100 mM SDS in 25 mM phosphate buffer (pH 2.5) with 22% (*v/v*) methanol *Capillary voltage:* -20 kV	UV: 200 nm	24 min	10-25 µg/L	[44]
Tobacco-specific N-nitrosamines	Human urine	*nmBGE*: 80 mM phosphate buffer (pH 2.5) with 10% (*v/v*) acetonitrile *HCB*: 109 mM phosphoric acid *HCB hydrodynamic injection:* equivalent to 13.3 mm *water plug injection*: equivalent to 1.3 mm *Sample electrokinetic injection:* 300 s (-10 kV) *Sample matrix:* water	Uncoated fused silica capillary of 48.5 cm (40 cm effective length) x 50 µm I.D. *mBGE:* 75 mM SDS in 80 mM phosphate buffer (pH 2.5) with 10% (*v/v*) acetonitrile *Capillary voltage:* -25 kV	UV: 240 nm	9 min	4-16 µg/L	[45]

Compounds	Samples	ISEI-sweeping conditions	MECC conditions	Detection	Analysis time	LODs	Ref.
Methadone and metabolites	Serum samples	*nmBGE*: 100 mM phosphate buffer (pH 4.0) with 20% (*v/v*) tetrahydrofuran *HCB*: 300 mM phosphate buffer (pH 4.0) *HCB hydrodynamic injection:* 30 s (10 psi) *Sample electrokinetic injection:* 500 s (+10 kV) *Sample matrix:* water	Uncoated fused silica capillary of 41 cm (30 cm effective length) x 50 µm I.D. *mBGE:* 100 mM SDS in 100 mM phosphate buffer (pH 4.0) with 22% (*v/v*) tetrahydrofuran *Capillary voltage:* -15 kV	UV: 214 nm	50 min	0.2-0.4 µg/L	[46]
Cotinine	Serum samples	*nmBGE*: 50 mM phosphoric buffer (pH 2.5) *HCB*: 75 mM phosphoric buffer (pH 2.5) *HCB hydrodynamic injection:* equivalent to 8 mm *water plug injection:* equivalent to 1 mm *Sample electrokinetic injection:* 180 s (+10 kV) *Sample matrix:* 0.1 mM phosphoric acid	Uncoated fused silica capillary (20 cm effective length) x 50 µm I.D. *mBGE:* 75 mM SDS in 50 mM phosphoric acid (pH 2.5) *Capillary voltage:* -12 kV	UV: 200 nm	4 min	0.2 µg/L	[47]
Ractopamine	Porcine meat	*nmBGE*: 55 mM phosphate buffer (pH 2.75) with 25% (*v/v*) methanol *HCB*: 125 mM phosphate buffer (pH 2.75) with 15% (*v/v*) methanol *HCB hydrodynamic injection:* 40 s (5 psi) *Sample electrokinetic injection:* 12 min (+9 kV) *Sample matrix:* water	Uncoated fused silica capillary of 50.2 cm (40 cm effective length) x 50 µm I.D. *mBGE:* 125 mM SDS in 55 mM phosphate buffer (pH 2.75) with 25% (*v/v*) methanol *Capillary voltage:* -25 kV	UV: 230 nm	18 min	3-5 µg/kg	[48]

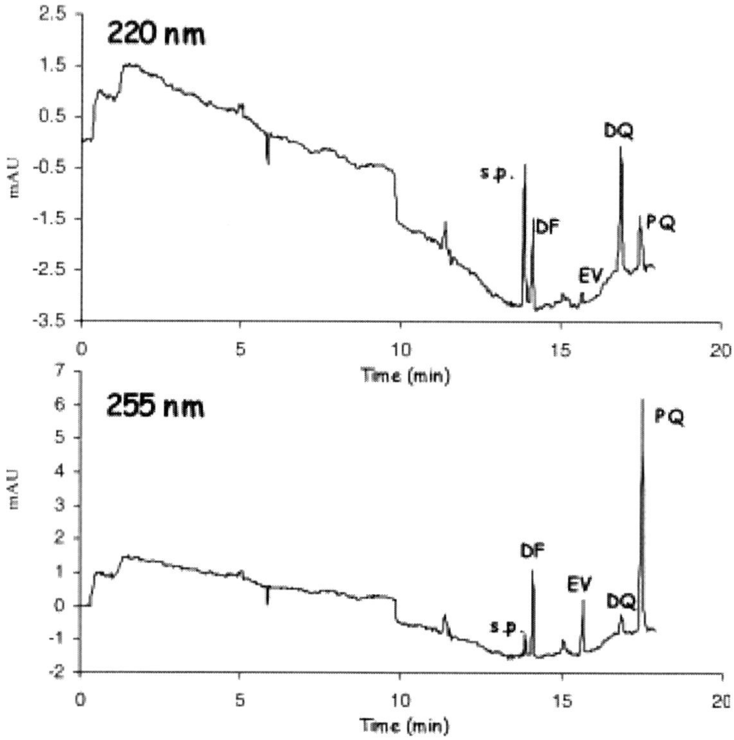

Figure 17. Electropherogram of a tap water analyzed by CSEI-sweeping-MECC. Sample concentration; 10 μg/L PQ, DQ and EV and 50 μg/L DF. Other conditions as in Figure 16b. Peak identification as in Figure 6. Reprinted with permission from reference [16]. Copyright (2002) Elsevier.

LODs in the range 0.5-3.3 μg/L were obtained, which are higher than those obtained when standards in purified water were used, due to the relatively high salinity of tap water (conductivity: 152.4 μS/cm) that produced a reduced field enhancement when the electrokinetic injection was used. However, the obtained values were similar to those previously reported by the same authors in a previous work using a combination of SPE and stacking with sample matrix removal in CZE for tap water [49].

When dealing with FASI procedures, such as in the case of ISEI-sweeping techniques, the water plug injected before the electrokinetic enhanced injection of the sample play an important role because it creates a higher electric field at the tip of the capillary, which improves stacking efficiency. Due to their low conductivity, acetonitrile and methanol were added into the sample solution by some workers to improve the concentration sensitivity. For instance, Zhu et al. [40] compared the use of pure water and water:acetonitrile (1:1 *v/v*) as high resistivity solvent on the analysis of phenoxy acid herbicides in water samples by CSEI-sweeping-MEKC, and the obtained results are shown in Figure 18. As can be seen, an important sensitivity enhancement is achieved by adding acetonitrile in the water plug previous to the FASI injection.

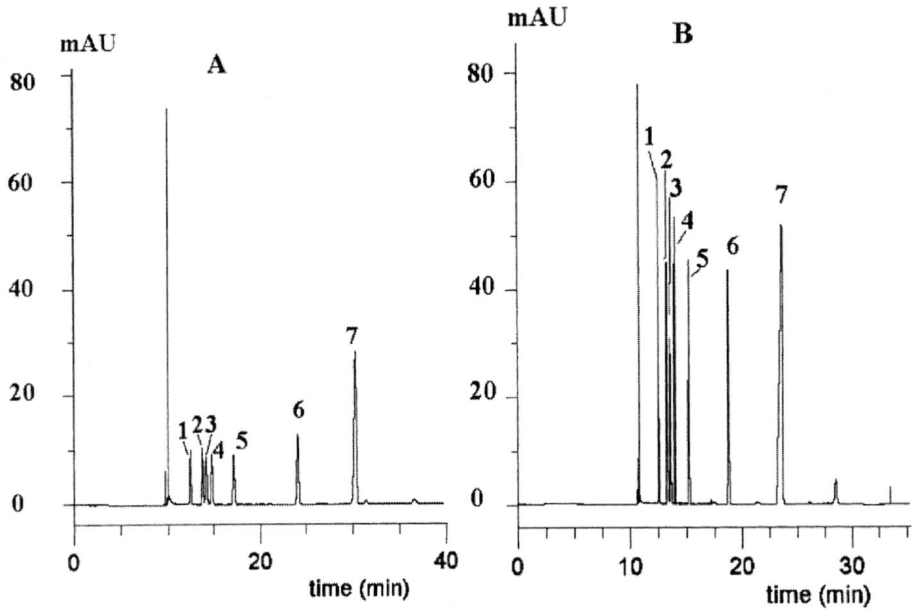

Figure 18. Effect of adding acetonitrile in the water plug on the stacking efficiency of CSEI-sweeping-MECC. Conditions: high resistivity solvent plug length, 3.31 cm; sample injection, 10 min; sample concentration, 100 ng/mL. High resistivity solvent: (A) water, (B) water:acetonitrile (1:1 v/v). Peak identification: 1, 4-(2,4-dichlorophenoxy) butyric acid; 2, 2,4,5-trichlorophenoxyacetic acid; 3, 2-(2,4,5-trichlorphenoxy)propionic acid; 4, 2-(2,4-dichlorophenoxy) propionic acid; 5, 2-(4-chloro-2-methylphenoxy) butyric acid; 6, 2,4-dichlorophenoxyacetic acid; and 7, 4-chlorophenoxyacetic acid. Reprinted with permission from reference [40]. Copyright (2002) American Chemical Society.

An interesting application of CSEI-sweeping-MECC is the one described by Lin et al. for the analysis of methamphetamine, ketamine, morphine and codeine in hair samples [43]. After pretreatment of hair samples with hydrochloric acid, neutralization and extraction with ethyl acetate, the extracts were evaporated and reconstituted in water previous to CSEI-sweeping-MECC analysis. Figure 19 shows the electropherogram of one hair sample under optimal conditions. Limits of detection down to 50 pg/mg hair for methamphetamine and ketamine, 100 pg/mg hair for codeine and 200 pg/mg hair for morphine were achieved.

The authors validated the proposed CSEI-sweeping-MECC method by analyzing the addict hair samples by liquid chromatography-mass spectrometry (LC-MS), showing good coincidence of results. Thus, CSEI-sweeping-MECC has proven to be feasible for application in detecting trace levels of abused drugs in forensic analysis.

Recently, Wang et al. [48] proposed a CSEI-sweeping-MECC method for the analysis of ractopamine in porcine meat. Limits of detection down to 3-5 µg/kg were achieved, which represented a 900-fold sensitivity enhancement in comparison to those observed with conventional CZE methods. The proposed method showed to be fast and suitable for serving as a routing tool for the examination of ractopamine in meat samples.

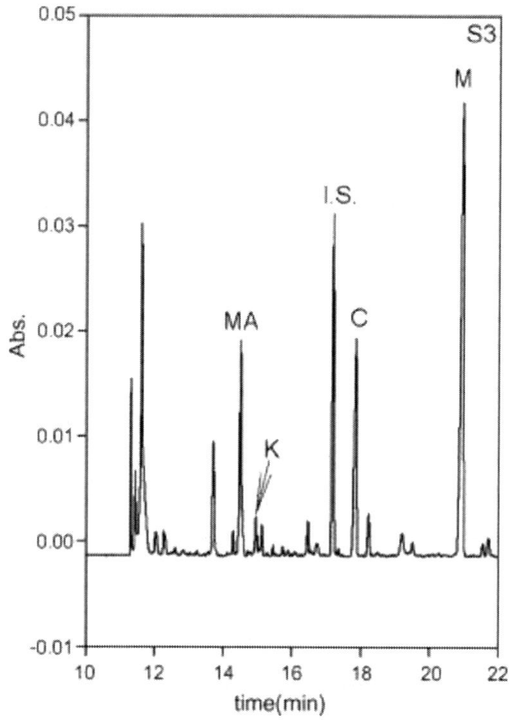

Figure 19. Electrophoerigram of one real hair sample obtained from an addict person under optimized conditions (see Table 4). Drug concentrations (mg of hair): methamphetamine (MA) 18.59 ng/mg; ketamine (K) 1.18 ng/mg; morphine (M) 43.95 ng/mg; and codeine (C) 20.68 ng/mg. Reprinted with permission from reference [43]. Copyright (2007) Elsevier.

2.4. Dynamic pH Junction-Sweeping

Dynamic pH junction is based on the creation of a pH discontinuity that is established by injecting the sample at a different pH than the BGE and can be used to concentrate weakly ionic analytes. Recently, combining dynamic pH junction and sweeping during MECC has been reported to lead to further improvements in sensitivity [50, 51]. This approach integrates the merits of both dynamic pH junction and sweeping and improves separation selectivity and sensitivity. Thus, the dynamic pH junction-sweeping method can be used to focus both weakly ionic and neutral analytes, as well as to improve the focusing performance for certain analytes as compared to either dynamic pH junction or sweeping formats alone. Britz-McKibbin et al. [50] first used dynamic pH junction-sweeping-MECC using laser-induced fluorescence for the determination of flavin derivatives with a picomolar detection limit. For example, as a way to enhance SDS partitioning and analyte sweeping, flavins may be dissolved in acidic phosphate buffer, pH 6, in order to reduce their negative charge (riboflavin (RF) is neutral) while retaining the selectivity of borate separation under alkaline conditions. Figure 20 depicts the RF focusing setup using a combined dynamic pH junction-sweeping technique when the sample electrolyte used is phosphate, whereas the BGE consists of borate pH 8.5 with SDS. Enhanced flavin focusing may be realized by using dynamic pH junction-sweeping format, since band narrowing is induced by several distinct processes (which may be additive), including buffer pH, borate complexation, and micelle partitioning.

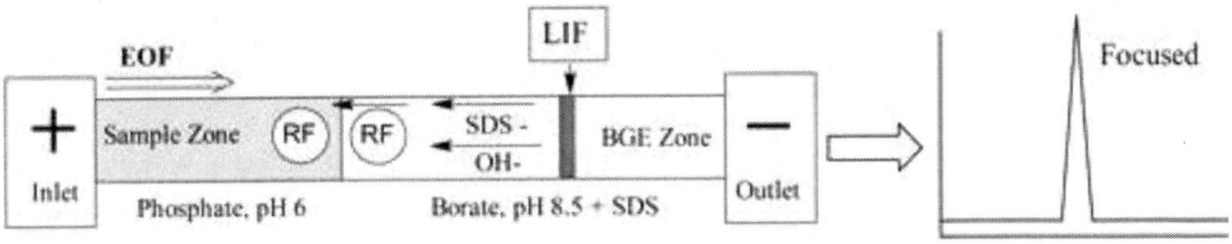

Figure 20. Schematic representation of dynamic pH junction-sweeping analyte focusing. Reprinted with permission from reference [50]. Copyright (2002) American Chemical Society.

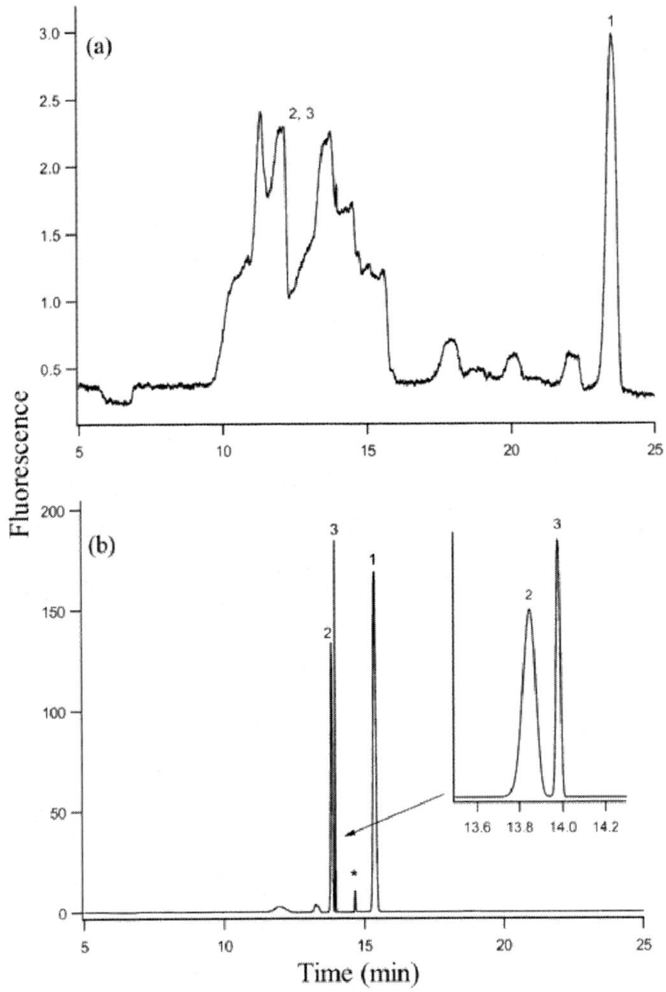

Figure 21. Comparison of flavin focusing using large injection plugs (8.2 cm) with (a) conventional sweeping and (b) dynamic pH junction-sweeping. BGE: 140 mM borate, 100 mM SDS, pH 8.5. Sample solutions contained 0.2 μM flavins dissolved in either (a) 140 mM borate, pH 8.5, and (b) 75 mM phosphate, pH 6.0. Peak identification: 1, riboflavin; 2, flavin mononucleotide; and 3, flavin adenine dinucleotide. *, system peak. Reprinted with permission from reference [50]. Copyright (2002) American Chemical Society.

This combined focusing approach is aimed at overcoming the often poor band-narrowing efficiency of conventional sweeping (using anionic micelles) and dynamic pH junction for hydrophilic and neutral analytes, respectively. As an example, Figure 21 shows two electropherograms comparing flavin focusing using large injection plugs (8.2 cm) with conventional sweeping and dynamic pH junction-sweeping. As can be seen, analyte focusing considerably improves when using dynamic pH junction for these weakly ionic compounds.

Table 5 summarizes several applications of dynamic pH junction-sweeping-MECC in food and bio-analytical applications [52-55].

For instance, Yu and Li [52] proposed a dynamic pH junction-sweeping-MECC method for the on-line preconcentration of toxic pyrrolizidine alkaloids in Chinese herbal medicine. For that purpose, a long plug of sample prepared in an acidic sample matrix (phosphate buffer pH 4.0) was introduced into the capillary. Under these conditions, pyrrolizidine alkaloids (PAs) undergo protonation since the sample matrix pH is bellow their pKa values. Upon application of separation voltage, hydroxide ions (OH^-) in the mBGE (micellar borate buffer at pH 9.1, above PAs pka values) with rapid mobility migrate into the sample zone, leading to an abrupt local pH increase at the front edge of the sample. At that moment, PA molecules with positive charge originally in the leading edge are suddenly deprived of protons and in the mean time, become neutral and solubilize in the SDS micelles. As a result, mobilities of PAs in the front edge experience a dramatic drop and then reverse from positive to negative, i.e., counter to the EOF (lower velocity), whereas charged PAs in the remaining sample zone still migrate to the detector with positive mobility (higher velocity). Consequently, original large sample plug is focused into sharp sample zones as the higher velocity PAs in the back section of sample compresses into the front edge section with lower velocity. Normal MECC separation starts after the compression process is finished. A 23.8- to 90.0-fold increase in sensitivity was achieved with the proposed method, obtaining LODs as low as 30 μg/L for the analyzed PAs.

Recently, an interesting on-line preconcentration approach was described by Rageh et al. [54] by combining dynamic pH junction-sweeping with large volume sample stacking conditions as three consecutive steps for on-line focusing in the sensitive quantitation of urinary nucleosides by MECC. For that purpose, a low conductivity aqueous sample matrix free from borate and a high conductivity BGE (containing borate, pH 9.25) were needed. Briefly, the method involved filling the capillary with BGE followed by an extremely large sample plug (up to 94% of the capillary length). Then, a negative voltage was applied, where EOF is toward the cathode (in the inlet position) and hydroxide and borate ions migrate toward the anode (in the outlet position). The focusing by dynamic pH junction starts when hydroxide ions deprotonate the nucleosides. Then, borate commences to sweep the nucleosides within the sample zone via complexation. At that moment, focusing by dynamic pH-junction ends, whereas the deprotonated analytes stacked at the sample matrix/BGE boundary. Nucleosides will continue to be accumulated by sweeping and currently the sample matrix is pumped out from the injection end. When the largest part of the sample matrix plug is removed, the polarity is switched to positive voltage and the separation by MECC starts.

This method allowed achieving LODs down to 10 μg/L, representing the lowest LOD reported so far for the analysis of nucleosides using CE techniques with UV detection, providing a comparable sensitivity to CE-MS techniques.

Table 5. Selection of dynamic pH junction-sweeping-MECC methods food and bio-analytical applications

Compounds	Samples	dynamic pH junction-sweeping conditions	MECC conditions	Detection	Analysis time	LODs	Ref.
Toxic pyrrolizidine alkaloids	Chinese herbal medicine	*sample matrix*: 10 mM phosphate buffer (pH 4.0) with 20% (*v/v*) methanol *Sample injection: 265* s (20.7 mbar)	Uncoated fused silica capillary of 60 cm (50 cm effective length) x 50 µm I.D. *mBGE:* 30 mM SDS in 20 mM borate (pH 9.1) with 20% (*v/v*) methanol *Capillary voltage:* 20 kV	UV: 220 nm	20 min	30 µg/L	[52]
Dipeptides	-	*sample matrix*: 25 mM sodium dihydrogen phosphate (pH 2.5) *Sample injection:* up to 39% capillary length	Uncoated fused silica capillary of 60 cm (50 cm effective length) x 50 µm I.D. *mBGE:* 21 mM SDS and 16 mM Brij35 in 100 mM borate (pH 9.0) *Capillary voltage:* 20 kV	Laser-induced fluorescence: 488 nm (excitation); 535 (emission)	14 min	1.0-5.0 pmol/L	[53]
Benzoic and sorbic acids	Food products	*sample matrix*: 2.5 mM phosphate buffer (pH 3.0) *Sample injection:* 360 s (with 20 cm height difference between capillary ends)	Uncoated fused silica capillary of 50 cm (40 cm effective length) x 75 µm I.D. *mBGE:* 40 mM SDS in 15 mM tetraborate (pH 9.2) with 0.1% (*v/v*) poly(ethylene oxide) *Capillary voltage:* 15 kV	UV: 230 nm	7 min	6.1-8.2 nmol/L	[54]
Urinary nucleosides	Urine	dynamic pH junction-sweeping with Large volume sample stacking *sample matrix*: water *Sample injection:* up to 94% of capillary length	Uncoated fused silica capillary of 64.8 cm (50 cm effective length) x 50 µm I.D. *mBGE:* 30 mM SDS in 20 mM borate (pH 9.1) with 20% (*v/v*) methanol *Capillary voltage:* -20 kV (for sample matrix removal); 20 kV (for separation)	UV: 257 nm	18 min	10 µg/L	[55]

Conclusion and Future Trends

Fundamentals aspects of micellar electrokinetic chromatography regarding theoretical principles (separation and composition of micellar solutions) have been addressed. MECC is becoming a very popular technique since it allows separation of neutral as well as ionic compounds with a simple instrumentation. An appropriate selection of micellar background electrolyte composition will allow analysts to achieve good electrophoretic separations of complex matrices under MECC.

As in the case of capillary zone electrophoresis, MECC is also becoming very popular in multiple application fields such as bio-analytical, food and environmental applications, and the number of publications on these topics is increasing. Although many scientists consider that an important handicap of MECC techniques is detection, due to the low amount of samples injected into the capillary and the short optical path length frequently employed (capillary internal diameter with on-column UV detection), this problem can easily be overcome today by employing on-line preconcentration methods. Many electrophoretic-based on-line preconcentration methods are available, and the fundamentals of some of them based on stacking and sweeping phenomena, such as sweeping, field-amplified sample injection (FASI), ion-exhaustive sample injection-sweeping (IESI-sweeping), and dynamic pH junction-sweeping have been presented and discussed. Examples of relevant applications in bio-analytical, food and environmental analysis of these on-line preconcentration methods have also been addressed. Huge sensitivity enhancements have been reported with some of these methods. For instance, up to 1,000,000-fold enhancement with IESI-sweeping-MECC.

Today, MECC, as well as other CE techniques, are becoming powerful tools for the practical and routine ultra-trace analysis by using on-line electrophoretic preconcentration methods. The combination of different on-line preconcentration methods will provide a variety of new approaches to improve sensitivity, making CE a very promising technique for future applications in many disciplines.

Acknowledgments

This work has been funded by the Spanish Ministry of Economy and Competitiveness under the project CTQ2012-30836, and from the Agency for Administration of University and Research Grants (Generalitat de Catalunya, Spain) under the project 2014 SGR-539.

References

[1] Terabe, S., Otsuka, K., Ichikawa, K., Tsuchiya, A. & Ando, T. (1984). Electrokinetic separations with micellar solutions and open-tubular capillaries. *Anal. Chem.*, *56*, 111-113.

[2] Schmitt-Kopplin, P. (2008). Capillary electrophoresis - Methods and Protocols. Totowa, NJ. Humana Press.

[3] Terabe, S., Otsuka, K. & Ando, T. (1985). Electrokinetic chromatography with micellar solution and open-tubular capillary. *Anal. Chem.*, *57*, 834-841.

[4] Terabe, S. (2009). Capillary separation: micellar electrokinetic chromatography. *Annu. Rev. Anal. Chem.*, *2*, 99-120.

[5] Ghowsi, K., Foley, J. P. & Gale, R. J. (1990). Micellar electrokinetic capillary chromatography theory based on electrochemical parameters: optimization for three modes of operation. *Anal. Chem.*, *62*, 2714-2721.

[6] Muijselaar, P. G., Otsuka, K. & Terabe, S. (1997). Micelles as pseudo-stationary phases in micellar electrokinetic chromatography. *J. Chromatogr. A*, *780*, 41-61.

[7] Otsuka, K. & Terabe, S. (2000). Enantiomer separation of drugs by micellar electrokinetic chromatography using chiral surfactants. *J. Chromatogr. A*, *875*, 163-178.

[8] Quirino, J.P. (2006). On-line sample enrichment in electrokinetic chromatography. In: Electrokinetic chromatography, Puell, U. (ed) *10*:207-31, Chichester, UK. Wiley. 539 pp.

[9] Simpson, S. L., Quirino, J. P. & Terabe, S. (2008). On-line sample preconcentration in capillary electrophoresis: Fundamentals and applications. *J. Chromatogr. A*, *1184*, 504-541.

[10] Chien, R. L. & Burgi, D. S. (1992). On-column sample concentration using field amplification in CZE. *Anal. Chem.*, *64*, 489A-496A.

[11] Kitagawa, F. & Otsuka, K. (2014). Recent applications of on-line sample preconcentration techniques in capillary electrophoresis. *J. Chromatogr. A*, *1335*, 43-60.

[12] Quirino, J. P. & Terabe, S. (1998). Exceeding 5000-fold concentration of dilute analytes in micellar electrokinetic chromatography. *Science*, *282*, 465-468.

[13] Quirino, J. P. & Terabe, S. (1999). Sweeping of Analyte Zones in Electrokinetic Chromatography. *Anal. Chem.*, *71*, 1638-1644.

[14] Quirino, J. P. & Terabe, S. (1999). Sweeping with an enhanced electric field of neutral analyte zones in electrokinetic chromatography. *J. High Resolut. Chromatogr.*, *22*, 367-372.

[15] Isoo, K., Otsuka, K. & Terabe, S. (2001). Application of sweeping to micellar electrokinetic chromatography-atmospheric pressure chemical ionization-mass spectrometric analysis of environmental pollutants. *Electrophoresis*, *22*, 3426-3432.

[16] Núñez, O., Kim, J. B., Moyano, E., Galceran, M. T. & Terabe, S. (2002). Analysis of the herbicides paraquat, diquat and difenzoquat in drinking water by micellar electrokinetic chromatography using sweeping and cation selective exhaustive injection. *J. Chromatogr. A*, *961*, 65-75.

[17] Tsai, I. L., Sun, S. W., Liao, H. W., Lin, S. C. & Kuo, C. H. (2009). Rapid analysis of melamine in infant formula by sweeping-micellar electrokinetic chromatography. *J. Chromatogr. A*, 1216, 8296-8303.

[18] Chiang, J. F., Hsiao, Y. T., Ko, W. K. & Wu, S. M. (2009). Analysis of multiple abused drugs and hypnotics in urine by sweeping CE. *Electrophoresis*, *30*, 2583-2589.

[19] Gao, W. H., Chen, G. P., Chen, Y. W., Li, N. N., Chen, T. F. & Hu, Z. D. (2011). Selective extraction of alkaloids in human urine by on-line single drop microextraction coupled with sweeping micellar electrokinetic chromatography. *J. Chromatogr. A*, *1218*, 5712-5717.

[20] Aranas, A. T., Guidote, A. M., Jr. Haddad, P. R. & Quirino, J. P. (2011). Sweeping-micellar electrokinetic chromatography for the simultaneous analysis of tricyclic antidepressant and Î²-blocker drugs in wastewater. *Talanta*, *85*, 86-90.

[21] Wu, W. C., Tsai, I. L., Sun, S. W. & Kuo, C. H. (2011). Using sweeping-micellar electrokinetic chromatography to determine melamine in food. *Food Chem.*, *128*, 783-789.

[22] Fang, H., Yang, F., Sun, J., Tian, Y., Zeng, Z. & Xu, Y. (2011). Centrifuge microextraction coupled with sweeping-MEKC to analyze trace steroid hormones in urine samples. *Talanta*, *85*, 2148-2153.

[23] Lin, S. C., Liu, H. Y., Lin, S. W., Yao, M., Wu, U. I., Kuo, H. P. & Kuo, C. H. (2012). Simultaneous determination of triazole antifungal drugs in human plasma by sweeping-micellar electrokinetic chromatography. *Anal. Bioanal. Chem.*, *404*, 217-228.

[24] Gao, W., Chen, Y., Chen, G., Xi, J., Chen, Y., Yang, J. & Xu, N. (2012). Trace analysis of three antihistamines in human urine by on-line single drop liquid-liquid-liquid microextraction coupled to sweeping micellar electrokinetic chromatography and its application to pharmacokinetic study. *J. Chromatogr. B: Anal. Technol. Biomed. Life Sci.*, *904*, 121-127.

[25] Maijo, I., Borrull, F., Aguilar, C. & Calull, M. (2012). Determination of anti-inflammatory drugs in river water by sweeping-micellar electrokinetic capillary chromatography. *J. Liq. Chromatogr. Relat. Technol.*, *35*, 2134-2147.

[26] El Hady, D. A. & Albishri, H. M. (2014). Alkyl imidazolium ionic liquid based sweeping-micellar electrokinetic chromatography for simultaneous determination of seven tea catechins in human plasma. *J. Chromatogr. B: Anal. Technol. Biomed. Life Sci.*, *969*, 224-229.

[27] Tsai, I. C., Su, C. Y., Hu, C. C. & Chiu, T. C. (2014). Simultaneous determination of whitening agents and parabens in cosmetic products by capillary electrophoresis with on-line sweeping enhancement. *Anal. Methods*, *6*, 7615-7620.

[28] World Health Organization, Toxicological and Health Aspects of Melamine and Cyanuric Acid, Report of a WHO expert Meeting In collaboration with FAO, supported by Health Canada, Health Canada, Ottawa, Canada 1-4 December 2008, World Health Organization, Geneva 2008. Available at: http://whqlibdoc.who.int/publications/2009/9789241597951_eng.pdf (accessed February 2, 2015).

[29] Lian, D. S., Zhao, S. J., Li, J. & Li, B. L. (2014). Progress in stacking techniques based on field amplification of capillary electrophoresis. *Anal. Bioanal. Chem.*, *406*, 6129-6150.

[30] Liu, Z., Sam, P., Sirimanne, S. R., McClure, P. C., Grainger, J. & Patterson, D. G., Jr. (1994). Field-amplified sample stacking in micellar electrokinetic chromatography for on-column sample concentration of neutral molecules. *J. Chromatogr. A*, *673*, 125-132.

[31] Quirino, J. P. & Terabe, S. (1997). Stacking of neutral analytes in micellar electrokinetic chromatography. *J. Capillary Electrophor.*, *4*, 233-245.

[32] Gallart-Ayala, H., Núñez, O., Moyano, E. & Galceran, M. T. (2010). Field-amplified sample injection-micellar electrokinetic capillary chromatography for the analysis of bisphenol A, bisphenol F, and their diglycidyl ethers and derivatives in canned soft drinks. *Electrophoresis*, *31*, 1550-1559.

[33] Zhu, L., Tu, C. & Lee, H. K. (2001). Liquid-phase microextraction of phenolic compounds combined with on-line preconcentration by field-amplified sample injection at low pH in micellar electrokinetic chromatography. *Anal. Chem.*, *73*, 5655-5660.

[34] Liu, L., Chen, X. & Hu, Z. (2005). Online concentration by field-amplified sample injection in acidic buffer for analysis of fangchinoline and tetrandrine in herbal medicine by flow injection-micellar electrokinetic capillary chromatography. *Electrophoresis*, *26*, 4456-4464.

[35] Fang, H., Yang, F., Sun, J., Zeng, Z. & Xu, Y. (2007). Pressure-assisted field-amplified sample injection with reverse migrating micelles for analyzing trace steroids in MEKC. *Electrophoresis*, *28*, 3697-3704.

[36] Wu, Y. W., Liu, J. F., Xiao, T. X., Han, D. Y., Zhang, H. L. & Pan, J. C. (2009). Field-amplified sample injection for the determination of albumin and transferrin in human urines by MEKC. *Electrophoresis*, 30, 668-673.

[37] Sun, H. & Wu, Y. (2013). Field-amplified sample injection for the determination of isonicotinamide and nicotinamide in whitening cosmetics and supplemented foodstuffs by MEKC. *Anal. Methods*, *5*, 5615-5621.

[38] Quirino, J. P. & Terabe, S. (2000). Approaching a million-fold sensitivity increase in capillary electrophoresis with direct ultraviolet detection: cation-selective exhaustive injection and sweeping. *Anal. Chem.*, *72*, 1023-1030.

[39] Quirino, J. P., Iwai, Y., Otsuka, K. & Terabe, S. (2000). Determination of environmentally relevant aromatic amines in the ppt levels by cation selective exhaustive injection-sweeping-micellar electrokinetic chromatography. *Electrophoresis, 21*, 2899-2903.

[40] Zhu, L., Tu, C. & Lcc, II. K. (2002). On-line concentration of acidic compounds by anion-selective exhaustive injection-sweeping-micellar electrokinetic chromatography. *Anal. Chem.*, *74*, 5820-5825.

[41] Chiang, H. Y. & Sheu, S. J. (2004). Analysis of ephedra-alkaloids using sweeping and cation-selective exhaustive injection and sweeping micellar electrokinetic chromatography methods. *Electrophoresis*, *25*, 670-676.

[42] Lin, Y. H., Li, J. H., Ko, W. K. & Wu, S. M. (2006). Direct and sensitive analysis of methamphetamine, ketamine, morphine and codeine in human urine by cation-selective exhaustive injection and sweeping micellar electrokinetic chromatography. *J. Chromatogr. A*, *1130*, 281-286.

[43] Lin, Y. H., Lee, M. R., Lee, R. J., Ko, W. K. & Wu, S. M. (2007). Hair analysis for methamphetamine, ketamine, morphine and codeine by cation-selective exhaustive injection and sweeping micellar electrokinetic chromatography. *J. Chromatogr. A*, *1145*, 234-240.

[44] Lin, Y. H., Chiang, J. F., Lee, M. R., Lee, R. J., Ko, W. K. & Wu, S. M. (2008). Cation-selective exhaustive injection and sweeping micellar electrokinetic chromatography for analysis of morphine and its four metabolites in human urine. *Electrophoresis*, *29*, 2340-2347.

[45] Yang, Y., Nie, H., Li, C., Bai, Y., Li, N., Liao, J. & Liu, H. (2010). On-line concentration and determination of tobacco-specific N-nitrosamines by cation-selective exhaustive injection-sweeping-micellar electrokinetic chromatography. *Talanta*, *82*, 1797-1801.

[46] Wang, C. C., Chen, C. C., Wang, S. J. & Wu, S. M. (2011). Cation-selective exhaustive injection and sweeping micellar electrokinetic chromatography for the analysis of methadone and its metabolites in serum of heroin addicts. *J. Chromatogr. A*, *1218*, 6832-6837.

[47] Xu, X. & Fan, Z. H. (2012). Concentration and determination of cotinine in serum by cation-selective exhaustive injection and sweeping micellar electrokinetic chromatography. *Electrophoresis*, *33*, 2570-2576.

[48] Wang, C. C., Lu, C. C., Chen, Y. L., Cheng, H. L. & Wu, S. M. (2013). Chemometric optimization of cation-selective exhaustive injection sweeping micellar electrokinetic chromatography for quantification of ractopamine in porcine meat. *J. Agric. Food Chem.*, *61*, 5914-5920.

[49] Núñez, O., Moyano, E. & Galceran, M. T. (2002). Solid-phase extraction and sample stacking-capillary electrophoresis for the determination of quaternary ammonium herbicides in drinking water. *J. Chromatogr. A*, *946*, 275-282.

[50] Britz-McKibbin, P., Otsuka, K. & Terabe, S. (2002). On-line focusing of flavin derivatives using dynamic pH junction-sweeping capillary electrophoresis with laser-induced fluorescence detection. *Anal. Chem.*, *74*, 3736-3743.

[51] Britz-McKibbin, P. & Terabe, S. (2002). High sensitivity analyses of metabolites in biological samples by capillary electrophoresis using dynamic pH junction-sweeping. *Chem. Rec.*, *2*, 397-404.

[52] Yu, L. & Li, S. F. Y. (2005). Dynamic pH junction-sweeping capillary electrophoresis for online preconcentration of toxic pyrrolizidine alkaloids in Chinese herbal medicine. *Electrophoresis*, *26*, 4360-4367.

[53] Chen, Y., Zhang, L., Cai, Z. & Chen, G. (2011). Dynamic pH junction-sweeping for on-line focusing of dipeptides in capillary electrophoresis with laser-induced fluorescence detection. *Analyst (Cambridge, U. K.)*, *136*, 1852-1858.

[54] Rageh, A. H., Kaltz, A. & Pyell, U. (2014). Determination of urinary nucleosides via borate complexation capillary electrophoresis combined with dynamic pH junction-sweeping-large volume sample stacking as three sequential steps for their on-line enrichment. *Anal. Bioanal. Chem.*, *406*, 5877-5895.

[55] Hsu, S. H., Hu, C. C. & Chiu, T. C. (2014). Online dynamic pH junction-sweeping for the determination of benzoic and sorbic acids in food products by capillary electrophoresis. *Anal. Bioanal. Chem.*, *406*, 635-641.

In: Capillary Electrophoresis (CE)
Editor: Christian Reed

ISBN: 978-1-63483-122-2

Chapter 6

CE-C⁴D FOR THE DETERMINATION OF CATIONS IN PARENTERAL NUTRITION SOLUTION

P. Paul, T. Gasca Lazaro, E. Adams and A. Van Schepdael*
KU Leuven-University of Leuven, Department of Pharmaceutical and Pharmacological Sciences, Pharmaceutical Analysis, Leuven, Belgium

ABSTRACT

The capillary electrophoretic (CE) analysis of inorganic ions in parenteral nutrition solution in association with capacitively coupled contactless conductivity detection (C^4D) is a simple, flexible, economic and eco-friendly method. The aim of this study was to improve the repeatability and linearity properties as well as the application of this validated method to estimate the quantity of each inorganic cation in commercial samples. The method is carried out on an uncoated fused silica capillary with 50 μm i.d. and 365 μm o.d. and 60 cm length of which 50 cm is the effective length. Before the actual analysis, the capillary is rinsed sequentially with 0.05 M H_3PO_4 for 10 minutes followed by water for 20 minutes. To ensure a stable baseline, an additional rinsing of the capillary by 0.1 M NaOH, water and background electrolyte (BGE) consisting of 8 mM of L-arginine and 5 mM of DL-malic acid has been performed. Both constant current (CC) and constant voltage (CV) CE separation show acceptable linearity ($R^2 > 0.995$) for all cations in concentration ranges up to 100 μg/mL. The CC separation mode gives lower migration time (MT), better resolution and peak integration than the CV mode. Repeatability of peak area for individual cations is increased further by employing the rinsing sequence in-between sample injections as well as by using lithium chloride (LiCl) as internal standard. Although the CC mode is found to improve repeatability of peak area, it exhibits more day-to-day variability. The %RSD of the MT and the relative peak area (RPA) of sodium and potassium are however always within the specified limit. The CV mode shows good repeatability for calcium and magnesium. The sample quantification by calibration curve shows out-of-the-limit values for all analytes due to marked matrix interference. The standard addition method, in the same way proved ineffective to approximate the actual quantity of analytes in parenteral nutrition (PN)

* E-mail: ann.vanschepdael@pharm.kuleuven.be.

solutions. Finally, a single point calibration technique proved fruitful in the assay of cations by the use of simulated standard solution.

1. Introduction

The analytical technique capillary electrophoresis (CE) is renowned for its simple and economic nature and has been established as a promising technique in the field of pharmaceutical assay, degradation studies and biological sciences. These features along with automation have extended its successful application to antibiotics [1] and drugs from different therapeutic classes [2].

It is suitable for most of the drugs in cationic or anionic form when coupled to a UV detector either in direct or indirect mode and additionally has established itself as potent alternative to conventional methods [3]. In line with this trend, CE analysis has also covered metallic components in the form of indirect detection [3]. However, CE-UV analysis of metal ions using indirect detection has some disadvantages, such as, poor sensitivity, higher cost and associated health hazards. Therefore, a direct approach for determination of metal ions in solution in a single run could be a good alternative to the existing techniques. Under these circumstances, capacitively coupled contactless conductivity detection (C^4D) could be appropriate because of its non-invasiveness, positioning flexibility and universality [4].

Therefore, CE-C^4D could be chosen to analyze all the ions simultaneously in a single run since it is a straightforward and economic technique for quantitative measurement of ions.

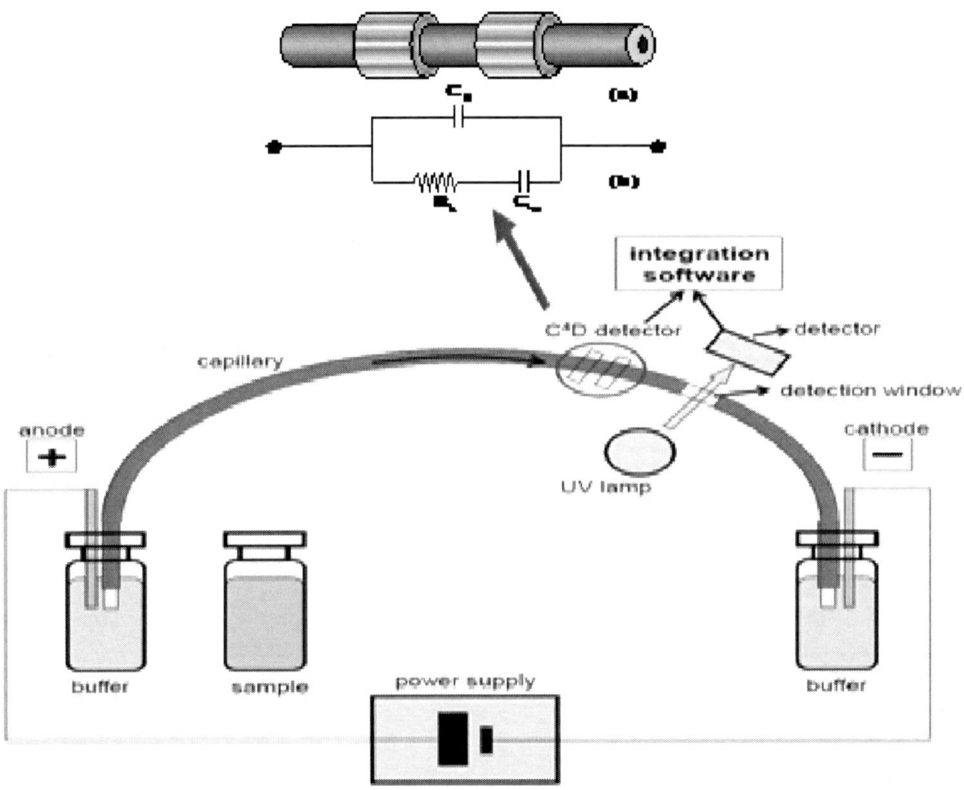

Figure 1.1. Schematic of a typical CE-C^4D instrumentation [7].

It finds extensive use after the publication of experimental results by Zemann et al. [5] and Fracassi et al. [6] in the 90s. C^4D enables data acquisition of isolated ionic compounds independently without any contact with the analyte. In principle, it measures the conductivity difference between stacked sample within the capillary and the background electrolyte (BGE), basically organic buffers. C^4D consists of two electrodes mounted axially along the capillary; with the electrodes being linked to a high voltage potential source. The entire instrumentation is illustrated schematically in Figure 1.1.

2. Materials and Method

2.1. Materials

The experiments of this chapter were intended to quantify inorganic cations in parenteral nutritional products. Therefore, in order to avoid unwanted leaching of those ions, especially sodium, proper grade of plastic recipients was used for the storage or preparation of reagents instead of glass.

Plastic bottles were purchased from Fisher Scientific (Loughborough, UK). Plastic vials were bought from Analis S.A. (Suarlée, Belgium). ChromafilXtra PET-45/25 filters were obtained from Macherey-Nagel (Düren, Germany) and syringes were obtained from Filterservice S.A. (Eupen, Belgium).

2.2. Chemicals and Reagents

The solvents and all reagents used in this project were of analytical grade. Methanol was obtained from Acros Organics (Geel, Belgium), sodium hydroxide was purchased from VWR Chemicals (Leuven, Belgium).

In order to prepare standard solution, the following salts were collected: sodium chloride (NaCl) was purchased from Fisher Scientific, calcium chloride ($CaCl_2$) was bought from Sigma-Aldrich (Germany), magnesium chloride anhydrous ($MgCl_2$) was purchased from Fluka (USA), potassium chloride (KCl) from Merck (Darmstadt, Germany) and lithium chloride (LiCl) was purchased from Acros Organics (New Jersey, USA). The sodium hydroxide (NaOH) and phosphoric acid (H_3PO_4) were obtained from VWR Chemicals (Belgium) and Chem-Lab NV (Belgium) respectively.

The buffer components malic acid (99%) and L-arginine were sourced from Janssen Chimica (Beerse, Belgium) and Applichem (Darmstadt, Germany) respectively.

Dilution of all standard and sample solutions was conducted by using ultra-pure Milli-Q water from a gradient purification module from Milli-Q (Millipore, France). All the standard, sample and BGE solutions were filtered through a 0.45 μm ChromafilXtra PET-45/25 filter before injection in order to remove any chance of unwanted capillary clogging.

Parenteral nutrition (PN) samples were obtained from the local university hospital pharmacy of UZ Leuven. Three types of samples were investigated: PN90, PN110 and PN170. These samples are basically a composite solution of inorganic and organic salts of Na^+, K^+, Ca^{2+} and Mg^{2+}, amino acids solution (Vaminolact®) and 50% glucose at varying concentrations. The composition of Vaminolact® can be found in Table 2.1.

2.3. Stock Solutions

Stock solutions of reference standard and sample solutions were prepared in order to save on time for the analysis. Different concentrations of standard and sample preparations were made through careful dilution using volumetric glass pipettes, previously rinsed with Milli-Q gradient water, dried and stored in appropriately cleaned plastic containers. For proper storage of the stock solutions, standard and sample solutions were tightly capped and wrapped-around with parafilm to ensure complete closure.

Table 2.1. Composition of Vaminolact®

L-Amino acid	Content (g)
Alanine	3.15
Arginine	2.05
Aspartic acid	2.05
Cystein	0.5
Glutamic acid	3.55
Glycine	1.05
Histidine	1.05
Isoleucine	1.55
Leucine	3.5
Lysine	2.8
Methionine	0.65
Phenyl alanine	1.35
Proline	2.8
Serine	1.9
Taurine	0.15
Threonine	1.8
Tryptophan	0.7
Tyrosine	0.25
Valine	1.8
Water	Quantity sufficient to 500 mL

Stock solutions of NaCl, KCl, LiCl, $CaCl_2$ and $MgCl_2$ were prepared at different concentration levels to suit different analytical operations. Usually the stock solution of each salt was prepared at a concentration level of 1 mg/mL. However, the assay by the standard addition method was conducted by using a more concentrated solution of all analytes, such as NaCl and KCl (20 mg/mL) and LiCl, $MgCl_2$ and $CaCl_2$ (10 mg/mL).

2.4. Buffer Solution

The buffer solution was prepared by dissolving accurately weighed amounts of L-arginine and DL-malic acid in a glass volumetric flask to get a concentration of 8 mM and 5 mM respectively. The buffer solution was stored in a thoroughly cleaned plastic container to avoid metallic leaching, at 4^{0}C in a refrigerator during not-in-use. Each time, the buffer temperature was brought to room temperature and used in experimentation after filtering through a 0.45 μm filter.

Buffer was prepared every week and during experimentation, the separation buffer was changed every three runs to avoid any flaws in response resulting from electrolytic pH change of the BGE.

2.5. Instrumentation and Electrophoretic Conditions

2.5.1. CE Instrumentation and Conditions

The entire work described in this chapter made use of a P/ACE MDQ instrument (Beckman Coulter, Inc. Fullerton, CA, USA) in conjunction with an eDAQ C^4D device (eDAQ, Denistone East, Australia) as signal recorder. The electrophoretic data recording and processing were accomplished by two licensed software packages- PowerChrom v2 (eDAQ) and 32 Karat™ 4.0 (Beckman Coulter). However, it is noteworthy that the former one was mainly dedicated for detector module control, data recording and processing whereas the latter software was used as electrophoresis controlling unit. The parameter inputs and necessary modulation for conducting electrophoresis were performed with this program.

As usual with most of the capillary electrophoresis making use of uncoated fused silica capillaries, the capillary dimension was set at 50 μm for internal diameter (i.d.) and 375 μm outer diameter (o.d.) having very thin, manufacturer defined (approximately 15 μm) polyamide coatings. Basically, this coating imparts structural integrity and stress-resistance to the fragile-in-nature fused silica capillary. For our research purpose, fused silica capillary was purchased from Polymicro Technologies (Phoenix, AZ, USA).

2.5.2. C^4D Instrumentation and Conditions

The parameter setting for C^4D, by and large depends on the nature and composition of the background electrolyte solution with desired pH value. Therefore, the same experiment, but with a different electrolyte composition will require a different parameter input in order to get an optimal signal-to-noise ratio and peak shape of the electropherogram. In our project, the eDAQ C^4D detector was set to 600 kHz input frequency with 60% gain (peak to peak amplitude of 60 V). The detector response was processed and analyzed using PowerChrom v2 software (eDAQ, Denistone East, Australia) while the entire operation of electrophoretic separation of ions and acquisition of detector response were based on synchronous operation of PowerChrom v2 and 32 Karat™ 4.0 software packages.

2.6. Experimentations

2.6.1. Electrophoresis

The instrumental parameters utilized in the beginning of this work can be found below, but the separation was also carried out under constant current (6.2 μA).

Capillary dimension: Uncoated fused silica capillary with total length (L_T) of 50 cm and effective length (L_E) of 30 cm (50 μm i.d.)
Capillary temperature: Maintained at 25°C with liquid coolant
Pre-conditioning: Between runs with running buffer for 5 min (pressure = 20 p.s.i.)

Injection: Hydrodynamic injection with a pressure of 0.5 p.s.i. for 5.0 s
Separation: Normal polarity, constant voltage (15 kV)
BGE: 8 mM L-arginine + 5 mM malic acid
Instrument: P/ACE MDQ (Beckman Coulter Inc. CA. USA)
Detection: C^4D (EDAQ, Denistone East, Australia)
Run time: 10 minutes

2.6.2. Rinsing Steps

For further optimization and capillary equilibration with the BGE, the capillary was rinsed at the beginning of the day with 0.1 M sodium hydroxide (NaOH) for three minutes. Another three and four minutes rinsing with water and BGE followed by application of electrophoresis voltage for ten minutes were carried out.

2.6.3. Constant Current Operation

In order to improve linearity and repeatability, the separation electric field was shifted from constant voltage (15 kV) to the corresponding constant current mode. A comparison of the results in terms of percentage relative standard deviation (%RSD) of migration time (MT) and relative peak area (RPA) from both separation modes was made to observe any benefits of one over the other. To this end, proper dilution of standard and sample stocks were injected six times following the running conditions displayed above with a buffer change every three injections, with a view to avoid buffer depletion.

2.6.4. Separation with Organic Solvents in BGE

Organic solvent addition into the BGE at a certain proportion, for example, methanol and acetonitrile 10 percent and 20 percent v/v respectively and EDTA were intended to improve the resolution and repeatability. During the course of this work, magnesium and calcium peaks appeared very close making integration of individual peaks improper which ultimately leads to variation in the peak areas. Numerous works to improve the selectivity have been reported that involve organic solvents or organic additives. Previously, work was undertaken to improve the resolution between magnesium and calcium by inclusion of chelating agents, for example, EDTA, to the buffer solution. However, these attempts proved ineffective by the fact that the resultant electropherogram showed disappearance of these two peaks in question.

2.6.5. Linearity Evaluation

In this part of method validation, five levels of standard solution were made by diluting 10, 20, 30, 40 and 50 μL of standard stock solution using a micropipette to 10.0 mL with Milli-Q water. This dilution provides a concentration of 20, 40, 60, 80 and 100 μg/mL of potassium and sodium cations and 10, 20, 30, 40 and 50 μg/mL of magnesium and calcium. Standard solutions from each concentration level were injected hydrodynamically at 0.5 p.s.i. for five seconds and electrophoretic separations were carried out at constant current (6.2 μA) for ten minutes with preliminary capillary rinsing conditions as mentioned in section 2.6.2. Moreover, an internal standard (LiCl) at definite volume was included in all standard solutions.

2.6.6. Precision Evaluation

Validation of a CE-based analytical method stipulates to undertake precision assessment in terms of migration time and peak areas [8, 9, 10].

An in house specification stipulates that the %RSD for CE based methods should be below 3% for RPA whereas MT %RSD should be lower than 2%. To conduct the repeatability evaluation, peak areas and MTs as well as %RSD value for both were determined at different concentration levels.

In CE, the MT of an analyte influences the actual peak area due to the fact that a slow moving analyte will spend more time in the detector window thereby exhibiting higher peak areas than the fast moving one. Therefore, in order to account for such phenomenon, the corrected peak area (CPA) for individual ions is utilized in the estimation of %RSD of analyte response. It was calculated by dividing the peak area by the respective migration time.

Moreover, due to significant instrumental variability of the injected amount in the hydrodynamic technique, LiCl was used as internal standard at the same concentration level to off-set injection to injection variability thereby improving the repeatability of the C^{4}D responses at each concentration level of analyte.

Five levels of standard solution, through proper dilution of standard stock solution, were made, all of which contained the same amount of LiCl as internal standard. The analyses were performed according to the protocols given in section 2.6.1, but at a constant current of 6.2 μA. Six injections from each concentration level of standard solution were made and %RSD was calculated.

In precision analyses, for samples PN90 and PN170, 15.0 mL of each was diluted to 500.0 mL with Milli-Q water while for PN110, 5.0 mL was diluted to 10.0 mL with the same solvent. Finally, %RSD values of RPA for each analyte in each individual sample were calculated.

2.6.7. Quantification by Calibration

Assay of analytes in a sample by this technique involved construction of a standard calibration curve for each analyte followed by interpolation of response data from the sample. In order to obtain a calibration curve, five concentrations of standard were made by diluting 10, 20, 30, 40 and 50 μL of stock solution to 10.0 mL with Milli-Q water to give concentrations of 10, 20, 30, 40, 50 μg/mL for magnesium and calcium and 20, 40, 60, 80 and 100 μg/mL for potassium and sodium. Similarly, the sample solution was diluted appropriately to obtain a concentration corresponding to a point that fits well into the range of the calibration plot. Moreover, all standard and sample solutions received 10 μL each of 10 mg/mL LiCl as internal standard. Afterwards, both the standard and sample solutions were analyzed by CE-C^{4}D.

2.6.8. Quantification by Standard Addition

A preferential approach for analyte quantification is standard addition when there is marked interference from the matrix of the sample [11]. Quantitative evaluation of samples by standard addition can be carried out in two ways:

a) Graphical approach
b) Quantification by the standard addition equation.

Analyses were undertaken following both techniques, but the latter approach is simple to conduct under the condition that standard-fortified samples do show linear response.

The graphical standard addition technique (a) was performed by transferring an aliquot (200 μL) of sample (PN90, PN170) into five volumetric flasks (10.0 mL) followed by addition of 10 μL of LiCl solution as internal standard. Into each of the flasks, 0, 10, 20, 30 and 40 μL of standard solutions were added. All the samples were then analyzed using the protocols described above. Afterwards, a plot of concentration against the resultant responses (RPA) of individual ions was constructed and extrapolation of the curve was made to obtain the concentration in the sample solution.

In the technique (b), 5.0 mL of sample were transferred to two 10.0 mL volumetric flasks. Both flasks were then filled up to the mark with Milli-Q water after adding 20 μL of standard solution in one flask and 10 μL of approximately 10 mg/mL stock solution of LiCl in both.

Calculation of the individual ion concentration was done by the following mathematical expression [10]:

$$\frac{I_x}{I_{s+x}} = \frac{C_x}{C_x + C_s}$$

I_x = Response of sample without standard
I_{s+x} = Response of analyte from sample with added standard
C_x = Final concentration of analyte in sample
C_s = Final concentration of standard added

2.6.9. Quantitative Analysis by Simulated Standard Solution

A standard solution of analytes as per the formulation protocols of the PN solution was made by accurately measuring all the components and dissolving them into water for injection. Meanwhile, a solution of Vaminolact® (Table 2.1) and 50% glucose solution was prepared and all solutions were mixed.

The same electrophoretic conditions as mentioned in section 2.6.1 were employed for assay of PN formulations except that the initial few runs were not used in content estimation in order to have consistent electropherograms resulting from adequate equilibration of BGE with the interior of the capillary. The resultant peak area for each analyte from the standard solution was normalized according to the purity claim by the reagent suppliers.

3. Results and Discussion

3.1. Detector Parameters and Their Optimization

When using CE-C^4D, the performance of a developed method largely depends on the sensitivity of the C^4D device. Conventionally, the CE separation method utilizes a buffer and an AC voltage on the input electrode for analyte detection. Depending on the nature and characteristics of the selected background electrolyte, identical detector settings will behave differently and become apparent in the resultant electropherogram. This voltage application stage in the C^4D electric module has tremendous consequences on the manifestation of the output signal in terms of peak shape and signal–to-noise ratio thereby affecting sensitivity. An intricate relationship between the peak shape and the voltage applied on the actuator electrode has been reported [4]. Application of an inappropriate voltage may result in unwanted phenomena such as peak overshooting which is characterized by a baseline dip in the leading edge or trailing edge of the peak [4]. This defective electropherogram severely limits accurate measurement of ionic contents of the analyte. Moreover, it has been proven that the detector response with significant sensitivity of an established CE-C^4D method will not deteriorate that much as long as the applied voltage is kept within a narrow span around the voltage yielding maximum sensitivity. Therefore, CE-C^4D based method development not only necessitates the optimization of the electrophoretic step, but also prudent selection of C^4D parameters at values that ensure the desired outcome. This process of detector optimization is called C^4D-profiling.

The best performance of the C^4D detector was demonstrated at an input frequency of 600 kHz and 60% gain with headstage "on" for the buffer composition of 8 mM L-arginine and 5 mM DL-malic acid. In this chapter, all the analyses were performed using the same detector configuration with similar buffer composition.

3.2. Impact of Organic Solvent on Resolution of Magnesium and Calcium

CE-C^4D analysis of PN110 and standard solution at a lower concentration corresponding to PN110 showed electropherograms with magnesium and calcium peaks partially overlapping and the calcium peak displayed tailing thereby making manual peak integration difficult and less precise (Figure 3.1).

Organic solvents alter the ionic mobility by changing the solvation radii [13, 14] and viscosity of the running buffer thereby contributing to the enhanced resolution of peaks. Moreover, inclusion of 20% ACN (v/v) in 100 mM Acetate/Tris BGE has been shown to offer resolution between K^+, Na^+, Mg^{2+} and Ca^{2+} higher than 1.5 [12].

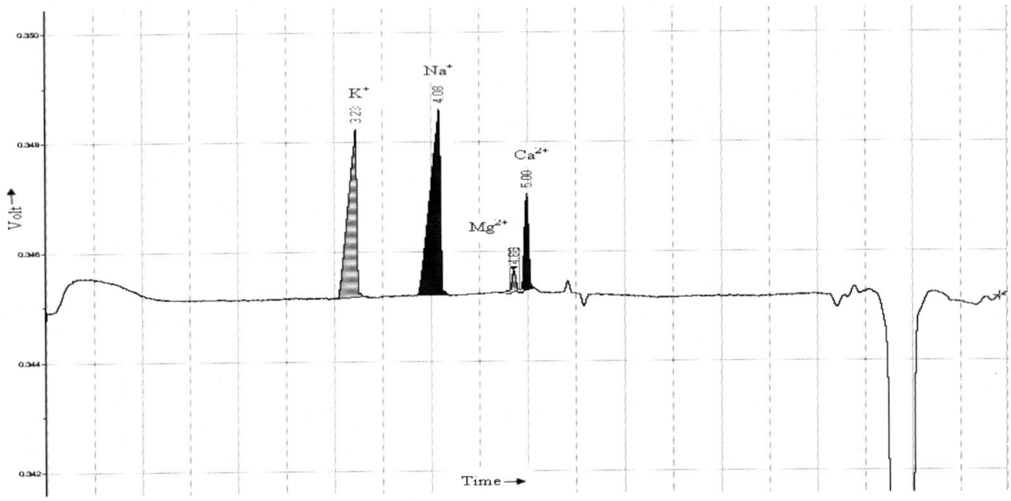

Figure 3.1. Electropherogram of PN110 obtained by using the conditions of section 2.6.1.

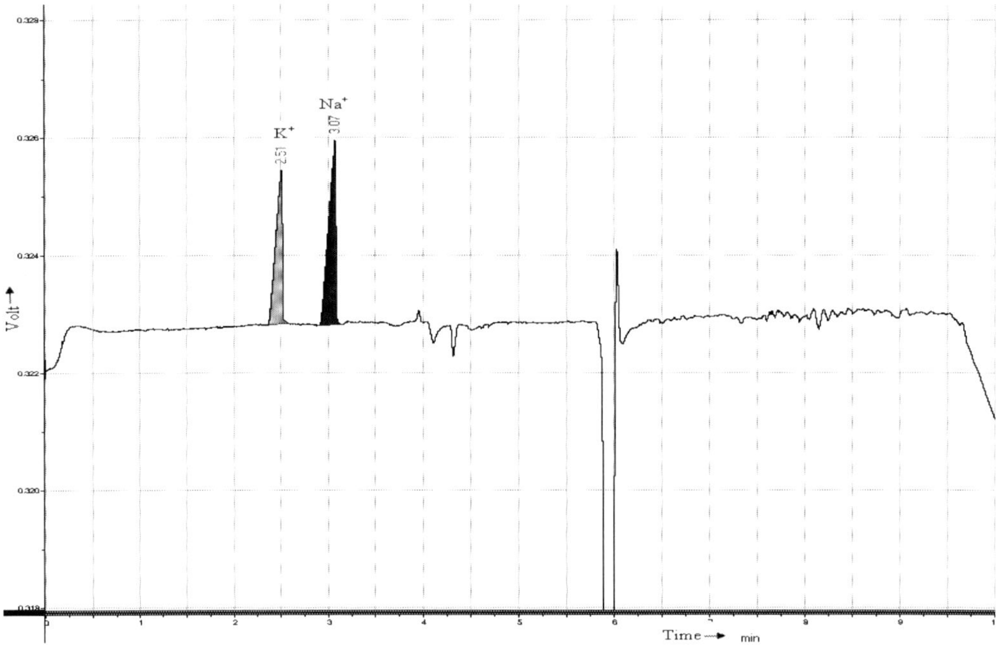

Figure 3.2. Electropherogram of PN110 with 20% ACN in BGE.

In addition, organic solvents with low permittivity constant favor ion pair formation leading to greater selectivity towards specific ion species and show good resolution [13].

Despite having evidence of improvement in selectivity by inclusion of organic solvents in the BGE, in this study the CE-C^4D analysis of these ions by the buffer system (L-arginine and DL-malic acid) is an exception to this which shows complete absence of Mg^{2+} and Ca^{2+} peaks (Figure 3.2). A complex interaction of the BGE in association with organic solvents like MeOH/ACN in our case might have changed the mobility of the ions in such a way that they may co-migrate with the neutral species or their conductance may be suppressed well below the detection limit under the C^4D conditions used.

3.3. Impact of Vaminolact® on Peak Areas

Vaminolact® is used as the amino acid source during the preparation of PN solutions.

In light of the difference in peak areas of similar concentrations of standard and PN solutions, a certain volume of the standard was also analyzed after spiking with a certain volume of Vaminolact® and 50% glucose. The results are summarized below (Table 3.1):

Despite the peak areas and corresponding CPAs of each analyte in the standard solution spiked with Vaminolact® and glucose show higher values than those without such addition, internal standard addition sufficiently reduces the variability resulting from the matrix interferences as reflected by the fact that the mean RPA of all analytes in both cases appears almost the same. The PN preparation contains amino acids at a certain proportion. These amino acids are charged species and may co-migrate or form a complex with the analyte thereby increasing the conductance. This enhancement in analyte conductance might lead to higher peak areas of the individual ions including the internal standard. Therefore, RPAs of each analyte with and without Vaminolact® were found relatively the same.

Table 3.1. Data from CE-C⁴D analysis of standard with and without being spiked with Vaminolact® and glucose

	With Vaminolact®+50% glucose				Without Vaminolact®+50% glucose			
Analyte	Mean RPA	%RSD	MT (min)	%RSD	Mean RPA	%RSD	MT (min)	%RSD
Potassium	4.84	4.1	3.01	< 0.1	4.30	4.3	2.94	0.2
Sodium	3.03	2.5	3.76	0.2	3.68	2.5	3.67	0.2
Magnesium	0.60	3.4	4.52	< 0.1	0.63	5.8	4.37	0.2
Calcium	0.44	8.7	4.64	< 0.1	0.51	7.3	4.46	0.2

3.4. Precision Study

3.4.1. Constant Current versus Constant Voltage

Although the popularity of CE is increasing over time, still this analytical technique is mainly considered to be an alternative and *complementary* technique to liquid chromatography (LC) [15] because of its simplicity, economic and eco-friendly nature.

However, CE is not yet used as widely as LC in analytical chemistry for several obvious reasons. Among them, the major one is lower precision as well as robustness in comparison with LC [16].

Besides, method transferability of a CE based technique is far more complex [15]. This is due to the large number of factors involved in a CE separation. Despite several reports of success in transferring CE methods [17, 18], inter-laboratory operations observed lack of precision and failed to fit in the system suitability limits as per protocol specification [18, 19].

According to Mayer [20] the precision of a CE method is determined by variability of MT and PA. Variability of MT is ascribed to errant electro-osmotic flow (EOF), analyte electrophoretic mobility, capillary re-equilibration and conditioning and capillary temperature while variability in injection volume, diffusion, wall interaction, peak integration, MT and fluctuation of capillary temperature are considered to be behind the inconsistency of peak areas.

A multitude of solutions has been reported to counter such obstacles in the CE technique [16, 17, 20, 21] among which are: use of internal standards, reduction of injection volume variability and use of optimal rinsing procedures to avoid wall interaction, maintain a constant mobility and a constant EOF. In addition, keeping the capillary temperature constant is very crucial in CE, fluctuation of which inevitably leads to less precise MTs and peak areas.

During optimization of this method, a constant EOF and reduction of variation in MT were achieved through undertaking several rinsing protocols, maintaining the pH of the BGE as constant as possible and keeping the temperature at a fixed value of 25°C by circulation of liquid coolant. In addition, inter-injection variability is also reduced by use of LiCl leading to amelioration of peak areas consistency.

Precision evaluation is an integral part of analytical method validation and is considered at three levels [22]: i) repeatability, ii) intermediate precision, and iii) reproducibility.

In this chapter, improvement of the intra-day and inter-day repeatability has been undertaken by application of both constant current (CC) and conventional constant voltage (CV) since several researchers have reported the application of the constant current condition as an effective means of improving the validation parameters [18] like precision.

3.4.2. Intra-Day Repeatability

The experimental results from CC and CV mode suggest that the analytical precision of this method is good in CV mode (Figures 3.3-3.6; Tables 3.3, 3.5, 3.7 and 3.9). However, apart from this experiment, CC separation employed in the linearity study and quantification studies by standard addition showed good %RSD values for all analytes (Table 3.14). CC may reduce variable heat generation inside the capillary thereby acting against axial and radial temperature fluctuation which ultimately leads to more precise MT, resolution and peak areas [15]. It is believed that the CC mode is preferred in the context of CE method transfer.

It has also been observed that the MT in CC mode is lower than in CV mode leading to a decreased run time in addition to amelioration of linearity and MT repeatability profile.

Table 3.2. %RSD of MT and RPA for PN90 in CC mode

Analyte Name	Mean MT (min)	%RSD	Mean RPA	%RSD
Potassium	3.03	0.3	11.37	14.7
Sodium	3.78	0.4	9.19	14.2
Magnesium	4.42	0.5	0.42	2.8
Calcium	4.54	0.5	2.50	7.6

Table 3.3. %RSD of MT and RPA for PN90 in CV mode

Analyte Name	Mean MT (min)	%RSD	Mean RPA	%RSD
Potassium	3.43	0.4	14.95	2.3
Sodium	4.30	0.3	12.04	1.9
Magnesium	5.05	0.4	0.54	3.2
Calcium	5.19	0.4	3.19	2.6

Table 3.4. %RSD of MT and RPA for PN110 in CC mode

Analyte Name	Mean MT (min)	%RSD	Mean RPA	%RSD
Potassium	3.00	0.2	4.323	3.5
Sodium	3.77	0.2	4.380	3.8
Magnesium	4.48	0.9	0.205	9.5
Calcium	4.59	0.3	0.957	5.1

Table 3.5. %RSD of MT and RPA for PN110 in CV mode

Analyte Name	Mean MT (min)	%RSD	Mean RPA	%RSD
Potassium	3.35	< 0.1	4.20	1.6
Sodium	4.23	0.1	4.24	0.8
Magnesium	5.05	0.1	0.21	5.7
Calcium	5.17	0.1	0.93	2.3

Table 3.6. %RSD of MT and RPA for PN170 in CC mode

Analyte Name	Mean MT (min)	%RSD	Mean RPA	%RSD
Potassium	3.06	0.2	12.720	2.3
Sodium	3.77	0.2	5.506	2.8
Magnesium	4.46	0.3	0.247	4.5
Calcium	4.57	0.3	1.257	5.4

Table 3.7. %RSD of MT and RPA for PN170 in CV mode

Analyte Name	Mean MT (min)	%RSD	Mean RPA	%RSD
Potassium	3.44	0.4	10.910	1.8
Sodium	4.25	0.4	4.729	2.6
Magnesium	5.05	0.5	0.225	5.6
Calcium	5.17	0.4	1.111	4.8

Table 3.8. %RSD of MT and RPA for Standard solution in CC mode

Analyte Name	Mean MT (min)	%RSD	Mean RPA	%RSD
Potassium	3.02	0.1	4.141	2.7
Sodium	3.79	0.1	3.358	3.1
Magnesium	4.54	0.1	0.601	7.1
Calcium	4.64	0.1	0.454	8.1

Table 3.9. %RSD of MT and RPA for Standard solution in CV mode

Analyte Name	Mean MT (min)	%RSD	Mean RPA	%RSD
Potassium	3.37	< 0.1	4.040	0.9
Sodium	4.24	< 0.1	3.278	1.4
Magnesium	5.09	< 0.1	0.562	1.6
Calcium	5.20	< 0.1	0.447	4.7

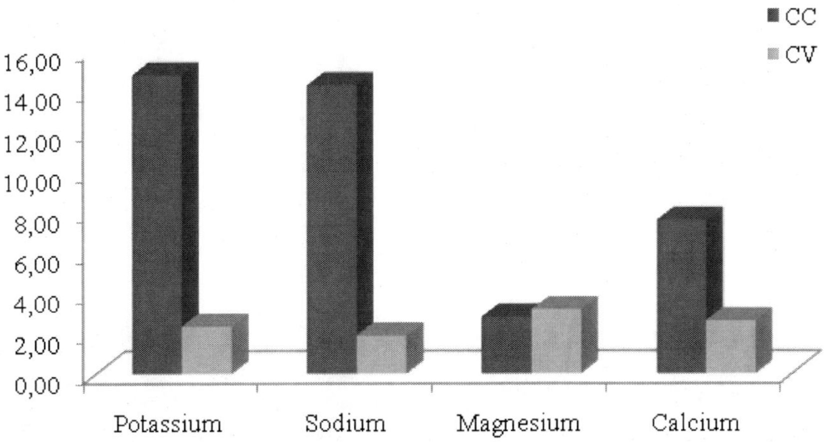

Figure 3.3. %RSD of analytes RPA in PN90 at CC versus CV.

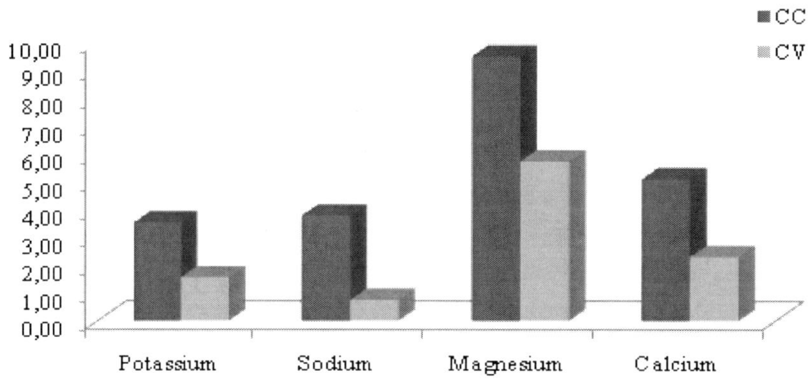

Figure 3.4. %RSD of analytes RPA in PN110 at CC versus CV.

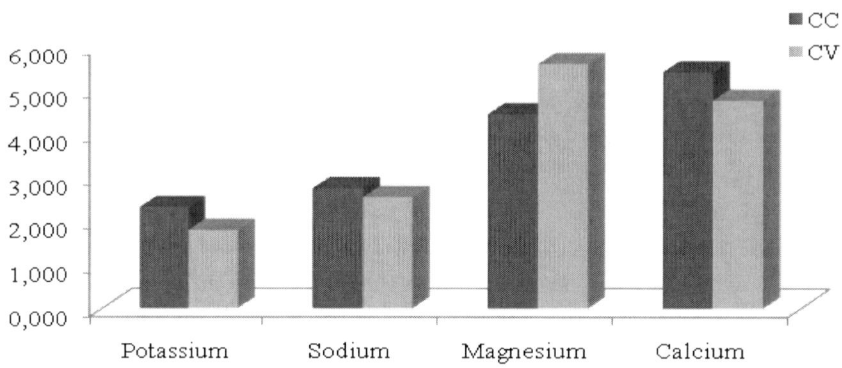

Figure 3.5. %RSD of analytes RPA in PN170 at CC versus CV.

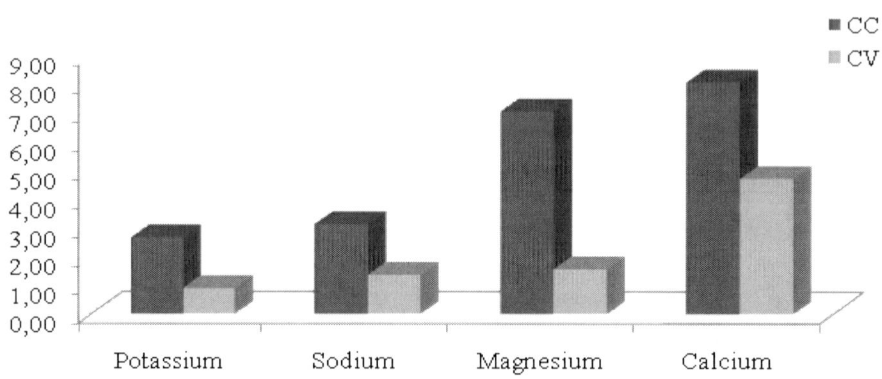

Figure 3.6. %RSD of analytes RPA in standard at CC versus CV.

3.4.3. Inter-Day Repeatability

During inter-day repeatability evaluation of this method in CC mode, a significant variance in terms of both MT and RPA from different days was observed despite keeping all the parameters and reagents the same. This work was done simultaneously with quantification by the equation-based standard addition method b) where a certain volume of sample (PN90) was diluted separately into two volumetric flasks by water or spiked with a definite volume of mixture of standard analytes. The sample afterward was subjected to identical CE-C^4D treatment on two consecutive days and the following tables (Tables 3.10-3.13) display the

%RSD of MT and RPA for sample PN90 and spiked PN90. It is quite obvious from the tabulated data that the %RSD of MT in unspiked PN90 is more than for the spiked PN90 sample on a day-to-day basis. The %RSD of RPA for spiked and unspiked PN90 samples exhibit different patterns. Unspiked %RSD of RPA is higher in day-1 as compared to day-2 while for spiked PN90 showed vice-versa. The results from the inter-day repeatability study showed a high degree of variability in terms of both MT and RPA. For the unspiked PN90, the inter-day %RSD of MT ranges from 2.4 to 26.8% and that of RPA from 4.0 to 11.5%. On the other hand, the spiked PN90 showed inter-day %RSD of MT and RPA ranging from 3.2 to 4.2% and 4.7 to 7.8% respectively.

In CC mode large %RSD in terms of inter-day MT and RPA may be attributed to the fact that in CC mode between-day variation in applied voltage may occur while keeping the current constant [15].

Table 3.10. %RSD of MT and RPA of all analytes on Day-1 in unspiked PN90

	Day-1			
Analyte	MT (min)	%RSD	RPA	Intra-day %RSD
Potassium	3.60	1.2	9.97	5.6
Sodium	4.75	1.5	9.83	4.7
Magnesium	5.88	1.8	0.41	4.6
Calcium	6.08	1.8	2.09	5.0

Table 3.11. %RSD of MT and RPA of all analytes on Day-2 in unspiked PN90

	Day-2			
Analyte	MT (min)	%RSD	RPA	Intra-day %RSD
Potassium	3.73	2.0	9.46	1.7
Sodium	4.97	2.7	9.57	2.8
Magnesium	6.12	3.5	0.48	10.3
Calcium	6.28	4.2	2.48	4.5

Table 3.12. %RSD of MT and RPA of all analytes on Day-1 in spiked PN90

	Day-1			
Analyte	MT (min)	%RSD	RPA	Intra-day %RSD
Potassium	3.70	0.6	12.13	1.7
Sodium	4.91	0.8	11.52	2.7
Magnesium	6.10	0.8	0.78	2.8
Calcium	6.31	0.9	2.29	3.3

Table 3.13. %RSD of MT and RPA of all analytes on Day-2 in spiked PN90

	Day-2			
Analyte	MT (min)	%RSD	RPA	Intra-day %RSD
Potassium	3.93	0.7	11.53	5.7
Sodium	5.30	0.9	10.98	5.5
Magnesium	6.59	1.1	0.83	6.2
Calcium	6.82	1.1	2.60	5.7

3.5. Linearity

Five levels of standard concentration containing all cations were made and analyzed using CE-C^4D at constant current mode for the construction of calibration curves and the obtained results are summarized below (Table 3.14).

It is quite apparent from the above Table 3.14 that the repeatability values at all concentration levels for all analyte ions are variable and for most of the cases are higher than the in house specified limit for peak areas. These results were always variable from day to day operations. These observations are in line with the outcome observed in the operation of CC versus CV separation mode executed on the same day. However, experimental results from most of the other working days in CC mode were found repeatable with %RSD values lying within the regulatory limits as reflected by the small %RSD values in graphical standard addition experiments for analyte quantification (Tables 3.24-3.26; Figures 3.15-3.17). Despite high %RSD, the construction of calibration plots of RPA against the corresponding concentration gives satisfactory results for linearity as displayed by the following figures:

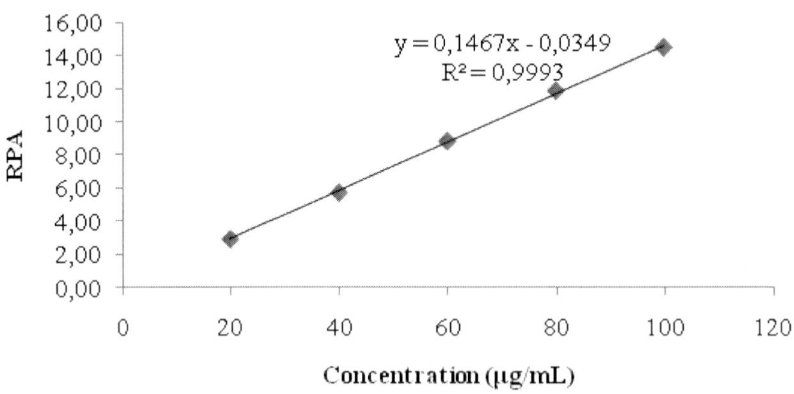

Figure 3.7. Calibration plot of potassium.

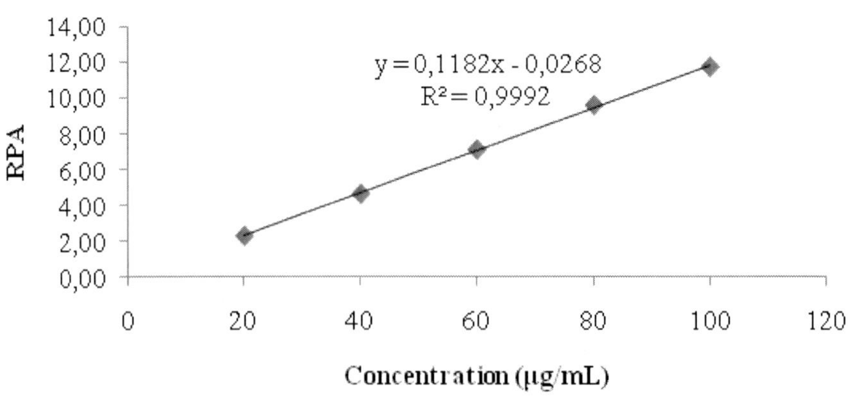

Figure 3.8. Calibration plot of sodium.

Table 3.14. Linearity data at five concentration (µg/mL) levels for all four analytes with %RSD of RPA

Level	Conc.	RPA K^+	%RSD	Conc.	RPA Na^+	%RSD	Conc.	RPA Mg^{2+}	%RSD	Conc.	RPA Ca^{2+}	%RSD
1	19.96	2.90	3.1	20.02	2.32	1.7	9.91	0.39	3.2	9.75	0.22	4.8
2	39.92	5.72	2.1	20.04	4.64	1.6	19.82	0.76	3.7	19.50	0.53	6.5
3	59.88	8.82	2.3	20.06	7.13	3.0	29.73	1.22	3.2	29.25	0.97	2.8
4	79.84	11.85	6.6	20.08	9.59	7.0	39.64	1.58	7.7	39.00	1.26	5.3
5	99.80	14.48	4.3	20.10	11.69	4.7	49.55	1.97	3.6	48.75	1.61	5.6

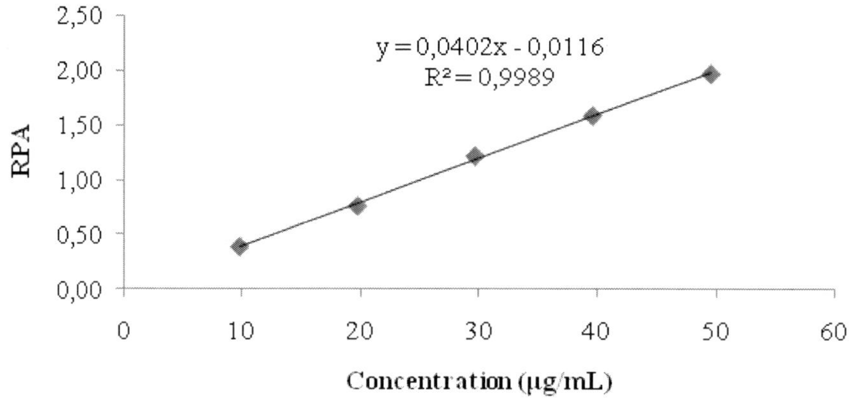

Figure 3.9. Calibration plot of magnesium.

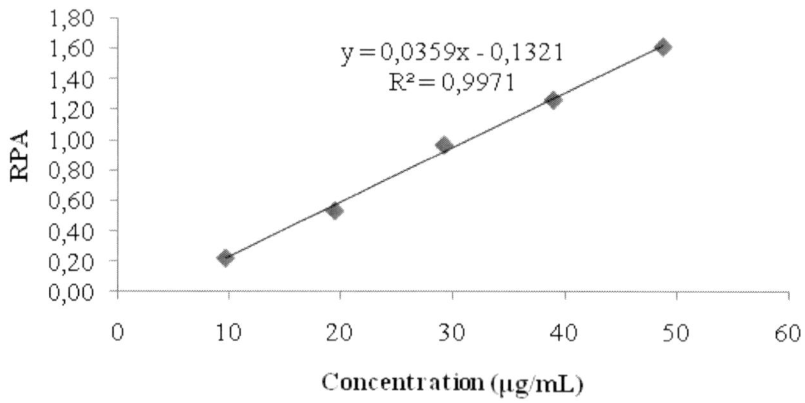

Figure 3.10. Calibration plot of calcium.

A R^2 value of over 0.995 is considered acceptable to demonstrate good linearity of the analytical method [11].

In CC mode the coefficients of determination for sodium, potassium, magnesium and calcium are 0.9992, 0.9993, 0.9989 and 0.9971, respectively, which demonstrates a good linearity profile for sodium and potassium over the concentration range of 20 - 100 µg/mL and for magnesium and calcium over the range of 10-50 µg/mL. This is an improvement with respect to the CV separation mode where the R^2 values for sodium and potassium were found relatively good but for calcium and magnesium these values were bad (far less than 0.995) indicating a poor linearity profile for those cations (data not shown). It has been observed that the detector response of Ca^{2+} and Mg^{2+} was smaller as compared to that of Na^+ and K^+. This might be due to strong interaction of these cations with the BGE components leading to decreased conductance for these ions. Moreover, the calcium peak showed marked tailing at low concentration leading to integration problems. At higher concentrations, overlap of the calcium peak with this of magnesium was observed which explains partly the large variability in the calcium peak areas. In addition, magnesium concentrations in all PN preparations were small and any dilution during analysis resulted in a very small peak closely adjacent to the calcium peak, thereby leading to inaccurate peak integration and hence poor RPA repeatability.

The low response for Ca^{2+} and Mg^{2+} could also be deduced from their lower calibration curve slopes. The slopes for Na^+ and K^+ were found higher than those of Ca^{2+} and Mg^{2+} (Figures 3.7-3.10).

3.6. Quantification by Calibration Curve

The quantitative analysis of samples PN90, PN110 and PN170 was tried using the developed and optimized CE-C4D method.

For the quantitative study, the calibration curves discussed in the linearity section were used for estimation of analytes in standard solution to cross-check the validity of the method and for estimation of analytes in PN90 and PN170.

It involved a standard solution and PN solutions with differing analyte strength in each preparation and results are listed in Table 3.15:

Table 3.15. Concentration of analytes in PNs and standard solution

Analyte Name	PN90 (μg/mL)	PN110 (μg/mL)	PN170 (μg/mL)	STD (μg/mL)
Potassium	1098.6	29.99	582.60	20.063
Sodium	1013.8	28.28	537.96	20.026
Magnesium	83.85	2.25	43.70	10.053
Calcium	837.6	22.50	436.85	10.115

Table 3.16. %Content of analytes for PN90

Analyte Name	RPA	%RSD	Estimated conc. (μg/mL)	Expected conc. (μg/mL)	% Content
Potassium	6.22	3.5	42.86	21.97	195.1
Sodium	6.39	2.8	54.34	20.28	268.0
Magnesium	0.27	3.5	6.89	1.68	410.8
Calcium	1.41	3.8	44.09	16.75	263.2

Table 3.17. %Content of analytes for PN170

Analyte Name	RPA	%RSD	Estimated conc. (μg/mL)	Expected conc. (μg/mL)	% Content
Potassium	3.22	3.2	22.30	11.65	191.4
Sodium	3.27	3.6	27.96	10.76	259.9
Magnesium	0.15	13.0	3.99	0.87	456.0
Calcium	0.64	6.8	22.13	8.74	253.3

The results found for the PNs' analyte contents seriously deviated from the anticipated results. The estimated percent contents for potassium and sodium ions were found approximately 2 and 2.5 times, respectively, greater than expected while for the magnesium and calcium ions the expected content deviates by a factor of approximately 4 and 2.5, respectively (Tables 3.16-3.17). However, for the standard solution, the percentage content of all analytes approached the anticipated value (Table 3.18).

Table 3.18. %Content of all analytes for standard solution

Analyte Name	RPA	%RSD	Estimated conc. (μg/mL)	Expected conc. (μg/mL)	%Content
Potassium	6.02	2.7	41.28	39.93	103.4
Sodium	4.86	2.9	41.41	40.03	103.5
Magnesium	0.75	4.1	18.56	19.70	94.2
Calcium	0.55	8.6	18.91	19.62	96.4

3.7. Quantification by Standard Addition

Quantitative analysis of analytes in samples PN90, PN110 and PN170 was carried out by standard addition in order to off-set the matrix interference and find an appropriate estimation of the analyte concentration. Both graphical and equation based estimation of analyte ions were carried out and the results from the former approach are given below (Table 3.19):

Table 3.19. Quantitative RPA data of PN90 from the graphical mode of standard addition (Concentrations in every case are given in μg/mL)

	Potassium		Sodium		Magnesium		Calcium	
Level	Conc.	RPA	Conc.	RPA	Conc.	RPA	Conc.	RPA
Level-0	0	6.46	0	6.52	0	0.31	0	1.65
Level-1	20.063	9.38	20.026	8.65	10.115	0.72	10.053	2.03
Level-2	40.126	12.02	40.052	10.94	20.23	1.15	20.106	2.45
Level-3	60.189	15.00	60.078	13.22	30.345	1.62	30.159	2.89
Level-4	80.252	17.57	80.104	15.40	40.46	2.00	40.212	3.29

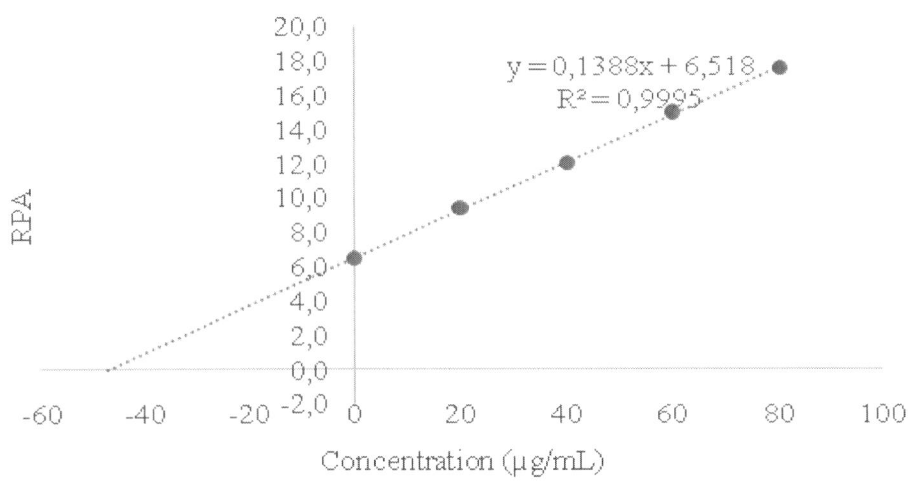

Figure 3.11. Standard addition curve for potassium in PN90.

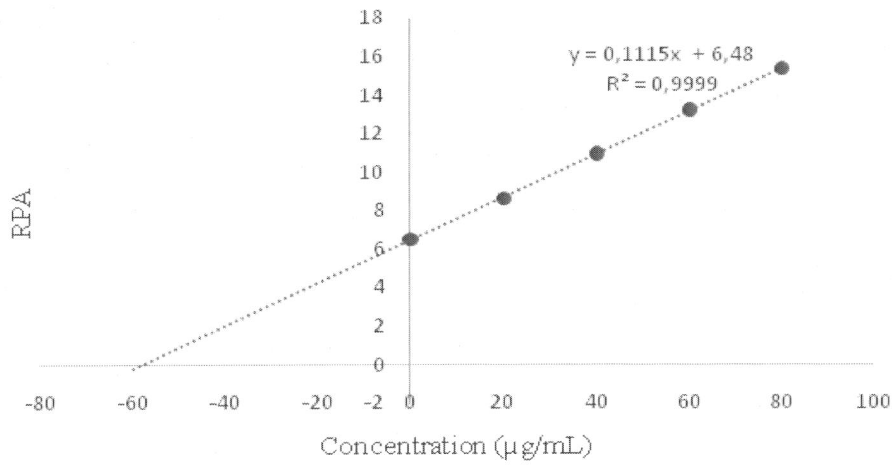

Figure 3.12. Standard addition curve for sodium in PN90.

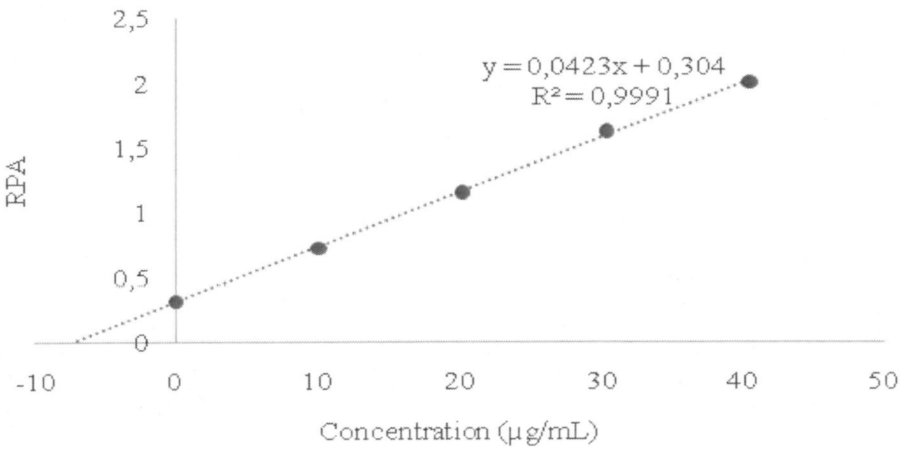

Figure 3.13. Standard addition curve for magnesium in PN90.

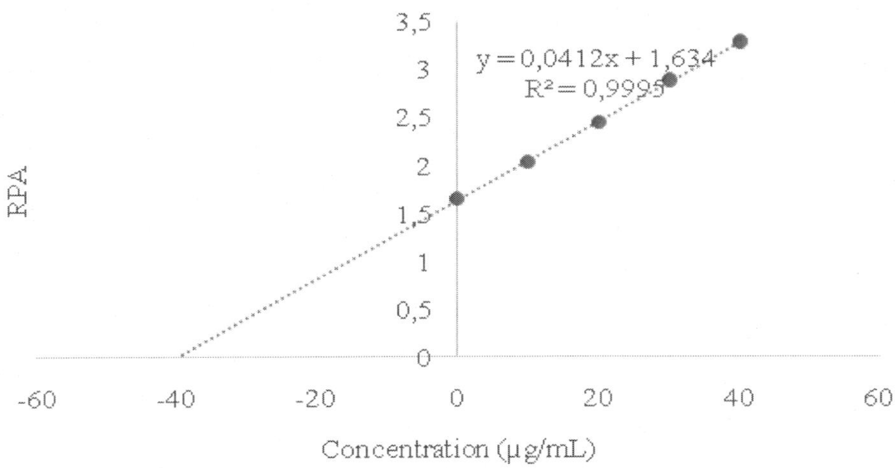

Figure 3.14. Standard addition curve for calcium in PN90.

In case of PN90 and PN110 the linearity profiles of the standard addition curve for K^+, Na^+, Mg^{2+} and Ca^{2+} were always found R^2= 0.999 (Figures 3.11-3.14 and Table 3.21), but with the PN170 the R^2 values for these cations were between 0.965 and 0.972 (Table 3.20).

Table 3.20. Standard addition curve equation for cations in PN170

Analyte	Straight line equation	R^2	Slope
Potassium	y = 0.183x + 2.639	0.972	0.183
Sodium	y = 0.151x + 2.777	0.967	0.151
Magnesium	y = 0.054x + 0.055	0.965	0.054
Calcium	y = 0.052x + 0.648	0.967	0.052

Table 3.21. Estimated quantity of all analytes in PN110 by standard addition method

Analyte	Equation	R^2	Estimated conc. (µg/mL)	Expected conc. (µg/mL)	% Content
Potassium	y = 0.035x + 1.000	0.999	28.57	29.988	95.3
Sodium	y = 0.028x + 1.181	0.999	42.18	28.277	149.2
Magnesium	y = 0.009x + 0.045	0.999	5	2.25	222.2
Calcium	y = 0.007x + 0.205	0.995	29.29	22.5	130.2

The amount of individual cations in the samples PN170 and PN90 was found higher than the specified limit (Tables 3.22 and 3.23):

Table 3.22. Estimated and expected amount of analytes in PN90 by standard addition method

Analyte Name	Estimated conc. (µg/mL)	Expected conc. (µg/mL)	% Content
Potassium	47.24	21.97	215.0
Sodium	58.34	20.28	287.7
Magnesium	7.23	1.68	431.1
Calcium	39.8	16.75	237.6

Table 3.23. Estimated and expected amount of analytes in PN170 by standard addition method

Analyte Name	Estimated conc. (µg/mL)	Expected conc. (µg/mL)	% Content
Potassium	14.4	11.65	123.6
Sodium	18.4	10.76	171.0
Magnesium	1.02	0.87	116.7
Calcium	12.46	8.74	142.6

Moreover, it is worth mentioning here that the %RSD values for all cations at each concentration level obtained from the data of the graphical standard addition method were found within the specified limit (Tables 3.24-3.26; Figures 3.15-3.17) with few exceptions which may be attributed to the unwanted CE or C^4D interference from the surrounding or from electrical glitches. Additionally, the C4D detector was found to be quite sensitive to external factors like mechanical jolt, static electricity.

Table 3.24. %RSD for all analytes from standard addition analysis of PN110

	Level-0	Level-1	Level-2	Level-3
Potassium	0.8	1.3	1.0	1.4
Sodium	1.7	1.0	1.1	2.0
Magnesium	8.1	2.4	1.6	1.1
Calcium	15.3	2.2	4.7	1.5

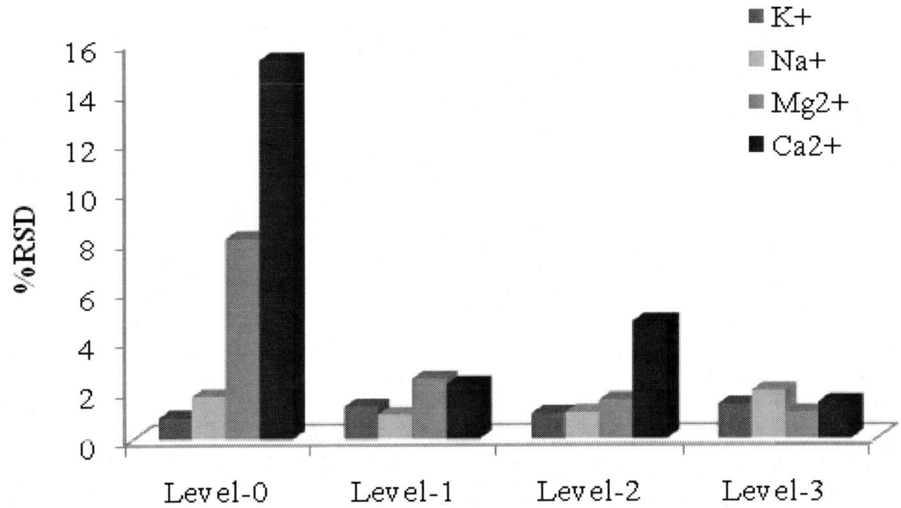

Figure 3.15. %RSD for all analytes from standard addition analysis of PN110.

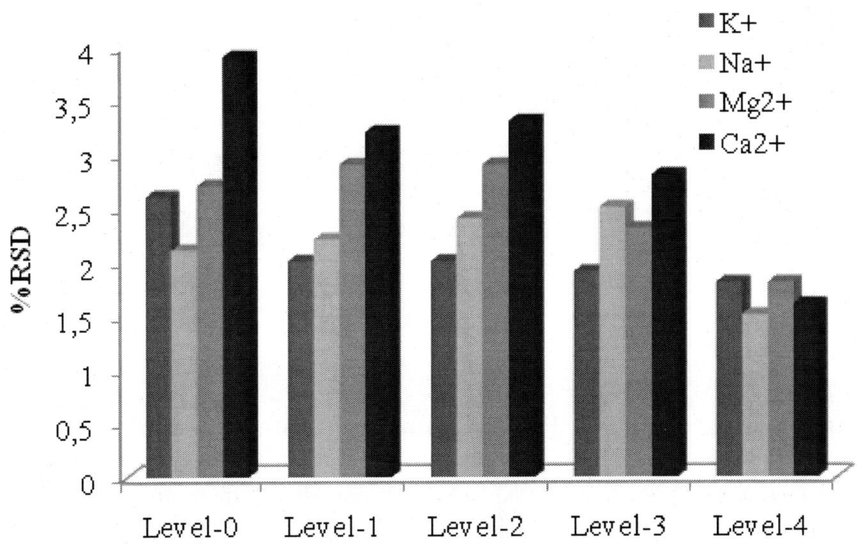

Figure 3.16. %RSD for all analytes from standard addition analysis of PN90.

Table 3.25. %RSD for all analytes from standard addition analysis of PN90

Analyte	Level-0	Level-1	Level-2	Level-3	Level-4
Potassium	2.6	2.0	2.0	1.9	1.8
Sodium	2.1	2.2	2.4	2.5	1.5
Magnesium	2.7	2.9	2.9	2.3	1.8
Calcium	3.9	3.2	3.3	2.8	1.6

Table 3.26. %RSD for all analytes from standard addition analysis of PN170

Analyte	Level-0	Level-1	Level-2	Level-3
Potassium	1.3	1.9	1.9	2.4
Sodium	1.2	1.8	1.6	2.0
Magnesium	9.4	3.1	2.9	2.6
Calcium	4.3	2.1	3.2	1.7

In either of the methods, the amounts of analytes calculated were always higher than what they are claimed to be. Moreover, even with the standard solution it gave erroneous results (data not shown here). In addition, the R^2 values of the standard addition curve for all cations are good (> 0.995) except for the PN170. During analysis of the PN170, more background noise was observed in few electropherograms that might lead to poor peak integration and hence large variability in calculated RPA.

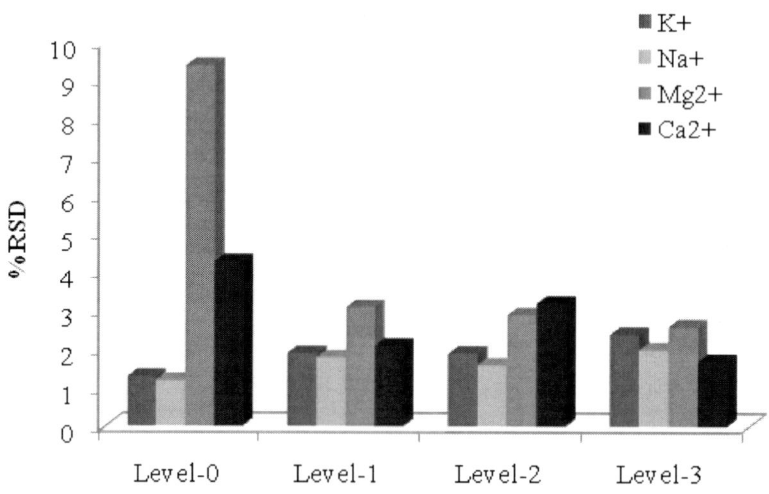

Figure 3.17. %RSD for all analytes from standard addition analysis of PN170.

3.8. Quantification by Single Point Calibration Using Simulated Standard

By this technique, the assay of PN90 was found reliable in terms of repeatability and estimated concentration of each analyte (Table 3.27). In contrast, the aqueous standard solution without amino acids displayed peak areas almost 1.5 times lower than that of sample solution.

Table 3.27. Measured response for sample and standard and content of analyte ions

	Sample		Standard		
Analyte	RPA	%RSD	RPA	%RSD	% Content
K^+	15.2	2.6	14.7	2.2	100.3
Na^+	16.1	2.7	14.9	2.0	102.4
Mg^{2+}	0.6	2.8	0.6	2.9	102.5
Ca^{2+}	3.1	3.4	2.9	3.3	100.7

The improvement in quantification by single point calibration using simulated standard solution was found effective and this outcome may be attributed to unpredictable influence of amino acids on the analytes' mobility and thereby on resultant peak area.

CONCLUSION

This CE-C^{4}D method for the assay of four inorganic cations can be used on routine basis as it has been demonstrated to be repeatable and effective. Moreover, the selectivity and sensitivity obtained with this method in addition to rapid analysis time and economy associated with it, can be a good alternative for the analysis of multiple metal ions in a single solution.

REFERENCES

[1] Flurer, C. L. (2003). Analysis of antibiotics by capillary electrophoresis. *Electrophoresis,* 24, 4116-4127.

[2] Carvalho, A. Z. Pauwels, J. De Greef, B. Vynckier, A. K. Yuqi, W. Hoogmartens, J. & Van Schepdael, A. (2006). Capillary electrophoresis method development for determination of impurities in sodium cysteamine phosphate sample. *Journal of Pharmaceutical and Biomedical analysis,* 42, 120-125.

[3] Altria, K. D. Kelly, M. A. & Clark, B. J. (1998). Current applications in the analysis of pharmaceuticals by capillary electrophoresis-I. *Trends in analytical chemistry,* 17, 204-214.

[4] Gillespi, E. Connolly, D. Macka, M. Hauser, P. & Paull, B. (2008). Development of a contactless conductivity detector cell for 1.6 mm O.D. (1/16th inch) HPLC tubing and micro-bore columns with on column detection. *Analyst,* 133, 1104-1110.

[5] Zemann, A. J. Schenell, E. Volgger, D. & Bonn, G. K. (1998). Contactless conductivity detection for capillary electrophoresis. *Analytical chemistry,* 70, 563-567.

[6] Fracassi da Silva, J. A. & do Lago, C. L. (1998). An oscillometric detector for capillary electrophoresis. Analytical Chemistry, 70, 4339-4343.

[7] El-Attug, M.N. (2011). PhD thesis, KU Leuven, Belgium.

[8] Altria, K. D. (1996). Capillary Electrophoresis Guidebook: Principles, Operation and Applications. (Vol. 52). Totowa, NJ: *Humana press.*

[9] Huber, L. (2007). Validation of Analytical Procedures. Agilent technologies. [2013 08 15]. Available from : http://www.chem.agilent. com/Library/primers/Public/5990-5140EN.pdf.

[10] Ermer, J. & Miller, J.H.M. (2005). Method validation in Pharmaceutical analysis: a guide to Best practice. (1st). Weinheim, Germany: *Wiley-VCH.*

[11] Harris, D. C. (2007). Quantitative Chemical Analysis: Statistics (7th). New York, NY: W. H. *Freeman and Company.*

[12] Nussbaumer, S. Fleury-Souverain, S. Bouchoud, L. Rudaz, S. Bonnabry, P. & Veuthey, J. (2010). Determination of potassium, sodium, calcium and magnesium in total parenteral nutrition formulations by capillary electrophoresis with contactless conductivity detection. *Journal of Pharmaceutical and Biomedical Analysis,* 53, 130-136.

[13] Varenne, A. & Descroix, S. (2008). Recent strategies to improve resolution in capillary electrophoresis-a review. *Analytica Chimica Acta,* 628, 9-23.

[14] Joyban, A. & Kenndler, E. (2006). Theoretical and empirical approaches to express the mobility of small ions in capillary electrophoresis. *Electrophoresis*, 27, 992-1005.

[15] De Cock, B. Dejaegher, B. Stiens, J. Mangelings, D. & Vander Heyden, Y. (2014). Precision evaluation of chiral capillary electrophoretic methods in the context of inter-instrumental transfer: Constant current versus constant voltage application. *Journal of Chromatography A,* 1353, 140-147.

[16] Faller, T. & Engelhardt, H. (1999). How to achieve higher repeatability and reproducibility in capillary electrophoresis. *Journal of Chromatography A,* 853, 83-94.

[17] Altria, K. D. & Luscombe, D. C. M. (1993). Application of capillary electrophoresis as a quantitative identity test for pharmaceuticals employing on-column standard addition. *Journal of Pharmaceutical and Biomedical Analysis,* 11, 415-420.

[18] Altria, K. D. & Fabre, H. (1995). Approaches to optimisation of precision in capillary electrophoresis. *Chromatographia,* 40, 313-320.

[19] Dehouck, P. Jagavarapu, P. K. R. Desmedt, A. Van Schepdael, A. & Hoogmartens, J. (2004). Intermediate precision study on a capillary electrophoretic method for chlortetracycline. *Electrophoresis*, 25, 3313-3321.

[20] Mayer, B. X. (2001). How to increase precision in capillary electrophoresis. *Journal of Chromatography A,* 907, 21.

[21] Petersen, N. J. & Hansen, S. H. (2012). Improving the reproducibility in capillary electrophoresis by incorporating current drift in mobility and peak area calculations. *Electrophoresis,* 33, 1021.

[22] ICH, International conference on harmonization of technical requirements for registration of pharmaceuticals for human use, in: Harmonized Tripartite Guideline: Validation of Analytical Procedures: Text and Methodology Q2 R1. (ICH 2013, http:/ich.org/fileadmin/public_Web_site/ICH_Products/Guidelines/Q2_R1/Step4/Q2_R__Guideline.pdf.). [2015 02 15].

In: Capillary Electrophoresis (CE)
Editor: Christian Reed
ISBN: 978-1-63483-122-2

Chapter 7

THEORETICAL PRINCIPLES AND APPLICATIONS OF HIGH PERFORMANCE CAPILLARY ELECTROPHORESIS

Ayyappa Bathinapatla, Suvardhan Kanchi*,
Myalowenkosi I. Sabela and Krishna Bisetty†
Department of Chemistry, Durban University of Technology,
Durban, South Africa

ABSTRACT

This book chapter is aimed at addressing the theoretical principles and applications of capillary electrophoresis (CE) for the separation of high intensity artificial sweeteners. Electrophoresis is a technique in which solutes are separated by their movement with different rates of migration in the presence of an electric field. Capillary electrophoresis emerged as a combination of the separation mechanism of electrophoresis and instrumental automation concepts in chromatography. Its separation mainly depends on the difference in the solutes migration in an electric field caused by the application of relatively high voltages, thus generating an electro-osmotic flow (EOF) within the narrow-bore capillaries filled with the background electrolyte. Currently capillary electrophoresis is a very powerful analytical technique with a major and outstanding importance in separations of compounds such as amino acids, chiral drugs, vitamins, pesticides etc., because of simpler method development, minimal sample volume requirements and lack of organic waste.

The main advantage of capillary electrophoresis over conventional techniques is the availability of the number of modes with different operating and separation characteristics include free zone electrophoresis and molecular weight based separations (capillary zone electrophoresis), micellar based separations (micellar electrokinetic chromatography), chiral separations (electrokinetic chromatography), isotachophoresis and isoelectrofocusing makes it a more versatile technique being able to analyse a wide range of analytes.

* E-mail: ksuvardhan@gmail.com.
† E-mail: bisettyk@dut.ac.za.

The ultimate goal of the analytical separations is to achieve low detection limits and CE is compatible with different external and internal detectors such as UV or photodiode array detector (DAD) similar to HPLC. CE also provides an indirect UV detection for analytes that do not absorb in the UV region. Besides the UV detection, CE provides five types of detection modes with special instrumental fittings such as Fluorescence, Laser-induced Fluorescence, Amperometry, Conductivity and Mass spectrometry. Infact, the lowest detection limits attained in the whole field of separations are for CE with laser induced fluorescence detection.

Regarding the applications of CE, the separation and determination of high intensity sweeteners were discussed in this chapter. The materials which show sweetness are divided into two types (i) nutritive sweeteners and (ii) non-nutritive sweeteners. The main nutritive sweeteners include glucose, crystalline fructose, dextrose, corn sweeteners, honey, lactose, maltose, invert sugars, concentrated fruit juice, refined sugars, high fructose corn syrup and various syrups. Non-nutritive sweeteners are sub-divided into two groups of artificial sweeteners and reduced polyols.

On the other hand, based on their generation; artificial sweeteners can further be divided into three types as (a) first generation artificial sweeteners which includes saccharin, cyclamate and glycyrrhizin (b) second generation artificial sweeteners are aspartame, acesulfame K, thaumatin and neohesperidinedihydrochalcone (c) neotame, sucralose, alitame and steviol glycosides falls under third generation artificial sweeteners. Artificial sweeteners are also classified into three types based on their synthesis and extraction: (i) synthetic (saccharin, cyclamate, aspartame, acesulfame K, neotame, sucralose, alitame) (ii) semi-synthetic (neohesperidinedihydrochalcone) and (iii) natural sweeteners (steviol glycosides, mogrosides and brazzein protein). Polyols are other groups of reduced-calorie sweeteners which provide bulk of the sweetness, but with fewer calories than sugars.

The commonly used polyols are: erythritol, mannitol, isomalt, lactitol, maltitol, xylitol, sorbitol and hydrogenated starch hydrolysates (HSH).

The studies revealed that capillary electrophoresis was successfully used for the separation of high intensity artificial sweeteners such as neotame, sucralose and steviol glycosides.

Additionally, the available methods for the other artificial sweeteners using capillary electrophoresis were reviewed besides the above indicated sweeteners.

1. Introduction to Capillary Electrophoresis

Electrophoresis is a technique in which solutes are separated by their movement with different rates of migration in an electric field. Depending on the type of electrophoresis the separation can be achieved by gel electrophoresis and capillary electrophoresis.

Capillary electrophoresis (CE) emerged by the combination of the separation mechanism of electrophoresis and instrumental automation concepts of chromatography. It is a versatile analytical technique with a major and outstanding importance in separations of compounds such as amino acids, chiral drugs, vitamins, pesticides, inorganic ions, organic acids, dyes, surfactants, peptides and proteins, carbohydrates, oligonucleotides, DNA restriction fragments, whole cells and even virus particles because of simpler method development, minimal sample volume requirements and lack of organic waste. Its separations depend on the difference in the solutes migration in an electric field caused by the application of relatively high voltages, thus generating an electro-osmotic flow (EOF) within the narrow-bore capillaries filled with BGE (Henk and Gerard, 2010).

1.1. Instrumentation

One key feature of CE is the overall simplicity of the instrumentation. The basic scheme of CE instrumentation consists of an auto sampler, two electrodes (the anode and a cathode), fused-silica capillary (20-100 mm I.D., 20-100 cm length) placed in buffer reservoirs. The electrodes are used to make electrical contact at high voltage power supply (up to 30 kV) operated in either positive or negative polarity. The sample is loaded into the capillary by replacing one of the reservoirs (usually at the anode) with a sample reservoir and applies either an electric field or an external pressure and then separation is performed as shown in Figure 1. Generally, the internal and external detectors such as UV/diode-array or fluorometric or electrochemical detector and mass spectrometer (MS) are coupled to the CE system which is present at the cathodic end (Henk and Gerard, 2010; McLaughlin et al., 1991).

1.2. Principle of Operation

The sample is introduced into the capillary from the anodic end by applying either hydrodynamic (external pressure) or electrokinetic (voltage) injection modes. With the buffer reservoir on each end, an electric field is applied through the capillary and separation depends on the migration of solutes against the field between anode and cathode. The solute migrations depend mainly on their sizes, degree of ionization, their charges as well as dielectric constant of the BGE.

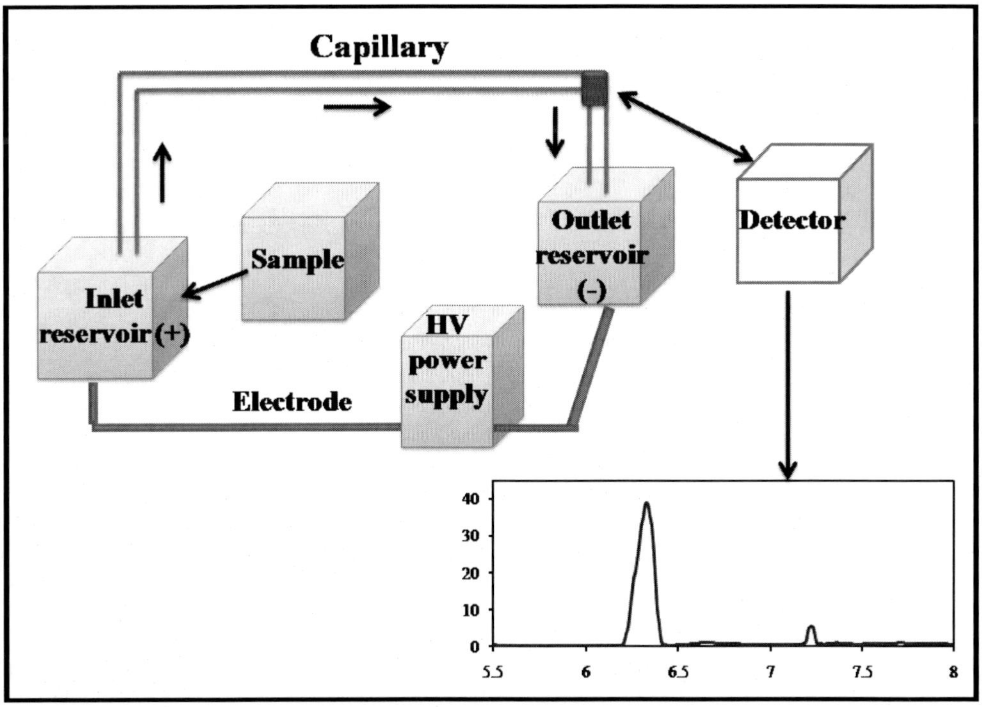

Figure 1. Basic components of capillary electrophoresis instrumentation.

As soon as the analytes are introduced into the capillary, voltage is applied and it enables the analyte molecules to migrate inside the capillary by the known phenomenon electrophoretic mobility and electro-osmotic flow (EOF). Finally, optical detection is done at the opposite end of the capillary which has an optical window aligned with the detector (Heiger, 2000).

1.3. Electro-Osmotic Flow

EOF is one of the fundamental processes based on electro-osmosis. This phenomenon is mainly generated from the surface charge of the capillary walls. Electro-osmotic flow is the bulk flow of the solute in the capillary and is consequence of the surface charge on the interior capillary wall. Cations migrate towards the negatively charged electrode (cathode), anions attracted by the positively charged electrode (anode) and neutrals are not attracted by either of the electrodes. Controlling the EOF can be achieved considerably by the efficiency and selectivity of the separation. The factors affecting the EOF are as follows: concentration/ionic strength of the BGE, electric field, pH, temperature and capillary coatings (e.g., silanol groups). The EOF enables the simultaneous analysis of cations, anions and neutral species in the same analysis.

Based on the pH of the BGE, change in the ionization capacity of silanol groups are observed in the inner walls of the capillary. The silanol groups (SiOH) produce hydrogen cations (H^+) into the BGE leaving the negative (SiO^-) groups on the inner walls of the capillary. Even at low pH the positive ions in the electrolyte, thus get attracted to the walls causes double ionization and forms a double layer which is known as zeta potential as shown in Figure 2. The ionization increases with increase in pH and same for the EOF. When the voltage is applied across the capillary the cations forming the diffuse double layer are attracted towards the cathode. Because they are solvated their movement drags the bulk solution within the capillary towards the cathode (Lukacs and Jorgenson, 1985). The magnitude of EOF can be defined by

$$u_{EOF} = (\frac{\varepsilon\zeta}{\eta})$$

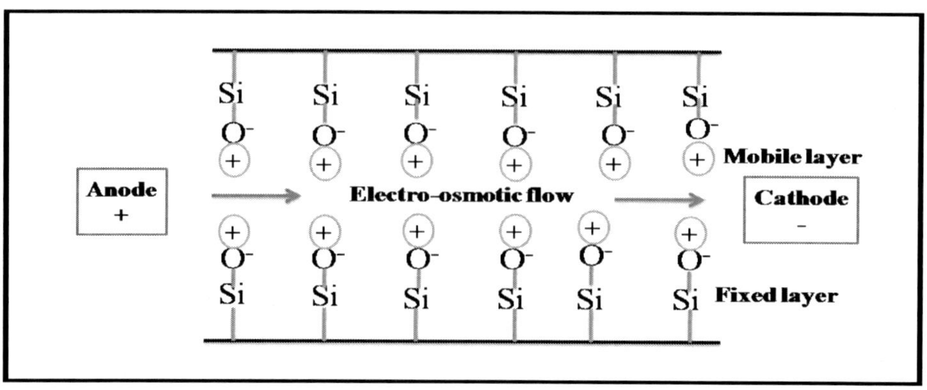

Figure 2. Development of electro-osmotic flow: Formation of negatively charged fused silica surfaces (SiO^-) and hydrated cations accumulating on surface.

where ε is the solution dielectric constant, ζ zeta potential and μ_{EOF} is EOF mobility.

The impact of pH on the analyte can also be substantial, particularly for complex zwitterionic compounds such as peptides. The charge on the compound is pH dependent and the selectivity of separation is affected substantially by pH. As a rule of thumb, select a pH that is at least two units above or below the pKa of the analyte to ensure complete ionization. At high alkaline pH the EOF may be so rapid that incomplete separation may occur (Introduction to capillary electrophoresis, Beckman coulter).

1.4. Electrophoretic Mobility

The CE efficiency, especially CZE mainly depends on the following fundamental principles of electrophoresis and electro-osmosis:

The electrophoretic mobility is determined by the electric force that the molecule experience, balanced by its frictional drag through the medium. This phenomenon can be described according to the equations shown below:

The electric force:

$$F_E = qE$$

From Stoke's law frictional force for spherical ion is:

$$F_F = -6\pi\eta rv$$

where q= Ion charge

η = Solution viscosity

r = Ion radius

v = Ion velocity

At transient point both electrical and frictional forces are equal

Hence,

$$qE = 6\pi\eta rv$$

and the ion velocity

$$(v) = \mu_e E$$

where μ_e = Electrophoretic mobility

E = Applied electric field

Finally,

$$qE = 6\pi\eta rv$$

$$qE = 6\pi\eta r\mu_e E$$

$$\text{Electrophoretic mobility } (\mu_e) = \frac{q}{6\eta r\pi} E$$

From this equation it is evident that small, highly charged species have high mobilities whereas large, minimally charged species have low mobilities, as shown in Figure 3 (Bird et al., 2001).

1.5. Analytical Parameters

(i) Migration Time

The time required for a solute to reach the detection point is called the "migration time", and is given by the quotient of migration distance and velocity. The apparent solute mobility can be calculated using equation shown below.

$$\mu_a = \frac{l}{tE} = \frac{lL}{tV}$$

where $\mu_a = \mu_e + \mu_{EOF}$

V = Applied voltage

l = Effective capillary length

L = Total capillary length

t = Migration time

E = Electric field

(ii) Dispersion

Peak dispersion σ^2, which result from molecular diffusion, takes place as the solute migrate through the capillary, is calculated using equation:

$$\sigma^2 = 2Dm = \frac{2DlL}{\mu_e v}$$

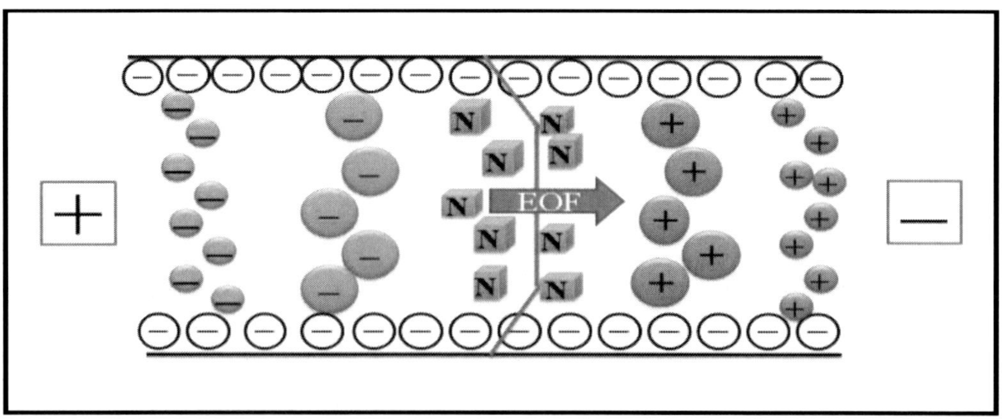

Figure 3. Differential solute migration superimposed on electro-osmotic flow in capillary zone electrophoresis.

where Dm = the solute's diffusion coefficient cm^2/s.

(iii) Efficiency

The separation efficiency in capillary electrophoresis can be calculated in terms of the number of theoretical plates and it is given by equation:

$$N = \frac{\mu V}{2D_m}$$

where: N = Number of theoretical plates

μ = Apparent mobility

D_m = Diffusion coefficient of the analyte

According to the above equation, the efficiency of separation is only limited by diffusion and is proportional to the strength of the electric field. In contrast to other separation techniques such as HPLC, the efficiency of capillary electrophoresis is typically much higher because of the absence of mass transfer between phases. In addition, the flat EOF driven system in CE does not significantly contribute to the band broadening than the characteristic of pressure driven flow in chromatography columns results in much efficiency and a number of theoretical plates.

(iv) Resolution

Achieving fair resolution among the sample components is the ultimate goal in separation science. Resolution is defined as the balancing of differential migration and the dispersive processes of the sample components. CE yields good separation of small molecules and resolution between two species can be calculated using equation shown below.

$$R = \frac{1}{4} N^{1/2} (\frac{\Delta\mu}{\mu^*})$$

$$\Delta\mu \ = \mu_2 - \mu_1$$

$$\mu^* = \mu_2 + \frac{\mu_1}{2}$$

With the substitution of number of theoretical plates (N) in the above equation gives:

$$R = (\frac{1}{4\sqrt{2}})(\Delta\mu)\left(\frac{v}{D(\mu^* + \mu_{EOF}}\right)^{1/2}$$

(V) Solute-Wall Interactions

Interaction between the solute and the capillary wall is unfavourable to CE. The peak tailing and total adsorption of the solute mainly depends on the level of interaction. The adsorption mainly caused by ionic interactions between negatively charged capillary walls and cationic solutes. In case of large peptides and proteins adsorption can occur due to the presence of numerous charges and hydrophobic moieties (Henk and Gerard, 2010; Heiger, 2000). The variance due to adsorption can be given by the equation:

$$\sigma^2 = \frac{k'V_{EOF}I}{(1+k')^2}\left(\frac{r^2k'}{4D}+\frac{2}{K_d}\right) \text{ k' = Capacity factor}$$

VEOF = Electro-osmotic flow velocity
D = Solute diffusion coefficient
l = Capillary effective length
Kd= First order dissociation constant
The variance is strongly dependent on the magnitude of the capacity factor.
For CZE method capacity factor is like in liquid chromatography

$$K' = \frac{t_r - t_0}{t_0}$$

where t_r= Elution time of retained solute
t_0= Elution time of an unretained solute
For EKC or MEKC method capacity factor is

$$K' = \frac{t_r - t_0}{t_0(1 - \frac{t_r}{t_m})}$$

where t_r= Elution time of retained solute
t_0= Elution time of an unretained solute
t_m= Elution time of pseudostationary phase

1.6. Modes of Operation

CE comprises of a family of techniques with different operating and separation characteristics, making it a more versatile technique being able to analyse a wide range of analytes.

The techniques are:

(i) Capillary Zone Electrophoresis (CZE)

CZE is the simplest mode in CE, where the capillary is filled with an electrolyte followed by injection of the sample at the inlet and electric field is applied. The basic principle of this mode is, analytes will migrate at different velocities (apparent mobility) due to their charges and sizes by applying an electric field. Hence, CZE separation is mainly governed by charge/size ratio with electrophoretic mobility which results in small and highly charged molecules migrate faster than larger and less charged. Neutral molecules cannot be separated because they migrate at the velocity of the EOF. CZE is widely employed in the separation of proteins and peptides but the problem with this mode is electrostatic binding of cationic substances to the walls of the capillary. This effect is observed in the case of proteins operating in a buffer that has a pH below the pKa of the analyte. The problem could be overcome by operating at least two pH units above the pKa of the protein. The use of treated capillaries is one of the several ways to reduce the wall binding (Introduction to capillary electrophoresis, Beckman coulter). Other applications for the CZE mode include the separation of inorganic anions and cations such as those normally separated by ion chromatography (Lauer and McManigill, 1986).

(ii) Micellar Electrokinetic Capillary Chromatography (MEKC)

MEKC is a hybrid form of electrophoresis and chromatography in which surfactants are added to the running buffer at concentrations that form micelles. It is widely used mode for industries including (bio) pharmaceutical, food, environmental and clinical industries.

The main strength of the MEKC is, the only electrophoretic technique that can be used for the separation of neutral solutes as well as charged species (Henk and Gerard, 2010).

MEKC principle of operation is based on the addition of surfactant to the background electrolyte example being SDS (anionic), CTAB and DTAB (cationic), CHAPS and CHAPSO (zwitterionic). At a concentration above the critical micelle, surfactant micelles are formed with hydrophobic tails oriented towards the centre and the charged heads oriented outside facing towards the buffer.

Depending on their charge, micelles travel either with the EOF or against the EOF and acts like a pseudo-stationary phase in chromatography as shown in Figure 4.

Those with a negative charge such as SDS travel against the EOF towards the anode. However, at neutral pH or basic pH the migration of micelles is slower than the EOF therefore, resulting in the net migration being towards the cathode favouring the direction of the EOF. As the solutes migrate through the column, partition between the micelles and the running buffer takes place through hydrophobic and electrostatic interactions (Terabe et al., 1984; Vindevogel and Sandra, 1992).

(iii) Capillary Gel Electrophoresis (CGE)

The principle of the CGE is identical to traditional slab or tube gel electrophoresis. It is mostly used for the separation of molecules such as protein and nucleic acids based on their size.

In order for the separation to be feasible the molecules have to be denatured using sodium dodecyl sulfate (SDS) and passed through a suitable polymer which acts as a molecular sieve making it easier for smaller molecules to migrate through the polymer as opposed to larger ones as shown in Figure 5. CGE is a very useful technique for separation of large biological molecules which have similar electrophoretic migration due to their similar charge-to-mass ratios which could not be varied and be resolved according to size without denaturing. This mode greatly applies to proteins and DNA analysis (Henk and Gerard, 2010; Lux et al., 1990).

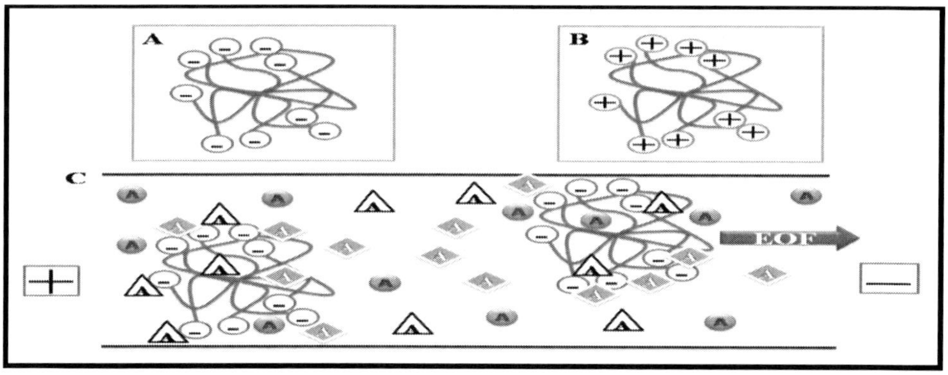

Figure 4. (A) negatively charged micelles (SDS) (B) Positively charged micelles (CTAB) and (C) Separation in MEKC (A=analyte).

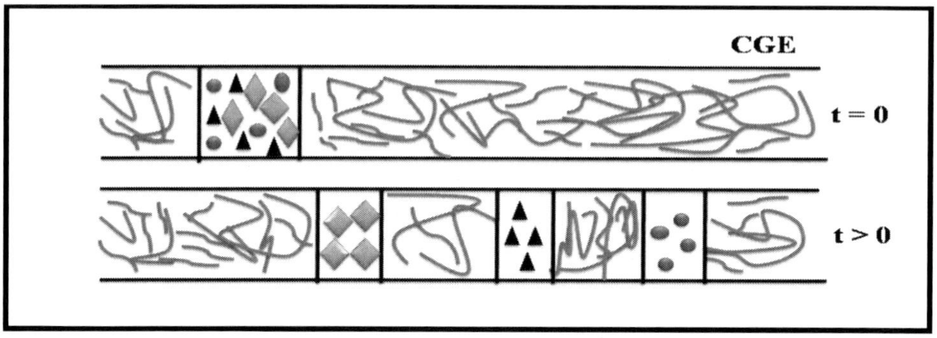

Figure 5. Size separation in CGE.

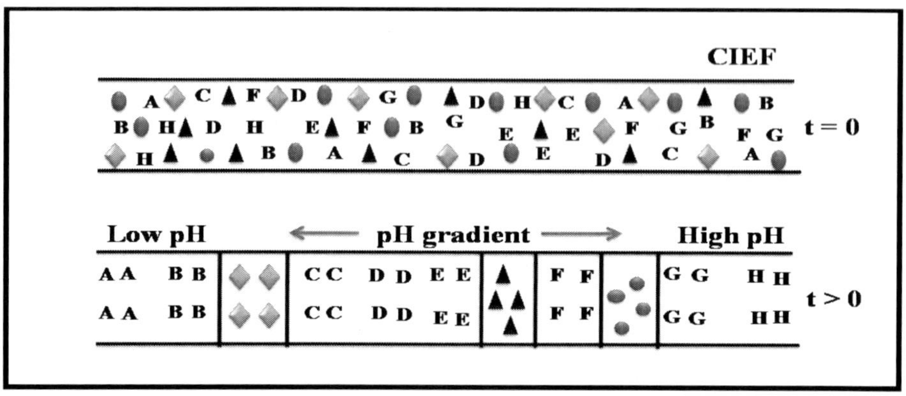

Figure 6. Separation in CIEF. A, B, C, D, E, F, G, H represent ampholyte molecules. Symbols , and represent solute molecules for example peptides and proteins.

(iv) Capillary Isoelectric Focusing (CIEF)

CIEF is referred to as a high resolution technique for the separation of ampholytes which are zwitterionic substances such as proteins, peptides and amino acids based on their isoelectric points (pI) rather than their apparent mobilities as shown in Figure 6.

CIEF employs ampholytes with both basic and acidic nature being able to have pI values that last the desired pH range between the anode and the cathode for the analysis. Its principle is based on the "focusing" method which is filled of the capillary with a mixture of ampholytes and solutes forming a pH gradient where the acidic and basic solutions are at the anode and cathode. When the electric field is applied the ampholytes and solutes are migrating through the capillary to the point where they reach their isoelectric points. Simply, if the analyte has a net charge that is positive it is mostly likely to migrate towards the cathode. At their isoelectric point (pI) migration stops and solute focused into a tight zone and it pass through the detection point by means of pressure or chemical means (Xu, 1996).

(v) Electrokinetic Chromatography (EKC)

EKC is best described by the cyclodextrin (CD) mediated mode, where enantiomers and diastereomers interacts differently with the CD and allows for their separation as shown in Figure 7.

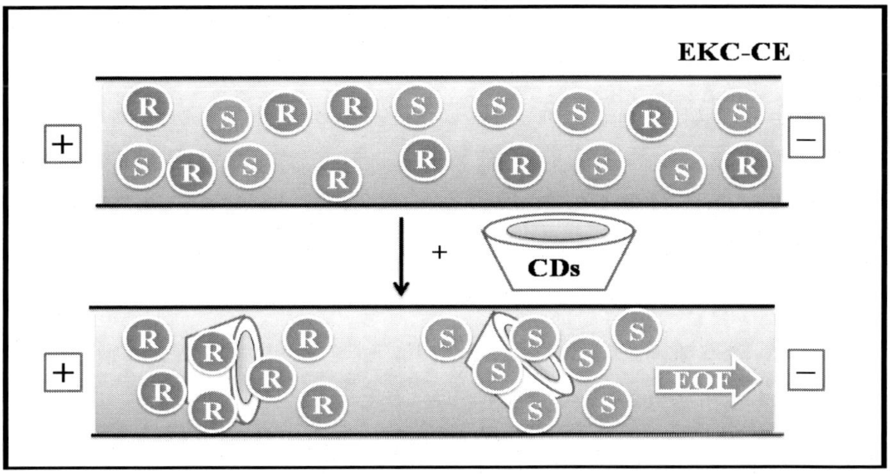

Figure 7. Separation mechanism for chiral compounds with cyclodextrin using EKC-CE method.

This approach has made a major impact in pharmaceutical industries for analysis of chiral drugs. Godel and Weinberger (Godel and Weinberger, 1995) explained this mode as a hybrid method of MEKC and CZE since it employs not only chiral micelles but non-micellar chiral selectors such as CDs. This method is, however preferred in the pharmaceutical industry as it is versatile compared to HPLC in enantio-separation since it's very difficult to separate enantiomers under normal CE and LC techniques. CDs are however the most widely used chiral selectors and simply added to the background electrolytes (Terabe, 1989). Native CDs are macrocyclic oligosaccharides formed from the enzymatic digestion of starch by bacteria.

These compounds are formed with 6, 7 or 8 glucopyranose units and are referred to as α-, β- or γ-CD respectively. The shape and size of the CDs are very important factor in chiral separation; generally CDs are torus-shaped and have a relatively hydrophobic internal cavity. The physical properties of the CDs are discussed in introduction section and briefly the formation of inclusion complexes with analytes is mainly depends on the size dimensions of interior hydrophobic cavity. It is also depends on the analyte size, if the analyte is too large, no complex is formed, if it is too small; the molecular contact with the CD may not be strong enough to impact the separations and this is the major limitation of CDs for chiral recognition.

The mechanism for chiral separation by CDs, which are mobility modifying BGE additive, is quite simple to understand. When a charged solute complexes with a neutral CD, its charge/mass ratio and thus its mobility decreases. Hence, the movement of free analytes will differ from the complexed species and the elution order of analytes depends on the degree of complexation. Differences in the equilibrium constants determine the ratio of free/complexed material. If the equilibrium constants are sufficiently different among the enantiomers, separation will occur. The general resolution for CE is shown in below equation:

$$R_s = 0.177\, \Delta\, \mu_{ep} \sqrt{\frac{EL}{(\mu_{ep} + \mu_{eo})Dm}}$$

R_s = resolution
$\Delta\mu$ = difference in mobility between the enantiomers
E = field strength
L = capillary length to detector
μ_{ep} = average mobility
μ_{eo} = electro-osmotic mobility
D_m = diffusion coefficient

Wren and later, Wren and Row have been derived equations to calculate $\Delta\mu$ and ultimately Rs. The mobility of the first enantiomer is:

$$\mu_a = \frac{\mu_1 + \mu_2 K_1[C]}{1 + K_1[C]}$$

For the second enantiomer the mobility is:

$$\mu_b = \frac{\mu_1 + \mu_2 K_2[C]}{1 + K_2[C]}$$

where:

μ_1 = the mobility of the uncomplexed solute
μ_2 = the mobility of the complexed solute
C = concentration of the chiral selector
K_1 and K_2 = the equilibrium constants

This shows that solute's apparent mobility is influenced by the proportion of time spent as complexed material. The difference in the apparent electrophoretic mobility of the two enantiomers $\Delta\mu$ is $\mu_a - \mu_b$ and can calculate using below equation:

$$\Delta\mu = \frac{[C](\mu_1 - \mu_2)(K_1 - K_2)}{1 + [C](K_1 + K_2) + K_1 K_2 [C^2]}$$

From the above equation, if $\mu_1 = \mu_2$ or $K_1 = K_2$, then $\Delta\mu = 0$. If [C] approaches zero or is very large, $\Delta\mu$ approaches zero as well. The greater affinity of the solute to the selector (large K), lower the optimal selector concentration. Therefore, both solute and type of the cyclodextrin selected impact the final result. The optimal concentration can be calculated using expression:

$$[C]_{opt} = \frac{1}{\sqrt{K_1 K_2}}$$

(vi) Capillary Electro Chromatography (CEC)

Electrochromatography is a term used to describe narrow bore packed column separations where the liquid mobile phase is driven not by hydraulic pressure as in HPLC but by electro-osmosis. An additional benefit of CEC compared to HPLC is the fact that the flow profile in a pressure driven system is parabolic, whereas in an electrically driven system it is plug-like and therefore much more efficient. Although Lecoq and Strain discussed the use of electro-osmotic flow in chromatography, Pretorius (Pretorius et al., 1974) first demonstrated the ability to use electro-osmotic flow in order to drive a mobile phase through a chromatography column. The advantages of using electro-osmosis to propel liquids through a packed bed are the same as for open capillaries i.e., reduced plate heights as a result of the plug flow profile and the ability to use smaller particles leading to higher peak efficiency than in pressure driven systems (HPLC). The driving force in CEC is electroosmotic flow and this is highly dependent on pH, the buffer concentration, the organic modifier and the type of stationary phase. The chemistry used to prepare the stationary phase can have a dramatic effect not only on the separation but also the speed of analysis, since the concentration of silanol groups present under the operating conditions largely determine the EOF. For conventional silica based stationary phases, the electro-osmotic flow drops off almost linearly between pH 10.0 and pH 2.0 often by as much as a factor of 3, and therefore; most CEC is performed above pH 8.0. The majority capillary electrochromatography have been performed on either C8 or C18 stationary phases. Non aqueous mobile phases in capillary electrochromatography was employed by Jorgenson and Lukacs (Jorgenson and Lukacs, 1981) where a mobile phase consisting of 100% acetonitrile electrically pumped through a capillary packed with 10 mm Partisil ODS-2 using a voltage of 30 kV for high efficiency separation of 9-anthracene molecule. Chiral CEC is a method where immobilizing the chiral cyclodextrin onto the surface of a fused silica capillary and then driving the mobile phase at applied voltage.

1.7. Instrumental Aspects

(i) Sample Injection

In order to maintain the high efficiency in CE only minute volumes (range up to nanoliters) of samples are loaded into the capillary. The two most commonly used are electrokinetic and hydrodynamic injection methods.

Electrokinetic injection is done by replacing the buffer vial at the injection end of the capillary with a sample vial by applying voltage for a certain period. In this type of injection the sample enters into the capillary through the pumping action and migration of the EOF. Electrokinetic injection is an important aspect in capillary gel electrophoresis, in the use of polymers as they mat be too viscous to be introduced via hydrodynamic injection, thus require voltage is the best alternative to migrate them into the capillary. Hydrodynamic injection is the most commonly used mode for sample introduction into the capillary. It could be performed in three ways namely, (i) by applying pressure at the injection point of the capillary, (ii) applying vacuum at the exit end of the separation capillary (iii) by the siphoning action which is described as the elevation of the sample vial relative to the exit vial. The quantity of sample loaded is nearly independent of the sample matrix with hydrodynamic injection (Huang et al., 1988) and these methods are not restricted to injection of samples as the vials(buffer, chiral selector, analyte etc.,) are interchangeable.

(ii) Capillary

The materials used for manufacturing the capillaries are Teflon and fused silica, at present the fused silica capillaries are widely in use for separations. The disadvantage of Teflon is difficult to obtain homogeneous inner diameters, exhibits sample adsorption problems and has poor heat transfer properties. Compared to Teflon, the fused silica has intrinsic properties; these include high temperature conductance and transparency over a wide range of an electromagnetic spectrum. Another advantage of using fused silica is easy to use for the manufacture of capillaries with small diameters of about a few micro metres as shown in Figure 8. From an analysis time perspective, capillaries with short effective lengths should be used. In CGE, 10 cm gel-filled capillaries and 50 to 70 cm effective length capillaries are used in CZE.

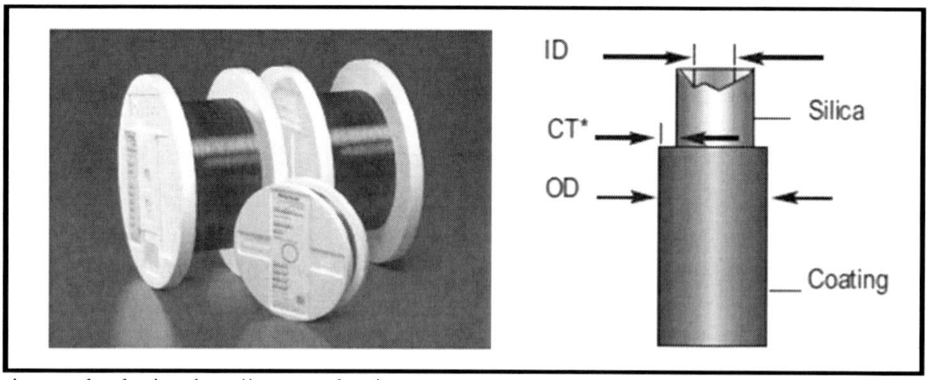

Polymicro technologies, http://www.polymicro.com.

Figure 8. Showing fused silica capillaries.

In most cases it is essential to regenerate the surface by preconditioning the capillary before analysis. Recently, wall coated capillaries gained much interest in CE analysis, because they provide good results in terms of analysis time and detection limits than conventional capillaries (Kok, 2000).

(iii) Capillary Conditioning

Maintaining a reproducible capillary surface is one of the most challenging aspects in CE. To achieve a good reproducibility, capillary conditioning is the important factor. The most commonly employed approach for reproducibility is to refresh the surface of capillary by deprotonation of the silanol groups and removal of the adsorbents and impurities. A typical wash method includes flushing a new capillary at 60 °C using the following sequence: first rinse with 20% methanol, then with1.0M NaOH, then with deionized water, and finally with the running buffer. At the beginning of each working day, the capillaries were conditioned by flushing for 10 min with 1.0 M NaOH, 5 min with deionized water and thereafter treated for

10 min with buffer solution. Other washing procedures can be employed with strong acids, organics such as methanol or DMSO or detergents (Introduction to capillary electrophoresis, Beckman Coulter).

(iv) High Voltage Power Supply

In CE a DC power supply is used to apply up to about 30 kV and current levels of 200 to 300 mA. Stable regulation of the voltage (± 0.1 %) is required to maintain high migration time reproducibility. The voltage power supply is able to reverse the polarity that can switch from the cathode to anode. Hence, there is no need to introduce the analyte in the cathodic end and also not necessary to move an on-line detector to the other end. It can provide high voltages up to 30 kV, which generate electro-osmosis and electrophoretic flow of the charged species and electrolytes through the capillary. For a good reproducibility of migration time the same voltage are to be applied for the entire analysis (Introduction to capillary electrophoresis, Beckman Coulter).

(v) Detector

A UV detector or photodiode array detector (DAD) is applicable in CE similar to HPLC. CE also provides an indirect UV detection for analytes that do not absorb in the UV region, in such cases a UV absorbing species (chromophore) is added to the buffer. Generally, in the analysis of peptides and carbohydrates (weak chromophores in UV range) an indirect UV method can be applied successfully. UV detection is widely used as a universal detector due to its collective detection nature. Besides the UV detection, CE provides nearly five types of detection modes with special instrumental fittings such as Fluorescence, Laser-induced Fluorescence, Amperomctry, Conductivity and Mass spectrometry. The limitations of each detection mode are presented in Table 1 (Ewing et al., 1989).

2. Artificial Sweeteners

In nature, number of food ingredients have sweetening features, a property that mainly varies with the change in food systems, temperature, physical state and the presence of other flavours. These food ingredients stimulate the sweet sensation by interacting with the sweet taste receptors in the mouth and throat.

The materials which show sweetness are divided into two types (i) nutritive sweeteners and (ii) non-nutritive sweeteners.

Nutritive sweeteners provide a sweet taste with the addition of energy and non-nutritive sweeteners provides a sweet taste without any addition of energy. The main nutritive sweeteners include glucose, crystalline fructose, dextrose, corn sweeteners, honey, lactose, maltose, invert sugars, concentrated fruit juice, refined sugars, high fructose corn syrup and various syrups. Non-nutritive sweeteners are sub-divided into two groups of artificial sweeteners and reduced polyols as shown in Figure 9.

On the other hand, based on their generation, artificial sweeteners can further be divided into three types as (a) first generation artificial sweeteners which includes saccharin, cyclamate and glycyrrhizin (b) second generation artificial sweeteners are aspartame, acesulfame K, thaumatin and neohesperidinedihydrochalcone (c) neotame, sucralose, alitame and steviol glycosides falls under third generation artificial sweeteners (Bahndorf and Kienle, 2004). Artificial sweeteners are also classified into three types based on their synthesis and extraction: (i) synthetic (saccharin, cyclamate, aspartame, acesulfame K, neotame, sucralose, alitame) (ii) semi-synthetic (neohesperidinedihydrochalcone) and (iii) natural sweeteners (steviol glycosides, mogrosides and brazzein protein) (Duffy and Anderson, 1998).

Table 1. Showing limitations of different detection modes in CE

Method	**Mass detection limits (moles)**	**Concentration detection (molar)/ 10 nL injection volume**	**Advantages/ disadvantages**
UV-Vis absorption	10^{-13}-10^{-15}	10^{-5}-10^{-8}	• Universal • Diode array offers spectral
Fluorescence	10^{-15}-10^{-11}	10^{-7}-10^{-9}	• Sensitive • Usually requires sample derivatization
Laser-induced	10^{-18}-10^{-20}	10^{-14}-10^{-16}	• Extremely sensitive fluorescence • Usually requires sample derivatization • Expensive
Amperometry	10^{-18}-10^{-19}	10^{-10}-10^{-11}	• Sensitive • Selective but useful only for electroactive analyses • Requires special electronics and capillary modification
Conductivity	10^{-15}-10^{-16}	10^{-7}-10^{-8}	• Universal • Requires special electronics and capillary modification
Mass spectrometry	10^{-16}-10^{-17}	10^{-8}-10^{-9}	• Sensitive and offers structural information • Interface between CE and MS complicated
Indirect UV, fluorescence amperometry	10-100, times less than direct methods	–	• Universal • lower sensitivity than direct method

Polyols are other groups of reduced-calorie sweeteners which provide bulk of the sweeteness, but with fewer calories than sugars. Polyols are used in a wide variety of food products, including chewing gums, confections, ice creams, toothpastes, mouth washes, pharmaceuticals and baked goods. The commonly used polyols are: erythritol, mannitol, isomalt, lactitol, maltitol, xylitol, sorbitol and hydrogenated starch hydrolysates (HSH) as shown in Figure 9 (Larry and Greenly, 2003).

The high consumption of nutritive sweeteners leads to increase in some chronic diseases like obesity, cardiovascular diseases (CVD), diabetes mellitus (type-II), dental caries, certain cancers and behavioural disorders (Shankar et al., 2013). Hence, many of the sweet lover's want the taste of sweetness without any addition of energy. In this regard food industries were introduced number of low-calorie artificial sweeteners in the food and beverage sectors. Artificial sweeteners and polyols can substitute to the nutritive sweeteners and therefore termed as macronutrient substitutes or sugar substitutes.

According to the Food Additives amendment to the Food, Drug and cosmetic Act, some sweeteners were considered as "Generally Recognized As Safe" (GRAS) and others were considered as additives (Fitch and Keim, 2012). Based on the 1958 amendment, Food and Drug Administration (FDA) states that United States of America must approve the safety of all the additives and sweeteners (Duffy and Anderson, 1998). The safety limit of sweeteners and food additives are expressed as the acceptable daily intake (ADI) and this concept is used by FDA and Joint Expert Committee of Food Additions (JECFA) of the United Nations Food and Agricultural Organization (UNFAO) and World Health Organization (WHO) (Duffy and Anderson, 1998).

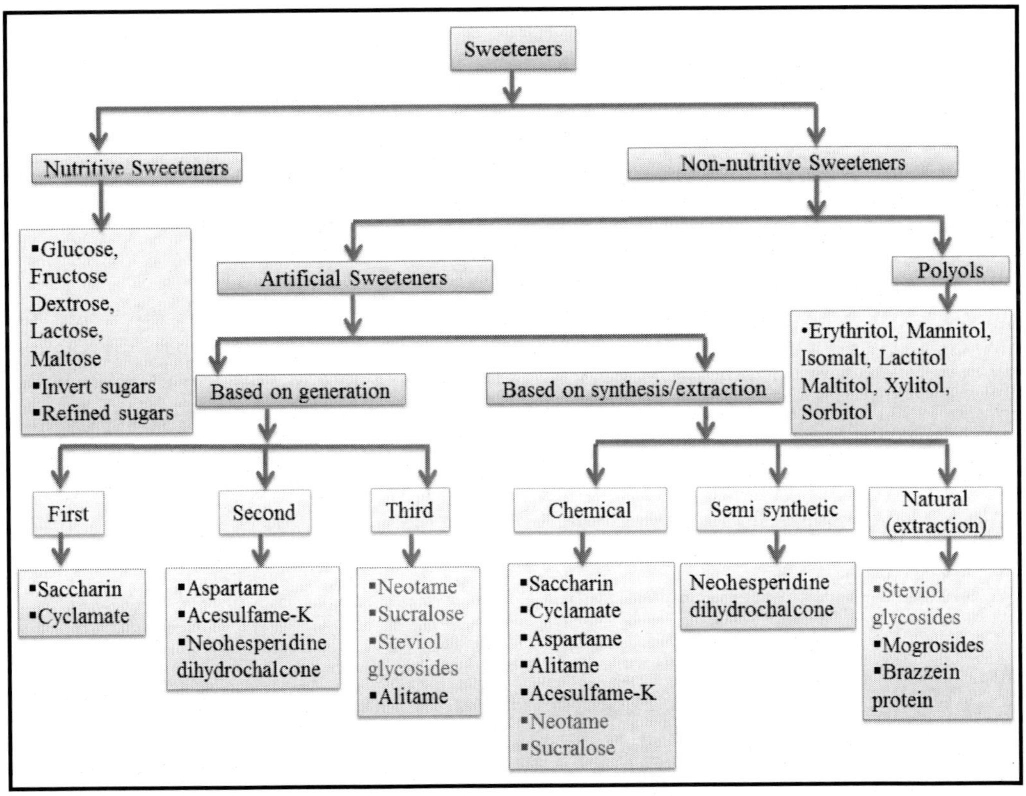

Figure 9. Classification of sweeteners.

2.1. Synthetic or Chemical (Artificial) Sweeteners

2.1.1. Saccharine (E954)

Saccharine (2H-1λ6,2-benzothiazol-1,1,3-trione) (Table 2) was first discovered by Remsen and Fahlberg at John Hopkins University in 1879 (Shankar et al., 2013). This non-caloric sweetener is 200 to 700 times sweeter than sugar and it was approved in nearly 90 countries. Saccharine is marketed under the brand names of Sweet'N Low, Sugar Twin and Necta Sweet. Due to excellent thermal stability and solubility of saccharine, it is used in the wide range of food products like soft drinks, baked goods, jams, canned fruit, candy, salad dressing, dessert toppings, chewing gum and in household products such as toothpaste, lip gloss, mouthwash, vitamins, and also in pharmaceuticals. Saccharin is a valuable alternative to sugars for people who suffer with diabetes because it does not undergo metabolic reaction in the gastrointestinal (GI) tract results in no effect on blood insulin levels. The Adequate Dietary Intake (ADI) for saccharin is set at 5 mg/kg body weight per day for adults and children. However, a survey of the usage pattern of saccharin in edible products in India reveals that 6 to10 year's age group exceeded the ADI by 54% and by adults 137% based on their life style. The extensive research was done between1970-80 and results indicated that high doses of saccharine in rats leads to bladder cancer. Since 1981 FDA passed a mandate that products containing saccharin carry a label warning about its potential as a human carcinogen. But recent studies made by National Institute for Environmental Health Sciences (NIEHS) suggest that for human beings saccharine is not carcinogen. Hence, FDA and National Toxicology Report on Carcinogens and labels no longer have to display on the saccharine containing food products. Saccharine is the oldest and most researched sweeteners and different analytical methods were developed for its determination individually or in combination with the other sweeteners.

2.1.2. Cyclamate (E952)

Cyclamate (sodium N-cyclohexylsulfamate) (Table 2) is first generation artificial sweetener with sweetness more than 30 times than sugar and approved in nearly 50 countries (Bopp et al., 1986). In 1966 first study was conducted on safety of cyclamate in animals and indicated that some intestinal bacteria can metabolize this sweetener as cyclohexylamine which is having chronic toxicity (Hellsten, 2010). Later in 1969 consumption of cyclamate in rats was studied and observed that it causes bladder cancer similar to saccharine. Hence, FDA banned this sweetener in 1969 and after in 1982 Cancer Assesment Committee of FDA reviewed carcinogenicity of cyclamate and they concluded that it is not carcinogenic and safe to use in foods (Duffy and Anderson, 1998).

2.1.3. Aspartame (E951)

Aspartame (methyl ester of L-aspartic acid and L-phenylalanine) (Table 2) was first discovered by James Schalatter, a chemist who was working on antiulcer drugs in 1965 (Shankar et al., 2013). This sweetener is ~250 times sweeter than sucrose and approved in 90 countries and in usage in nearly 6000 food products under the commercial name of Equal, NutraSweet, and Nutra Taste. Aspartame is used as a sweetener in many products which includes chewing gum, diet soda, dry drink mixtures, yogurt and pudding, instant tea and coffee. Industrially, aspartame can be synthesize by amino acids like aspartic acid, phenylalanine and methanol. Due to the presence of amino acids aspartame can produce energy nearly 4 kcal g^{-1}. According to the FDA, the acceptable daily intake of aspartame by humans is 50 mg kg^{-1} body weight, for both adults and children (Learn about cancer, 2012). The safety studies of aspartame reveal that, enzymes called "peptidases" which is used to break the peptide bond can rupture aspartame into two amino acids results in the formation of phenylalanine and aspartic acid. People, who suffer with genetic disorder like phenylketonuria, better to avoid usage of aspartame due to failure of converting, metabolize phenylalanine to tyrosine and this cause to brain damage (FDA statement on European aspartame study, 2013). Because of this reason, FDA suggested that, the products which are having aspartame as sweetener must have a label stating the containment of phenylalanine (Duffy and Anderson, 2012). Common side effects with regular usage of aspartame are brain tumours, systemic lupus, multiple sclerosis, and methanol toxicity causing to blindness, spasms, shooting pains, seizures, headaches, depression, anxiety, memory loss, birth defects, leukemia and death. In 2006 and 2007, the European Ramazzini Foundation (ERF) of Oncology and Environmental Sciences published two studies on aspartame toxicity. These results explained that intake of aspartame in rat's causes cancer, lymphomas and leukaemia (Shankar et al., 2013). These results were opposed by European Food Safety Authority (EFSA) in March 2009, and found that aspartame is not genotoxic or potential carcinogenic (Kroger et al., 2006).

2.1.4. AcesulfameK (E950)

AcesulfameK (potassium 6-methyl-2,2-dioxo-2H-1,2λ6,3-oxathiazin-4-olate) (Table 2) was discovered by a food company Hoechst in 1967 (Claub and Jensen, 1970). This sweetener is nearly 200 times sweeter than sucrose and approved nearly in 100 countries in more than 5000 food products. Due to its exceptional thermal stability it can be used in baking and cooking. It is well known sweetener under the brand names of Sunette, Sweet One and Swiss Sweet. In 1998, FDA was approved for the use of this sweetener in soft drinks and beverages but previously it was only allowed to use in foods such as sugar free baked goods, chewing gum and gelatine desserts. The ADI for acesulfameK is 15 mg kg^{-1} body weight (Kroger et al., 2006). Besides saccharine and cyclamate, acesulfameK is also used in the preparation of sweetener blends in the combination of aspartame and sucralose. Such combinations are not only providing a "more sugar-liketaste" but also decrease the total amount of sweetener used. The safety studies elucidate that acesulfameK is not a carcinogenic molecule because it excretes by kidneys in unchanged form after passes through the body. But one of the by-products of acesulfameK is acetoacetamide in body, which is a toxic molecule at high dosages. However, very low amounts of acesulfameK is used in the foods and the resultant bye product acetoacetamide is also in small amount which is not hazardous to the human body.

2.1.5. Alitame

Alitame ((3S)-3-amino-4-[[(1R)-1-methyl-2-oxo-2-[(2,2,4,4-tetramethyl-3-thietanyl) amino]ethyl] amino]-4-oxobutanoic acid) (Table 2) was discovered by chemists at the Pfizer pharmaceutical in 1979 (Ellis, 1995). This sweetener was approved in Australia, New Zealand, Mexico, PR China and the EU but not approved by the FDA which means it can't use in United States (Kroger et al., 2006). Alitame is nearly 2000 and 10 times sweeter than sucrose and aspartame respectively. The brand name of alitame is "Aclame" and due to its thermal and acidic stability, it is used in the wide range of foods and beverages, including bakery wares, water-based flavoured drinks, dairy-based drinks, desserts, cream, edible ices, jams, confectionery and some dietetic foods. Industrially alitame is in the form of dipeptide ester which can be synthesized using amino acids like L-aspartic acid and D-alanine with N-substituted tetramethylthietanyl molecule. The main advantage of alitame over aspartame is, 7-22 % of alitame is unabsorbed and excreted through the faeces and the remaining (78 to 93%) is hydrolysed to aspartic acid and alanine amide which is highly suitable for the people who suffer with phenylketonuria. Further, the formed aspartic acid is metabolized normally, and the alanine amide is excreted in the urine as a sulfoxide isomer, sulfone or conjugated with glucoronic acid (Chattopadhyay et al., 2011). Essentially, this shows that alitame is not hazardous and goes through normal processes in the body, even though it is metabolized to some degree.

2.1.6. Neotame

In 1996 Nofri and Tinti reported neotame as a non-nutritive artificial sweetener with an *N*-substituted aspartame derivative, (*N*-[*N*-(3,3-dimethylbutyl)-L-α-aspartyl]-L-phenylalanine-1-methyl ester) and with a dipeptide bond (Kroger et al., 2006) (Table 2). During year 2002 and 2013 neotame was approved by the United States Food and Drug Administration as an artificial sweetener (U.S. Food and Drug Administration, news releases, May 19, 2013). On the industrial scale, neotame which contains all the elements of aspartame can be prepared by the reductive alkylation of aspartame with a 3,3-dimethylbutyl group. In contrast to aspartame, neotame has less side effects and the mechanism of neotame safety compared to aspartame is generally due to the enzyme "Peptidases" which is used to break the peptide bonds in dipeptides (Fisher, 1989). In neotame the bond between the aspartic acid and the phenylalanine groups are effectively blocked by the presence of the 3,3-dimethylbutyl moiety, thus reducing the availability of phenylalanine, thereby eliminating concerns for those who suffer from phenylketonuria (Nofre and Tinti, 2000). The safety of the neotame has been investigated and the results indicated that neotame is not carcinogenic, genotoxic, teratogenic or associated with any reproductive toxicity (Scientific opinion, European Food Safety Authority Journal 2007). However 3,3-dimethylbutyraldehyde, a highly flammable component used in the synthesis of neotame, may cause minor side effects like irritation to the skin, eyes, respiratory and reproductive systems after prolonged consumption of the neotame (Mayhew et al., 2003). Due to the presence of amino acids and organic groups, neotame exhibits high sweeteness nearly 10,000 and 40 times sweeter than sugar and aspartame respectively (Prakash et al., 1999).

Neotame has two chiral canters at the C3 and C5 positions, hence it can form four diastereomers namely L,L; L,D; D,D and D,L neotame, and their sweetness is attributed to the presence of well-oriented hydrophobic groups in L,L-diastereomer (Prakash et al., 1999).

Neotame has been approved in more than 35 countries around the world including USA, Canada, Mexico, Argentina, Brazil, Russia, Australia, China, Philippines, Indonesia, Japan, Nigeria and South Africa (Hu et al., 2013). The Acceptable Daily Intake (ADI) for neotame has been set at 0-2 mg kg^{-1} body weight by the Joint Expert Committe for Food Additives in 2003 as well as by the European Food Safety Authority in 2007 (Fitch and Keim, 2012). The Centre for Science in the Public Interest indicates that neotame is still not being used as a sweetener throughout the world, yet it has a wide potential application as a third generation dipeptide sweetener (Hu et al., 2013; Tomasik, 2004). Owing to its low cost, safety and high intensity sweetness the demand and importance of neotame as a sweetener in foods has gained widespread recognition by the food industries. Accordingly, the growing interest on the use of neotame in food and beverages has prompted the need to develop a simple, accurate and reliable method for the determination of neotame. However, some natural food components in complicated food matrices will interfere with the determination of the analyte (Alghamdi et al., 2005; He et al., 2012; Vianna-Soares et al., 2002).

2.1.7. Sucralose (E955)

The high intensity artificial sweetener sucralose, also known as splenda or sucraplus[1,6-dichloro-1,6-dideoxy-β-D-fructofuranosyl-4-chloro-4-deoxy-α-D-galacto-pyranoside (CAS RN: 56038-13-2 and E955)] which is generally using as a sweetener and flavor enhancer in foods and beverages. It was discovered in 1976 by the *Tate and Lyle* company and it industrial preparation includes selective replacement of three hydroxyl groups with three chlorin atoms (Knight, 1994) (Table 2). The presence of two chlorine atoms on the four membered carbon ring (fructose portion) enhances the hydrophobic nature of the five membered carbon ring (galactose portion) which is present on the opposite side of the sucralose molecule. Due to this orientation, the sweeting strength of sucralose increased to 650 times when compared to the sucrose (Jenner and Smithson, 1989). Sucralose is exceptionally stable over a wide temperature and pH ranges, high intensive sweet taste, exceptional stability and excellent solubility characteristics, it has been intoduced into the food market over 3500 different food products through out the world. Sucralose was first approved to use in Canada in 1991 followed by Australia in 1993, New Zealand in 1996, the United States in 1998, and the European Union in 2004 (Schiffman and Rother, 2013). By 2008, it had been approved in over 80 countries, including Mexico, Brazil, China, India, Japan and Turkey (McNeil, 2009). In 2006, the *Food and Drug Administration* (FAD) amended the regulations for foods to include sucralose as a non-nutritive sweetener in food products. The acceptable daily intake (ADI) of sucralose is 0-15mg kg^{-1} body weight by the Joint FAO/WHO Expert Group on Food Additives (JECFA) in 1990 (Schiffman and Rother, 2013). The European Parliament and Council Directive in 2003 has proposed the maximum usable doses of sucralose in different food products as:beverages-300mg L^{-1}, yoghurts-350mg kg^{-1}, candy-200mg kg^{-1}, energy-reduced beer-10mg L^{-1} and in breath-freshening microsweets-2400mg kg^{-1} (European parliament and council Directive, 2003). Sucralose has been discussed as a possible human health hazard, mostly in public media because of its chlorinated structure. In a sub-chronic toxicity test, with 2.8-6.4 g kg^{-1} body weight per day of sucralose administered to rats in the diet, showed several ill-effects (Motwani et al., 2011; Roderoet al., 2009) such as increase in blindness, mineralization of pelvic area and epithelial hyperplasia in rats (Calza et al., 2013). In human body sucralose is hardly absorbed and almost 85 % excreted after metabolism and studies in human beings have shown that this sweetener did not pose carcinogenic, reproductive or neurological risk (Roberts et al., 2000).

The detection of sucralose and other carbohydrates like fructose, glucose and sucrose is a challenging task owing to its: (i) unavailability of the charged functions and (ii) lack of absorption of strong chromophoric nature in the UV region. Therefore, separation of non-absorbing nuetral molecules need a careful procedure with the suitable electrolyte systems. Survey of the literatures revealed that, most of the HPLC or CE separations of nuetral carbohydrates have been achieved either by (i) borate complexation, where separation was achieved based on the differences in the electrophoretic mobilities of the complexed solutes.

Where the degree of complexation variesto variety of sugars with borate, but this method required very high concentrations of analyte (nearly 5000 μg g^{-1}) to achieve well defined analyte peaks (ii) changing the nuetral molecule to charged molecule, this methods is appropriate to determine sugars and artificial sweeteners with strong absorbing buffers like 3,5 dinitro benzoic acid (DNBA) using capillary electrophoresis indirect UV method (McCourt et al., 2005; Stroka et al., 2003) and *p*-nitrobenzoyl chloride with HPLC-pre coloumn derivitization method (Nojiri et al., 2002) (iii) complex formation *via* SN_2 mechanism (a bimolecular nucliophilic substitution reaction), less hindered chloromethyl groups in sucralose easily undergoes substitution reaction in the presence of nucliophiles (electron rich species) (Motwani et al., 2011).

2.2. Semi Synthetic Sweeteners

2.2.1. NeohesperidineDihydrochalcone (E959)

NeohesperidineDihydrochalcone (NHDC)(1-[4-[[(2S,3R,4S,5S,6R)-4,5-Dihydroxy-6-(hydroxymethyl)-3-[[(2S,3R,4R,5R,6S)-3,4,5trihydroxy-6-methyl-2-tetrahydropyranyl]oxy]-2-tetrahydropyranyl]oxy]-2,6-dihydroxyphenyl]-3-(3-hydroxy-4-methoxyphenyl)propan-1-one), is a semi-synthetic sweetener which was first prepared by Horowitz and Gentili in 1963 (Amin et al., 2013). NHDC is nearly 670 and 4 times sweeter than sucrose and aspartame respectively. Industrially it can be synthesized from neohesperidin with alkaline hydrogenation (treatment with potassium hydroxide or another strong base) followed by catalytic hydrogenation (Table 2). The starting material of NHDC, neohesperidin is a bitter taste compound which can be isolated from the peel and pulp of orange, grapefruit, and other citrus fruit. Hence, NHDC also exhibits bitter after taste and to mask this bitter nature generally it is always used in the form of blends in combination of saccharin, cyclamate and acesulfameK. In the European Union, NHDC is commonly used as a non-nutritive sweetener and artificial sweetener, but this sweetener not yet approved by the FDA. In aqueous solutions 0.0045 % of NHDC is approximately equisweetto 5 % sucrose. The safety studies of NHDC demonstrate that about 750 mg neohesperidindihydrochalcone per kg body weight per day didn't show any effect in rats (Lina et al., 1990). Few analytical methods were developed for the determination of NHDC in foods products.

2.3. Natural Sweeteners

2.3.1. Steviol Glycosides (Diterpene Glycosides)

Stevia rebaudiana Bertoni, an herbaceous perennial shrub, belonging to the Asteraceae family also known as ''Sweet-Leaf'', "Sweet weed", "Honey leaf" and "Sweet-Herb" has attracted economic and scientific interest due to the non-nutritive sweetness and the therapeutic properties of its leaf (Midmore, 2002). From hundreds of years in Paraguay and Brazil stevia leaves have been using as "sweet treat" to prepare local teas and medicines. Japan and Korea, are the largest consumers of stevia extract consuming about 200 and 115 tons respectively, on an annual basis. In Japan, stevia replaces the artificial sweeteners like aspartame which were around since the 1970s. The stevia sweeteners are approximately 300 times sweeter than sugar (Liu et al., 1997; Geuns, 2010). Lately, the use of stevia has been approved by the Food and Drug Association in South Africa with the recent promulgation (Foodstuffs, Cosmetics and Disinfectants Act, 1972, 10th September 2012) of the new sweetener regulations (Regulations relating to the use of sweeteners in foodstuffs. Foodstuffs, cosmetics and disinfectants act (1972)). The regulations made by Joint FAO/WHO Expert Committee on Food Additives (JECFA) for steviol glycosides, requiring a purity level at least 95% of the seven well known steviol glycosides (Liu et al., 1997). The ADI for steviol glycosides by JECFA expressed from 2 to 4 mg kg^{-1} bodyweight (Geuns, 2010). Recently reported that stevia leaves contain more than 35 ent-kaurene-type diterpene glycosides, the most abundant of which are rebaudioside A (Reb A) and stevioside (Stv) (Zimmermann et al., 2011) (Table 2). Traditionally, the dry weight percentages of glycosides present in the leaves were reported as Stv ranging from 5 to 10 %, Reb A from 2 to 4 % and with a lower percentage reported for rebaudioside C (Reb C). On the other hand, the relative sweetness of the Stv ranges from 60 to 70 % and between 110 to 270 times sweeter than sugar, while Reb A ranges from 30 to 40 % and between 180 to 400 times sweeter than sugar, resulting in these two compounds being the sweetest compounds amongst the remaining glycosides (Midmore, 2002). The quality of the taste also varies among the compounds; Stv has slight bitterness and astringency after taste in addition to sweetness, while RebA has more pure sweetness than Stv, without any bitterness after taste comparatively similar to that of sucrose (Woelwer-Rieck, 2012; Tadhani et al., 2007). Hence, in most of the commercially available real stevia samples, preferably Reb A is using as sweetening component due to its exceptional stability and superior sweetness (Mauri et al., 1996). Apart from these sweetening properties, other health benefits of steviolglycosides includes antihypertensive, antihyperglycemic and anti-human rotavirus activities (Tadhani et al., 2007).

Table 2. Properties and structures of sweeteners

Sweetener	Properties of sweeteners	Structure[a]
Saccharine E-No: E-954 [a]CAS NO: 81-07-2	Formulae $C_7H_5NO_3S$ [a]Molecular weight 183.18 [a]pKa1.60 Log Kow0.910 [b]W.S (g L^{-1})4 [d]M.U.D (mg L^{-1})80[e] Sweeteness300-500	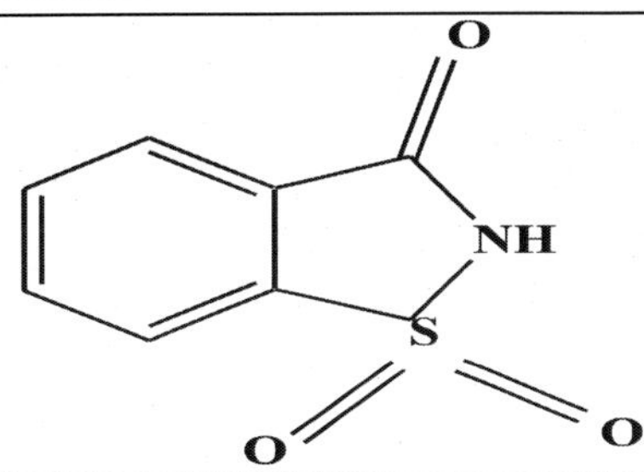
Cyclamate E-No: E-952 [a]CAS NO: 139-05-9	Formulae $C_6H_{12}NO_3SNa$ [a]Molecular weight 201.22 [a]pKa-8.66[c] Log Kow-2.63 [b]W.S (g L^{-1})1000 [d]M.U.D (mg L^{-1})250 Sweeteness30	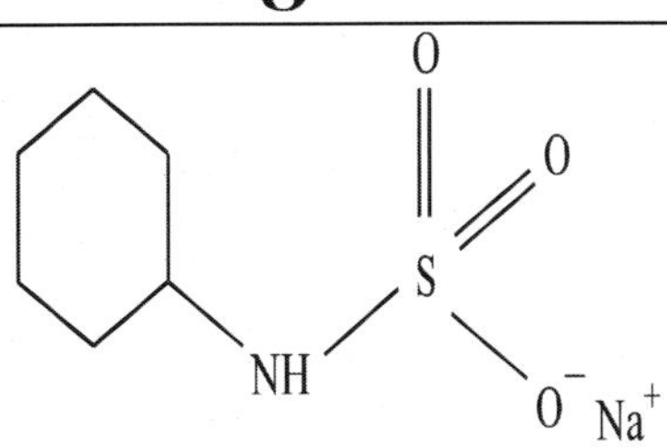

Table 2. (Continued)

Sweetener	Properties of sweeteners	Structurea
Aspartame E-No: E-951 [a]CAS NO: 22839-47-0	Formulae $C_{14}H_{18}N_2O_5$ [a]Molecular weight 294.30 [a]pKa[c]3.21, 5.0, 7.7 Log Kow0.542 [b]W.S (g L^{-1})10 [d]M.U.D (mg L^{-1})600 Sweeteness180-200	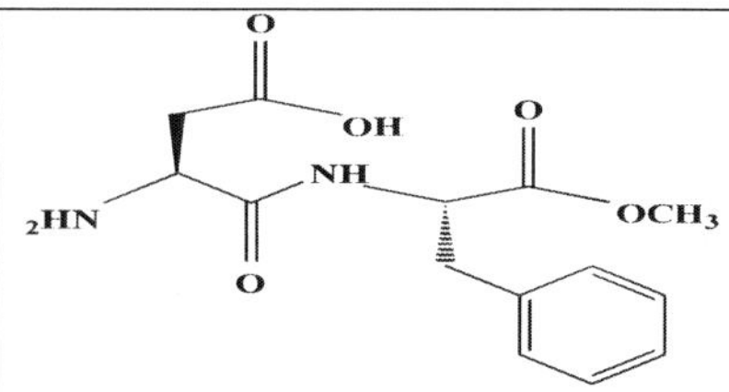
Acesulfame- K E-No: E-950 [a]CAS NO: 55589-62-3	Formulae $C_4H_4KNO_4S$ [a]Molecular weight 201.24 [a]pKa~2 Log Kow-0.31 [b]W.S (g L^{-1})270 [d]M.U.D (mg L^{-1})350 Sweeteness200	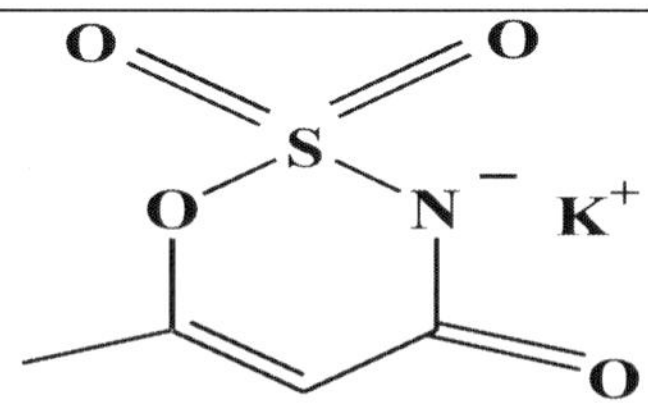
Alitame E-No: E-956 [a]CAS NO: 80863-62-3	Formulae $C_{14}H_{25}N_3O_4S$ [a]Molecular weight 331.431 [a]pKa[c]3.44, 8.23 Log Kow- [b]W.S (g L^{-1})0.18 [d]M.U.D (mg L^{-1}) Sweeteness2000	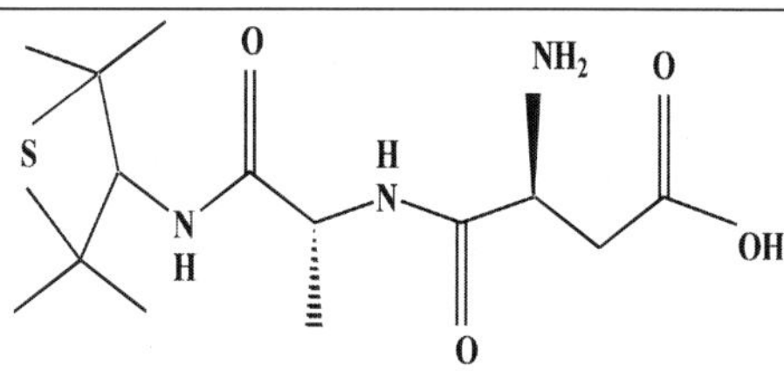
Neotame E-No: E-961 [a]CAS NO: 165450-17-9	Formulae $C_{20}H_{30}N_2O_5$ [a]Molecular weight 378.46 [a]pKa[c]3.68, 5.5, 8.1 Log Kow3.834 [b]W.S (g L^{-1})12.6 [d]M.U.D (mg L^{-1})20 Sweeteness10000	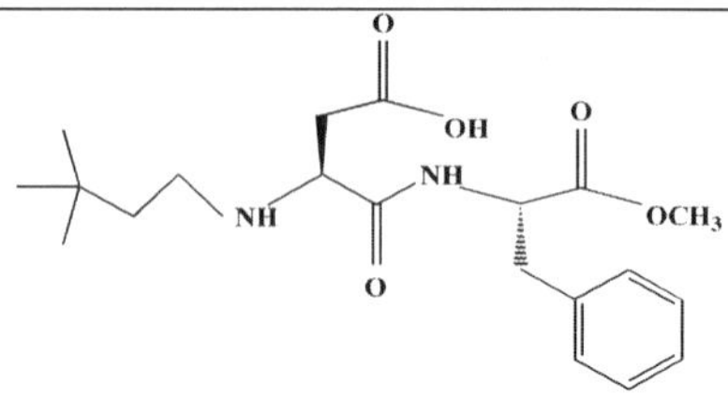
Sucralose E-No: E-955 [a]CAS NO: 56038-13-2	Formulae $C_{12}H_{19}C_{13}O_8$ [a]Molecular weight 397.63 [a]pKa11.8 Log Kow-0.49, -0.51, -1.0 [b]W.S (g L^{-1})282 [d]M.U.D (mg L^{-1})300 Sweeteness600	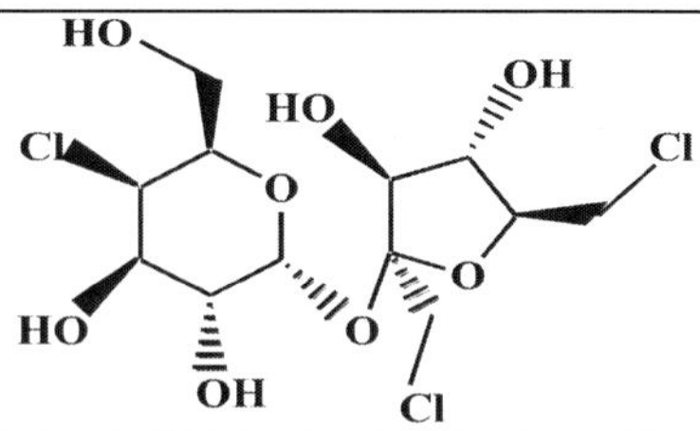
Neohesperidine dihydrochalcone E-No: E-959 [a]CAS NO: 20702-77-6	Formulae $C_{26}H_{36}O_{15}$ [a]Molecular weight 612.58 [a]pKa6.85 Log Kow0.205 [b]W.S (g L^{-1}) 0.4 -0.5 [d]M.U.D (mg L^{-1}) 30 Sweeteness1900	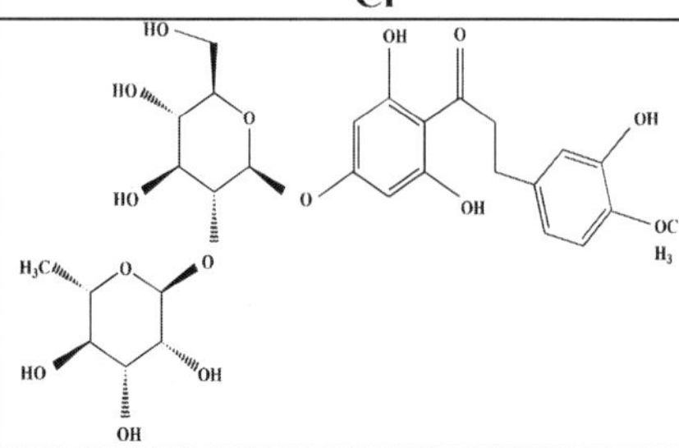

Sweetener	Properties of sweeteners	Structure[a]
Rebaudioside A E-No: E-960 [a]CAS NO:58543-16-1	Formulae $C_{44}H_{70}O_{23}$ [a]Molecular weight 967.01 [a]pKa8.0 Log Kow [b]W.S (g L^{-1}) 80 [d]M.U.D (mg L^{-1}) Sweeteness250-400	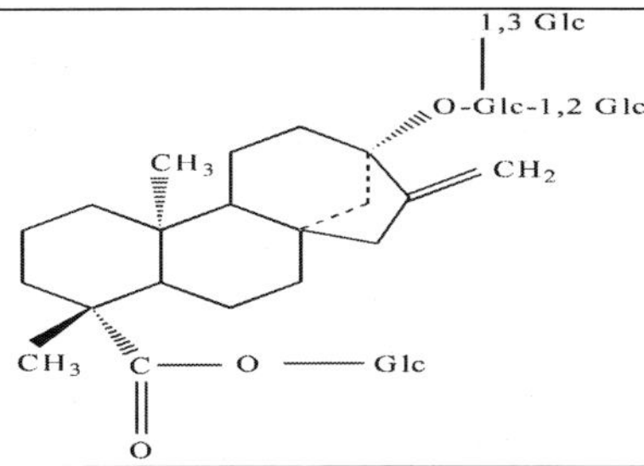
Stevioside E-No: E-960 [a]CAS NO: 57817-89-7	Formulae $C_{38}H_{60}O_{18}$ [a]Molecular weight 804.87 [a]pKa8.4 Log Kow- [b]W.S (g L^{-1}) 13 [d]M.U.D (mg L^{-1}) Sweeteness200-250	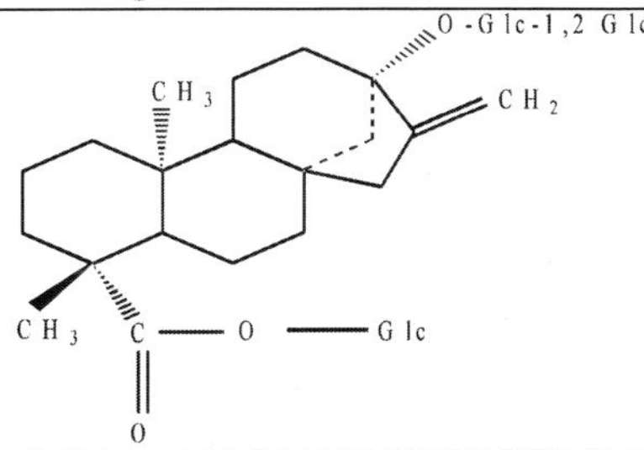

W. S = Water solubility.
M. U. D = Maximum usable dosase.
[a] Data from SciFinder Scholar Database (Calculated using Advanced Chemistry Development (ACD/Labs) Software VII. 02 (©1994–2011 ACD/Labs)):)): http://www.cas.org/products/ sfacad/10-4-2015.
[b] Experimental values, from database of physicochemical properties. Syracuse Research Corporation: http://www.syrres.com/esc/physdemo.htm10-42015.
[c] Protonated form.
[d] Maximun usable dose (MUD) authorized in EU legislation for use in non-alcoholic drinks. European Commission, Directive 94/35, 1994; European Commission, Directive 96/83, 1996;European Commission, Directive 2003/115, 2003; European Commission, Directive 2006/52, 2006 and European Commission, Directive 2009/163, 2009).
[e] 'Gaseosa': non-alcoholic water based drink with added carbon dioxide, sweeteners and flavourings, 100 mg L^{-1}.

On the other hand, the reported drawbacks for the impure stevia glycosides include hypotension, diuresis, natriuresis and kaliuresis (Mauri et al., 1996; Melis, 1992a, 1992b). The composition of the stevia components in the leaves is highly dependent on the nature of the soil, climate and the methods used for extraction and purification (Kuznesof, 2007).

2.3.2. Mogrosides (Triterpene Glycosides)

Mogrosides, which are cucurbitane-type triterpene glycosides, extracted from the fruit of Siraitiagrosvenorii (Luo-Han-Guo) and belongs to the family of Cucurbitaceae. The mogrosides are medically active compounds and used as pulmonary demulcent and emollient for treatment of dry cough, sore throat, dire thirst and constipation (Committee of National Pharmacopoeia, 2010). The reported pharmacological effects are conducive to human healthcare, including antitumor, anti-inflammation, anti-oxidative, anti-obesity and insulin-secretion stimulation (Takasaki et al., 2003; Di et al., 2011).

Besides the theurepatic nature, mogrosides has significant properties of high intensity sweetness and low calories, which make them to serve as a substitute for sugar in food, especially for obese and diabetic patients (Kasai et al., 1989). The major triterpenoids in Luo-

Han-Guo includes mogroside III, mogroside IV, siamenoside I, mogroside V and 11-oxomogroside V (Venkata et al., 2011).

2.3.3. Brazzein

Brazzein is a sweet protein, which was isolated from the African plant Pentadiplandrabrazzeana in 1989 (Van der Wel et al., 1989) and first reported in the scientific literature in 1994 by Ming (Ming et al., 1994). Nature Research Ingredients (NRI) is a company that is working to develop brazzein as a commercial product.

Brazzein is unlikely to be used as a sole sweetener or in combination of one or more other sweeteners because of its slow onset and lingering sweetness. It is up to 2000 times as sweet as sucrose on a weight basis, remarkably heat stable and water-soluble. Due to its protein nature it is expected that it would be digested just as other dietary proteins without any side effects (Walters, 2013).

3. Determination of Individual Sweeteners by Capillary Electrophoresis

3.1. Steviol Glycosides

Mauri (Mauri et al., 1996) reported the determination of diterpene glycosides from Stevia rebaudiana leaves using capillary electrophoresis. The optimum conditions for the analyses were: 20 mM sodium tetraborate buffer, pH 8.3, and 30 mM sodium dodecyl sulfate. The effect of the organic solvent (methanol) was studied on the resolution of three steviol glycosides and found that absolute amount of 1.6 nL per injected sample was optimal. Rebaudioside A and steviolbioside were isolated by semi-preparative high performance liquid chromatography (HPLC), and their structure was assessed by mass spectrometry. The separation of four steviol glycosides including stevioside, rebaudioside A, rebaudioside C and dulcoside A was reported by Liuand Li (Liuand Li, 1995) using capillary electrophoresis and high performance liquid chromatography. A comparative study was conducted between results obtained from capillary electrophoretic method and HPLC method. The individual steviol glycosides were obtained by HPLC fraction collection, and peaks in the electropherograms of the sweetener samples from Chinese refining factories were identified by comparing with those of individual steviol glycosides.

A simple subcritical fluid extraction (SubFE) method was developed for the extraction of four steviol glycosides including stevioside, rebaudioside A, rebaudioside C and dulcoside A by Liu (Liu et al., 1997). At optimum extraction conditions, the extraction efficiency of more than 88 % was obtained using methanol as a modifier. Further CE method was used for the analysis of stevioside among the four steviol glycosides. Recently, Bathinapatla (Bathinapatla et al., 2015) developed an EKC-CE method for the simultaneous separation and determination of Stevioside and Rebaudioside A in real stevia samples. The obtained results using 30-mM heptakis-(2,3,6-tri-o-methyl betacyclodextrin) as a separating agent, suggest that at optimum experimental conditions the detection limits of $2.017 \times 10^{-5}$and 7.386×10^{-5} M and relative standard deviations ($n = 5$) of 1.10 and 1.17 were obtained for rebaudioside A and stevioside, respectively. In addition, the molecular docking studies explained to a certain extent why the separation was successful. The calculated binding free energy results for the rebaudioside A and stevioside complexes formed with the separating agent showed that although both ligands penetrated deeply into the hydrophobic cavity of the separating agent, the presence of additional hydrogen bonding in the case of stevioside is probably responsible for its stronger binding affinity than that of rebaudioside A.

3.2. Neotame

A CZE method combined with solid phase extraction was developed for the determination of neotame in non-alcoholic beverages. The optimum separation conditions were 20 mmol L^{-1} sodium borate buffer, pH 8.0, 25 kV applied voltage, 5 s hydrodynamic injection at 30 mbar and ultraviolet detection at 191 nm. The calibration curve showed good linearity ($R^2 = 1.000$) in the range 0.5–100 μg mL^{-1}, and the limit of detection was 0.118 μg mL^{-1}. The method was successfully applied to the determination of neotame in two kinds of beverage with migration time less than 5 min, relative standard deviation (n = 3) less than 2% and recoveries ranging from 90 to 95 % (Hu et al., 2013). Recently, Bathinapatla (Bathinapatla et al., 2014) developed an electrokinetic chromatographic method for the chiral separation of neotame diastereomers (LL and DD) using heptakis(2,3,6-tri-o-methyl)betacyclodextrin as a chiral separating agent. The optimum conditions were 50mM phosphate buffer, pH 5.5, applied voltage 20 kV, cassette temperature of 30 ^{0}C, and a 4 s sample injection time. The calibration curve showed good linearity ($R^2 > 0.99$) with recoveries for both diastereomers, ranging from 95.66–99.00 % and the limits of detection for LL-neotame and DD-neotame were 0.01857 and 0.08214 mM, respectively. The developed method showed analytical precision with relative standard deviations (n = 5) of 1.20 % and 1.17 % with respect to migration time and peak area, respectively. Furthermore, thermodynamic parameters were also calculated according to the van't Hoff equations and the results coincide with the elution order of two compounds. In addition, molecular docking studies were performed to elucidate the mechanism of the separation. A large difference in the interaction energies of the corresponding diastereomers upon docking to the selector molecule elucidate the large resolution (*Rs*: 3.79) obtained experimentally. The results also revealed that both electrostatic and hydrophobic interactions played a significant role in stabilizing their inclusion complexes and consequently supported the elution order based on their differential stabilities.

3.3. Sucralose

An indirect UV-CZE method was developed for the separation and determination of high intensity sweetener sucralose (1,6-dichloro-1,6-dideoxy-β-D-fructofuranosyl-4-chloro-4-deoxy- α–Dgalactopyranose) using 3,5-dinitrobenzoic acid as a buffer at pH12.1. The method allowed determination of sucralose in low-calorie soft drinks, without any sample clean-up over a linear range of 42–1000 mg L^{-1} ($R^2 = 0.9991$). The limits of detection and determination were 28 and 42 mg L^{-1}, respectively (Stroka et al., 2003). The repeatability for a mean concentration of 100 mg L^{-1} was 4.2% for the signal area and 3.6% for the migration time, which is superior to HPLC methods described in the literature for determination of sucralose in beverages. Further, an indirect UV-CZE method was optimised chemo metrically for the qualification and quantification of sucralose in various food materials was reported by McCourt (McCourt et al., 2005). The optimum experimental conditions were 3,5-dinitrobenzoic acid (3 mM)/sodium hydroxide (20 mM) as a BGE at pH 12.1, a potential of 0.11 kV cm^{-1} and a temperature of 22 °C with detection wavelength 238 nm. The authors found that during the optimisation process two principal factors, capillary temperature and the electric field strength were affecting the resolution of sucralose. At the optimum experimental conditions, the detection limit of sucralose was > 30 mg kg^{-1}, with a linearity range of 50–500 mg kg^{-1} found in carbonated beverages, yoghurts and hard-boiled candy.

3.4. Aspartame

A CZE method was developed for the determination of aspartame in different food products at the optimum experimental conditions: 30 mM phosphate (phosphoric acid) and 19 mM Tris, pH 2.14, 30 kV applied voltage, detection at 211 nm and injection for 3 s at 12.5 cm Hg vacuum. A linear calibration curve was established using the concentrations ranging from 25-150 μg mL^{-1} and used for quantitative determinations of aspartame in typical food and beverage products. Six commercial samples are analysed and one diet cola with a known aspartame concentration gives an R.S.D. of 2.6 % from the manufacturer's value which is better than the HPLC determination (R.S.D = 7.0 %) (Pesek and Matyska, 1997). Additionally, capillary electrochromatography (CEC) method also tested for the analysis of aspartame using modified capillary by attachment of a diol moiety, 7-octene-l,2-diol. But the main problem with this method was quantitative determination of aspartame because peak area determinations were not reproducible enough to construct a good linear calibration curve in the same concentration rangeused for the CE experiments (Pesek and Matyska, 1997).

A comparative study was accomplished between HPLC and CE methods for the determination of aspartame (α-L-aspartyl-L-phenylalaninemethyl ester) LL-α-APM and several decomposition products namely LL-β-aspartame (LL-β-APM), L-α-aspartyl-L-phenylalanine (α-AP), L-β-aspartyl-L-phenylalanine (β-AP)and diketopiperazine (DKP) by Aboul-Enein and Bakr (Aboul-Enein and Bakr, 1997). The optimum conditions for CZE method were, 1:1 ratio of 25mM phosphate/25mM Borate Buffers, pH.9.0, applied voltage 15 kV and injection mode hydrostatic for 20 s. A linear regression analysis was carried out using the HPLC method over the range of 5-100 μg mL^{-1} for all the compounds while a higher linear range of 250-4000 μg mL^{-1} was obtained by CZE method. The separation efficiency (*N*) of all compounds was higher in CZE method but limit of detections were lower than HPLC method.

A simple and sensitive capillary zone electrophoresis (CZE) method was developed for the simultaneous determination of aspartame and strontium ranelate (antiosteoporetic drug) in pharmaceutical formulation for the treatment of postmenopausal osteoporosis (Carvalho et al., 2014). The optimum separation conditions were: borate buffer 50 mmol L^{-1} at pH 9.4 (BGE), applied potential of 30 kV, temperature set to 35 ^{0}C and hydrodynamic injection time of 10 s at a pressure of 50 mbar and detection wavelengths for ranelate and aspartame were 235 and 198 nm, respectively. The separation was carried out into a fused-silica capillary column (55 cm total length × 75 μm ID) and took less than 8 min. For both analytes, the method showed linear range from 1 to 40 μg mL^{-1}, with satisfactory detectability (limits of detection of 0.3

and 0.2 μgmL^{-1} for aspartame and ranelate, respectively). In addition, acceptable accuracy, good repeatability and intermediate precision (RSD = 2.6%) were obtained. The feasibility of the method was verified with recovery tests of analytes in the pharmaceutical sample. Recoveries varied from 85 ± 5% to 111 ± 2%, indicating the usefulness and effectiveness of the proposed method.

3.5. Cyclamate

Determination of cyclamate in low joule cordials and other low joule foods by capillary zone electrophoresis (CZE) with indirect ultraviolet (UV) detection at 254 nm was reported by (Thompson et al., 1995). The separation was performed using uncoated fused-silica capillary column with an electrolyte consisting of 1 mM hexadecyltrimethylammonium hydroxide, 10 mM sodiumbenzoate and α-hydroxyisobutyric acid as an internal standard. Cyclamate and sorbate are well separated from the other components in the foods in less than 5 min at the optimum separation conditions. The levels of cyclamate determined by CZE were in good agreement with those determined by the Association of Analytical Communities (AOAC) gravimetric method. A method for the determination of cyclamate in food was developed using solid-phase extraction (SPE) and capillary electrophoresis (CE) with indirect ultraviolet (UV) detection. Separation was performed on a fused-silica capillary using 1 mmol L^{-1} hexadecyltrimethylammonium bromide and 10 mmol L^{-1} potassium sorbate as a running buffer. Detection and reference wavelengths of cyclamate determined with a UV detector were 300 and 254 nm, respectively. The calibration curves for cyclamate showed good linearity in the range of 2–1000 μg mL^{-1} and the limits of detection in beverage, fruit in syrup, jam, pickles and confectionary are sample dependent and ranged from 5–10 μg g^{-1}. The recovery of cyclamate added at a level of 200 μg g^{-1} to various kinds of foods was 93.3–108.3 % and the relative standard deviation was less than 4.9 % (n = 3). Cyclamate was detected in one *waume*, two pickles, and two sunflower seeds. The quantitative values determined with CE correlated to those from high-performance liquid chromatography (HPLC) (the detected values of cyclamate in a sunflower seed measured by CE and HPLC were 3.40 g kg^{-1} and 3.51 g kg^{-1}, respectively) (Horie et al., 2007).

3.6. Neohesperidindihydrochalcone (NHDC)

The quantitative analysis of neohesperidindihydrochalcone in foodstuffs by capillary zone electrophoresis has been investigated. The best separation was obtained with a running buffer of 100 mM borate, pH 8.3. The method is linear between 2.9-42 mg L^{-1}, a correlation coefficient 0.9996; limit of detection 1.75 μg L^{-1}. Intra and inter-day precisions were 1.6 % RSD (n = 10) and 2.8 % RSD (n = 10), respectively. The proposed CE method has running time of 4.8 min which is faster than the reported HPLC methods running times of 9-20 min (Ruiz et al., 2000).

4. Determination of Mixture of Sweeteners (Blends) by Capillary Electrophoresis

A capillary zone electrophoretic (CZE) method was developed for the determination of aspartame in combination with caffeine and benzoic acid in diet cola soft drinks and in artificial sweetening powders. The optimum experimental conditions were, ionic strength of sodium phosphate buffer was 0.025 at pH 11, running voltage 15 kV and the hydrostatic injection was performed for 30 s. The results obtained from the CZE method was compared with the previously developed HPLC method in terms of repeatability, reproducibility, accuracy, linearity, separation efficiency and sensitivity. Authors were found that the separation efficiency of CZE was 65-110 times higher than that of HPLC; on the other hand, 10-20 times lower detection limits were obtained in HPLC (Jimidar et al., 1993). A micellar electrokinetic chromatography method was developed for the simultaneous analysis of artificial sweeteners and food additives. The mixture, comprising saccharin, aspartame, acesulfame K and propyl gallate, octyl gallate, dodecyl gallate, butylated hydroxyanisole, butylated hydroxytoluene, tertiary butylhydroquinone, *p*-hydroxybenzoic acid methyl ester, *p*-hydroxybenzoic acid ethyl ester, benzoic acid, sorbic acid. The separation was not resolved using single surfactant micellar systems consisting of sodium dodecyl sulfate (SDS), sodium cholate (SC) or sodium deoxycholate (SDC). The separation of these additives using mixed micellar systems, involving SDS/SC, SDS/SDC and SC/SDC, was investigated. Organic solvents were added to the mixed micellar phases to optimise the separation. The mixture was successfully separated using a 20 mM borate buffer with 35 mM SC, 15 mM SDS and 10 % methanol added at pH 9.3. Under the optimum separation conditions recoveries 100.86 % with RSD of 3.3 % was achieved for aspartame in jam samples (Boyce, 1999).

A micellar electrokinetic chromatography method was developed for the simultaneous determination of artificial sweetener (aspartame, saccharin, acesulfame K), preservatives (caffeine, sorbic acid, benzoic acid) and colours (brilliant blue FCF, green S, sunset yellow FCF, quinoline yellow, carmoisine, ponceau 4R, black PN) in carbonated soft drinks. The running buffer consists of 20 mM carbonate buffer with 62 mM sodium dodecyl sulfate (SDS) as the micellar phase at pH 9.5 and wavelength used 200 nm for sensitive determination. Under the optimum experimental conditions, in the presence of SDS a fair resolution between all additives was successfully achieved within a 15-min run-time in soft drinks. When applied to retail soft drink samples, this method allowed the reliable determination of additives with a limit of quantification of 0.01 mg mL^{-1} (Frazier et al., 2000). A micellar electrokinetic capillary method for the simultaneous determination of the sweeteners dulcin, aspartame, saccharin, and acesulfame K and the preservatives sorbic acid; benzoic acid; sodium dehydroacetate; and methyl, ethyl, propyl, isopropyl, butyl, and isobutyl-*p*-hydroxybenzoate in preserved fruits is developed (Lin et al., 2000). These additives are ion-paired and extracted using sonication followed by solid-phase extraction from the sample. Separation is achieved using a 57-cm fused-silica capillary with a buffer comprised of 0.05 M sodium deoxycholate, 0.02M borate-phosphate buffer (pH 8.6), and 5 % acetonitrile, and the wavelength for detection is 214 nm. The average recovery rate for all sweeteners and preservatives is approximately 90 % with good reproducibility, and the detection limits range from 10 to 25 $\mu g\ g^{-1}$. Fifty preserved fruit samples are analysed for the content of sweeteners and preservatives. The sweeteners found in 28 samples were aspartame

(0.17–11.59 g kg^{-1}) or saccharin (0.09–5.64 g kg^{-1}). Benzoic acid (0.02–1.72 g kg^{-1}) and sorbic acid (0.27–1.15 g kg^{-1}) were found as preservatives in 29 samples.

A capillary zone electrophoresis (CZE)method was used to validate the PLS-2 model UV–visible spectrophotometry in the analysis of sodium saccharin and aspartame in commercial non-caloric sweeteners (Cantarelli et al., 2008). Calibration plots were constructed for both saccharine and aspartame standards by UV–visible spectrophotometry with partial least-squares (PLS-2) model. Salicylic acid was used as an internal standard to evaluate the adjustment of the real samples to the PLS model. The concentration of analytes in the commercial samples was evaluated using the obtained model by UV spectral data. The result from validation studies in all cases a relative error of less than 11 % between the PLS-2 and the CZE methods. A new method for the rapid separation and sensitive determination of sulfanilamide artificial sweeteners, including saccharin sodium, acesulfame potassium and sodium cyclamate, by capillary electrophoresis with conductivity detection was developed (Jiang et al., 2009). Three analytes were well separated within 11 min in a fused-silica capillary under the optimal experimental conditions: running buffer: 15 mmol L^{-1} Tris-10 mmol L^{-1} H_3BO_3-0.2 mmol L^{-1} EDTA, electro-osmotic flow(EOF) inhibitor:0.2% tetraethylenepentamine, separation voltage:15 kV, electrokinetic injection:10 kV×10 s. The linear response ranges were 0.8-120,1.1-120,1.5-120 μmol L^{-1} with the LODs of 0.3,0.4,0.6 μmol L^{-1} for saccharin sodium, acesulfame potassium and sodium cyclamate, respectively. The relative standard deviations for the intra-and inter-day precisions were below 4.0%.

Capillary electrophoresis (CE) with capacitively coupled contactless conductivity detection (CE-C^4D) was used for the simultaneous determination of aspartame, cyclamate, saccharin and acesulfameK (Bergamo et al., 2011). A complete separation of all the analytes were attained less than 6 min under the optimum experimental conditions: 100 mmol L^{-1} TRIS and 10 mmol L^{-1} L-histidine (His)as BGE, Separation voltage 30 kV; gravity injection for 30 s at a height of 100 mm; silica capillary with 75 μm inner diameter and 70 cm length. C^4D operated at 450 kHz and 5.0 V peak amplitude. The limits of detection (LOD) were 4.2, 2.5, 1.5, 1.4 mg L^{-1} and quantification (LOQs) were 14.1, 8.2, 4.9, 4.7 mg L^{-1} for aspartame, cyclamate, saccharine and acesulfame K, respectively. The obtained detection limits were better than those obtained by CE with photometric detection. Recoveries ranging from 94% to 108 % were obtained for samples spiked with standard solutions of the sweeteners. The relative standard deviation (RSD) for the analysis of the samples with the CE-C4D method varied in the range of 1.5-6.5 %.

CE-C^4D with hydrodynamic pumping was developed for the determination of common sweeteners aspartame, cyclamate, saccharin and acesulfame K (Stojkovic et al., 2013). In order to obtain the best compromise between separation efficiency and analysis time hydrodynamic pumping was imposed during the electrophoresis run employing a sequential injection manifold based on a syringe pump. The analyses were carried out in an aqueous running buffer consisting of 150 mM 2-(cyclohexylamino)ethanesulfonic acid and 400 mM tris (hydroxymethyl) aminomethane at pH 9.1. The use of hydrodynamic pumping allowed easy optimization, either for fast separations (separation time of 190 s) or low detection limits (6.5 mol L^{-1}, 5.0 mol L^{-1}, 4.0 mol L^{-1} and 3.8 mol L^{-1} for aspartame, cyclamate, saccharin and acesulfame K respectively). The conditions for fast separations not only led to higher limits of detection but also to a narrower dynamic range. However, the settings can be changed readily between separations if needed. The four compounds were determined successfully in food samples.

References

Aboul-Enein, H. Y., and S. A. Bakr. Comparative Study of the Separation and Determination of Aspartame and Its Decomposition Products in Bulk Material and Diet Soft Drinks by Hplc and Ce. *Journal of Liquid Chromatography and Related Technologies,* 20 (1997)1437-1444.

Alghamdi, A. H., A. F. Alghamdi, and A. A. Alwarthan. Determination of content levels of some food additives in beverages consumed in riyadh city. *Journal of the King Saud University,* 18 (2005) 99-109.

Amin, E. K., B. Larijani, and A. E. Habibi, Neohesperidindihydrochalcone: Presentation of a small molecule activator of mammalian alpha-amylase as an allosteric effector. *FEBS Letters,* 587 (2013) 652–658.

Bahndorf, D., and U. Kienle. World market of sugar and sweeteners. International Association for Stevia Research, Hirschstraße 12, D- 70771 Leinfelden-Echterdingen, April 2004.

Bathinapatla, A., S. Kanchi, P. Singh, M. I. Sabela, M. Dovey and K. Bisetty. Analytical evaluation of steviol glycosides by capillary electrophoresis supported with molecular docking studies. *Journal of the Iranian Chemical Society,* 12 (2015)127–136.

Bathinapatla, A., S. Kanchi, P. Singh, M. I. Sabela and K. Bisetty. Determination of Neotame by High-Performance Capillary Electrophoresis Using ß-cyclodextrin as a Chiral Selector. *Analytical Letters,* 47 (2014) 2795-2812.

Bergamo, A. B., J. A. F. Silva, and D. P. Jesus. Simultaneous determination of aspartame, cyclamate, saccharin and acesulfame K in soft drinks and tabletop sweetener formulations by capillary electrophoresis with capacitively coupled contactless conductivity detection. *Food Chemistry,* 124 (2011) 1714–1717.

Bird, R. B., W. E. Stewart, and E. N. Lightfoot. Transport Phenomena. John Wiley & Sons Inc., 2nd edition, 2001.

Bopp, B. A., R. C. Sonders, and J. W. Kesterson. Toxicological aspects of cyclamate and cyclohexylamine. *Critical Reviews of Toxicology,* 16 (1986) 213–306.

Boyce, M. C., Simultaneous determination of antioxidants, preservatives andsweeteners permitted as additives in food by mixed micellar electrokinetic chromatography. *Journal of Chromatography A,* 847 (1999) 369–375.

Calza, P., V. A. Sakkas, C. Medana, A. D. Vlachou, F. Dal Bello, and T. A. Albanis. Chromometricassesment and investigation of mechanism involved in photo-Fenton and TiO_2 photocatalytic degradation of the artificial sweetener sucralose in aqueous media. *Applied catalysis B: Environmental,* 129 (2013) 71-79.

Carvalho, R. C., A. D. P. Netto, and F. F. C. Marques. Simultaneous determination of strontium ranelate and aspartame inpharmaceutical formulation for the treatment of postmenopausal osteoporosis by capillary zone electrophoresis. *Microchemical Journal,* 117 (2014) 214–219.

Cantarelli, M. A., R. G. Pellerano, E. J. Marchevsky and J. M. Camina. Simultaneous Determination of Saccharin and Aspartame in Commercial Noncaloric Sweeteners Using the PLS-2 Multivariate Calibration Method and Validation by Capillary Electrophoresis. *Journal of Agricultural Food Chemistry,* 56 (2008) 9345–9349.

Chattopadhyay, S., U. Raychaudhuri, and R. Chakraborty. Artificial sweeteners–a review. *Journal of Food Science and Technology,* (2011) DOI 10.1007/s13197-011-0571-1.

Claub, K., and H. Jensen. Cycloadditionen von halogensulfonylisocyanatenan acetylene. *Tetrahedron Letters,* 11 (1970) 119–122.

Committee of National Pharmacopoeia, Pharmacopoeia of P.R. China, Chemical Industry Press, Beijing, 197 (2010).

Di, R., M. T. Huang, and C. T. Ho. Anti-inflammatory activities of mogrosides from Momordicagrosvenori in murine macrophages and a murine ear edema model. *Journal of Agricultural and Food Chemistry,* 13 (2011) 7474–7481.

Duffy, V. B., and G. H. Anderson. Position of the American dietetic association: Use of nutritive and non-nutritive sweeteners. *Journal of the American Dietetic Association*, 98 (5) (1998) 580-587.

EC 2003 European parliament and council Directive 2003/115/EC: Amendment to European parliament and council Directive 94/35/EC on sweeteners for use in foodstuffs. Off JL114:15-17, L237:3-12.

Ellis, J. W., Overview of Sweeteners. *Journal of Chemical Education,* 72 (8) (1995) 671–675.

Ewing, A. G., R. A. Wallingford, and T. M. Olefirowicz. Capillary Electrophoresis. *Analytical Chemistry,* 61 (1989) 292A–303A.

FDA statement on European aspartame study. CFSAN/ Office of Food Additive Safety. Atwww.fda.gov/Food/FoodingredientsPackaging/ FoodAdditives/ucm208580.

Fisher, R. S., Aspartame, neurotoxicity and seizures: A review. *Journal of Epilepsy,* 2(2) (1989) 55-64.

Fitch, C., and K. S. Keim. Position of the academy of nutrition and dietetics: Use of nutritive and nonnutritive sweeteners. *Journal of the Academic Nutrition Diet*, 112 (2012) 739-758.

Frazier, R. A., E. L. Inns, N. Dossi, J. M. Ames, and H. E. Nursten. Development of a capillary electrophoresis method for the simultaneous analysis of artificial sweeteners, preservatives and colours in soft drinks. *Journal of Chromatography A,* 876 (2000) 213–220.

Geuns, J. M. C., Stevia and Steviol Glycosides (Euprint Ed., Heverlee, 2010).

Godel, H., and R. Weinberger. Chiral Recognition by Capillary Electrophoresis with Cyclodextrins. Application Note Hewlett-Packard Company, Publication Number 12-5963-5502E, 1995.

He, Y. C., S. Fang, and X. J. Xu. Simultanious determination of acesulfame-K, aspartame and stevioside in sweetener blends by ultraviolet spectroscopy with variable selection by sipls algorithm. *Macedonian Journal of Chemistry and Chemical Engineering,* 31 (2012) 17–28.

Heiger, D., High performance capillary electrophoresis: An introduction, Agilent Technologies, Publication Number 5968-9963E, (2000).

Hellsten, I., Implicit media frames: Automated analysis of public debate on artificialsweeteners. *Public Understanding of Science,*19 (2010) 590-608.

Henk, H. L., and P. R. Gerard. High performance capillary electrophoresis-agilent technologies. Publication number 5990-3777EN, (2010).

Horie, M., F. Ishikawa, M. Oishi, T. Shindo, A. Yasui, and K. Ito. Rapid determination of cyclamate in foods by solid-phase extraction and capillary electrophoresis. *Journal of Chromatography A,* 1154 (2007) 423–428.

Hu, F., L. Xu, F. Luan, H. Liu, and Y. Gao. Determination of neotame in non-alcoholic beverage by capillary zone electrophoresis. *Journal of the Science of Food and Agriculture*, 93 (2013) 3334–3338.

Huang, X., M. J. Gordon, and R. N. Zare. Bias in quantitative capillary zone electrophoresis caused by electrokinetic sample injection. *Analytical Chemistry,* 60 (1988) 375–377.

Introduction to Capillary electrophoresis, Beckman Coulter, Hand book. Available at:https://www.google.co.in/?gws_rd=ssl#q=Introduction+to+Capillary+electrophoresis %2C+Beckman+Coulter%2C+Hand+book_23-March2015.

Jenner, M. R., and A. Smithson. Physicochemical properties of the sweetener sucralose. *Journal of Food Science*, 54 (1989) 1646-1649.

Jiang, Y. X., R. X. Wei, G. Z. Yang, X. D. Luo, X. Y. Li, T. Y. Xie. Separation and Determination of Sulfanilamide Artificial Sweeteners in Beverages by Capillary Electrophoresis with Conductivity Detection. *Journal of Instrumental analysis*, 7, 2009.

Jimidar, M., and T. P. Hamoir. Comparison of capillary zone electrophoresis with high-performance liquid chromatography for the determination of additives in foodstuffs. *Journal of Chromatography A*, Volume 636, Issue 1, 23 April 1993, 179–186.

Kasai, R., R. L. Nie, K. Nashi, and K. Ohtani. Sweet cucurbitane glycosides from fruits of Siraitiasiamensis (chi-zi-luo-han-guo), a Chinese folk medicine. *Agricultural and Biological Chemistry,* 53 (1989) 3347–3349.

Kok, W., Capillary electrophoresis: instrumentation and operation. *Chromatographia,* 4 (2000) Online ISBN 978-3-322-83133-0.

Knight, I., The development and applications of sucralose, a new high-intensity sweetener. *Canadian Journal of Physiology and Pharmacology,* 72 (1994) 435-439.

Kroger, M., K. Meister, and R. Kava. Low-calorie sweeteners and other sugar Substitutes: A review of the safety issues. *Comprehensive Reviews in Food science and Food safety,* 5 (2006) 35-47.

Kuznesof, P. M., The 68th JECFA Steviolglycosides (CTA) (2007) pp. 1–8.

Larry, W., and D. C. Greenly. A doctor's guide to sweeteners. *Clinical Tearout,* 2 (2) 2003.

Lauer, H. H., and D. McManigill. Capillary zone electrophoresis of proteins in untreated fused silica tubing. *Analytical Chemistry,* 58 (1986) 166–170.

Learn about cancer. Aspartame. Last Revised: 05/28/2014. Available at http://www.cancer.org/Cancer/CancerCauses/OtherCarcinogens/AtHome/aspartame.

Lin, Y. H., S. S. Chou, F. Sheu, and Y. T. Shyu. Simultaneous Determination of Sweeteners and Preservatives in Preserved Fruits by Micellar Electrokinetic Capillary Chromatography. *Journal of Chromatographic Science,* Vol. 38, August 2000.

Lina, B. A. R., H. C. Dreef-van der Meulen, and D. C. Leegwater. Subchronic (13-week) oral toxicity of neohesperidinedihydrochalcone in rats. *Food and Chemical Toxicology,* 28 (7) (1990) 507-513.

Liu, J., C. P. Ong, and S. F. Y. Li. Subcritical fluid extraction of stevia sweeteners from stevia rebaudiana. *Journal of Chromatographic Science,* 35 (1997) 446-450.

Lukacs, K. D., and J. W. Jorgenson. Capillary zone electrophoresis: Effect of physical parameters on separation efficiency and quantitation. *Journal of High Resolution Chromatography,* 8 (1985) 407–411.

Lux, J. A., H. F. Yin, and G. Schomburg. A simple method for the production of gel-filled capillaries for capillary gel electrophoresis. *Journal of High Resolution Chromatography,* 13, (1990) 436-437.

Mauri, P., G. Catalano, C. Gardana, and P. Pietta. Analysis of Stevia glycosides by capillary electrophoresis. *Electrophoresis,* 17 (1996) 367-371.

Mayhew, D. A., C. P. Comer, and W. W. Stargel. Food consumption and body weight changes with neotame, a new sweetener with intense taste: differentiating effects of palatability from toxicity in dietary safety studies. *Regulation of Toxicology and Pharmacology,* 38 (2003) 124–143.

McCourt, J., J. Stroka, and E. Anklam. Experimental design-based development and single laboratory validation of a capillary zone electrophoresis method for the determination of the artificial sweetener sucralose in food matrices. *Analytical Bioanalytiocal Chemistry,* 382 (2005) 1269-1278.

McLaughlin, G., R. Palmieri, and K. Anderson. Benefits of automation in the separation of biomolecules by high performance capillary electrophoresis, Techniques in Protein Chemistry II, Academic Press, (1991) 3-19.

McNeil, Strong clinical database that supports safety of splenda sweetener products. McNeil Nutritionals, LLC. Retrieved 2009-12-30 (2009). http://www.splenda.com/

Melis, M. S., Renal excretion of stevioside in rats. *Journal of Natural Products,* 55 (1992) 688-690.

Melis, M. S., Stevioside effect on renal function of normal and hypertensive rats. *Journal of Ethnopharmacology,* 36 (1992) 213-217.

Midmore, D. J., and A. H. Rank. A new rural industry–stevia–to replace imported chemical sweeteners. A report for Rural Industries Research and Development Corporation RIRDC Web, Publication No. W02/022 RIRDC Project No. UCQ-16A (2002).

Ming, D., and H. G. Brazzein. A new high-potency thermostable sweet protein from Pentadiplandrabrazzeana B. *FEBS Letters,* 355 (1994) 106-108.

Motwani, H. V., S. Qiu, B. T. Golding, H. Kylin, and M. Tornqvist. Co(I)balamin reacts with sucralose to afford an alkylcobalamin: relevance to in vivo cobalamin and sucralose interaction. *Food and Chemical Toxicology,* 49 (2011) 750-757.

Nojiri, S., M. Nakazato, Y. Kasuya, I. Takano, M. Oishi, K. Yasuda, and S. Suzuki. Determination of sucralose in foods by HPLC using pre-column derivatization. *Journal of the Food Hygenic Society of Japan,* 43 (2002) 289-294.

Pesek, J. J., and M. T. Matyska. Determination of aspartame by high-performance capillary electrophoresis. *Journal of Chromatography A,* 781 (1997) 423-428.

Prakash, I., I. Bishay, and S. Schroeder. Neotame: Synthesis, stereochemistry and sweetness. *Synthetic Communications,* 29 (1999) 4461-4467.

Pretorius, V., B. J. Hopkins and J. D. Schieke. Electro-osmosis: A new concept for high-speed liquid chromatography. *Journal of Chromatography A,* 99 (1974) 23-30.

Regulations relating to the use of sweeteners in foodstuffs. Foodstuffs, cosmetics and disinfectants act (1972), http://www. doh.gov.za/docs/foodcontrol/additives/2012/regr 733.

Roberts, A., A. G. Renwick, J. Sims, and D. J. Snodin. Sucralose metabolism and pharmacokinetics in man. *Food and Chemical Toxicalogy,* 38 (2000) 31-41.

Rodero, A. B., L. D. Rodero, and R. Azoubel. Toxicity of sucralose in humans: A review. *International Journal of Morphology,* 27 (2009) 239-244.

Ruiz, T. P., C. M. Lozano, V. Tomfis, A. Sanz, and E. Bravo. Quantitative Assay for NeohesperidinDihydrochalconein Foodstuffs by Capillary Electrophoresis. *Chromatographia*, 51 (2000).

Schiffman, S. S., and K. I. Rother. Sucralose, a synthetic organochlorine sweetener: Overview of biological issues. *Journal of Toxicology and Environmental Health B*, 16 (2013) 399–451.

Scientific opinion of the Panel on Food Additives, Flavourings, Processing Aids and Materials in Contact with Food on a request from European Commission on neotame as a sweetener and flavor enhancer. *EFSAJ*, 581 (2007)1–3.

Shankar, P., S. Ahuja, and K. Sriram. Non-nutritive sweeteners: Review and update. *Nutrition,* 29 (2013) 1293–1299.

Stojkovic, M., T. D. Mai, P. C. Hauser. Determination of artificial sweeteners by capillary electrophoresis with contactless conductivity detection optimized by hydrodynamic pumping. *Analytica Chimica Acta,* 787 (2013) 254–259.

Stroka, J., N. Dossi, and E. Anklam. Determination of the artificial sweetener sucralose by capillary electrophoresis. *Food Additives and Contaminents,* 20 (2003) 524-527.

Takasaki, M., T. Konoshima, Y. Murata, M. Sugiura, H. Nishino, H. Tokuda, K. Matsumoto, R. Kasai, and K. Yamasak. Anticarcinogenic activity of natural sweeteners, cucurbitane glycosides, from Momordicagrosvenori. *Cancer Letters,* 198 (2003) 37–42.

Tadhani, M. B., V. H. Patel, and R. Subhash. In vitro antioxidant activities of Stevia rebaudiana leaves and callus. *Journal of Food Composition and Analysis,* 20 (2007) 323-329.

Terabe, S., K. Otsuka, K. Ichikawa, A. Tsuchiya, and T. Ando. Electrokinetic separations with micellar solutions and open-tubular capillaries. *Analytical Chemistry,* 56 (1984) 111-116.

Terabe, S., Electrokinetic chromatography: an interface between electrophoresis and chromatography. *Trends in Analytical Chemistry,* 8 (1989) 129–134.

Thompson, C. O., V. C. Trenerry, and B. Kemmery. Determination of cyclamate in low joule foods by capillary zone electrophoresis with indirect ultraviolet detection. *Journal of Chromatography A,* 704 (1995) 203-210.

Tomasik, P., Chemical and functional properties of food saccharides. CRC Press LLC, 397, 2004.

United States Food and Drug Administration. 2013. Department of Health and Human Services, Code of Federal Regulations Title 21 (21CFR172.829) Vol. 3, Revised as of April,2013.

Van der Wel, H., G. Larson, A. Hladik, C. M. Hladik, G. Hellekant, and D. Glaser. Isolation and characterisation of Pentadin, the sweet principle of Pentadiplandra-Brazzeana Baillon. *Chemical Senses,* 14 (1989) 75-79.

Venkata, S. P. C., and P. Indra. Cucurbitane glycosides from Siraitiagrosvenorii. *Journal of Carbohydrate Chemistry,* 30 (2011) 16–26.

Vianna-Soares, C. D., and J. L. S. Martins. Saccharin analysis in pharmaceutical and cosmetic preparations by derivative ultraviolet spectrophotometry. *Brazilian Journal of Pharmaceutical Sciences,* 38 (2002) 471-478.

Vindevogel, J., and P. Sandra, Introduction to Micellar Electrokinetic Chromatography. Hüthig Buch Verlag GmbH: Heidelberg, 1992, 100.

Walters, Sweeteners: Discovery, Molecular Design and Chemoreception, 2013.

Woelwer-Rieck, U., The leaves of Stevia rebaudiana (Bertoni), their constituents and the analyses thereof: A review. *Journal of Agricultural and Food Chemistry,* 60 (2012) 886-895.

Xu, Y., Tutorial: Capillary electrophoresis. *The Chemical Educator,* 1 (1996) 1-14.

Zimmermann, B. F., U. Woelwer-Rieck, and M. Papagiannopoulos. Separation of steviol glycosides by hydrophilic liquid interaction chromatography. *Food Analytical Methods,* 5 (2011) 266-271.

In: Capillary Electrophoresis (CE)
Editor: Christian Reed

ISBN: 978-1-63483-122-2

Chapter 8

CAPILLARY ZONE ELECTROPHORESIS WITH LASER INDUCED FLUORESCENCE (CZE-LIFD): A METHOD TO EXPLORE THE PHYSIOLOGICAL AND PATHOLOGICAL ROLES OF MONO AND POLYAMINES

Luis R. Betancourt [*]***, Pedro V. Rada, Maria J. Gallardo, Mike T. Contreras and Luis F. Hernandez***
Laboratory of Behavioral Physiology, School of Medicine,
University of Los Andes, Merida, Venezuela

ABSTRACT

Monoamines are chemicals containing an amine group and they possess enormous biological importance. They include most of the amino acids, the catecholamines, the indoleamines among the most important molecules. Polyamines are aliphatic chains containing multiple amine groups that generally originate from the amino acid arginine. They include citrulline, agmatine, ornithine, putrescine, spermine, spermidine and cadaverine. In general, they are concentrated in the micromolar to picomolar range. They participate in proliferation, differentiation, development, and cell signaling. Due to the lack of highly sensitive analytical techniques, most of the studies on mono and polyamines have been confined to tissue homogenates and very few studies have been carried out in extracellular fluids such as plasma, cerebral spinal fluid (CSF), or microdialysates of several tissues. The development of analytical techniques based on Capillary Zone Electrophoresis and Laser Induced Fluorescence Detection (CZE-LIFD) has been crucial to opening fields of studies in the aforementioned extracellular fluids and the physiological, as well as pathological role of polyamines. In the last two decades we have successfully applied CZE-LIFD to the study of meningitis, preeclampsia, the mechanism of memory circuits of the brain, schizophrenia, Parkinson's disease (PD) and

[*] Corresponding author: Luis Betancourt, MD., Laboratory of Behavioral Physiology, School of Medicine, University of Los Andes, Merida, Venezuela, e-mail: luisbeta@ula.ve

neuro-development. For such goals we have developed analytical techniques based on CZE-LIFD capable of detecting down to 2 nanomolar concentrations of glutamine, glutamate, arginine, agmatine, citrulline and putrescine in extracellular fluids. In CSF of meningitis-stricken children we found low glutamine levels, particularly when the etiological agent was *Haemophylus influenzae*. These levels increased to normal during the convalescence of the patient. This finding suggests that *H. influenzae* uses large amounts of glutamine probably because it lacks the first two enzymes of the Krebs cycle. In patients suffering preeclampsia low levels of arginine and high levels of agmatine in CSF and plasma were found. These results suggest that arginine might be an essential amino acid in preeclampsia patients and that it might be of therapeutic value. By means of brain microdialysis, 90 nanomolar concentration of agmatine were found in the *stratum radiatum* of the hippocampus in rats. The agmatine in the extracellular fluid of the hippocampus was nerve impulse and calcium dependent, suggesting an exocytotic origin and possible involvement in memory processes. Injecting agmatine by reverse microdialysis in the striatum it was found that extracellular dopamine increased, suggesting a role for agmatine in the control of automatic movements and a role in schizophrenia. Lately, we developed a method to measure putrescine and found that PD patients have higher levels of putrescine both in red cells and plasma from blood, providing a biological marker for PD and suggesting a role of putrescine and other polyamines in the degeneration of *substantia nigra* dopaminergic neurons, which is the hallmark of PD. Recently we found low levels of arginine and citrulline and a lack of correlation between arginine and citrulline in the plasma of preterm babies, as compared with fully developed neonates. These findings suggest that arginine and citrulline might be essential amino acids in premature babies; that they should be supplemented in their diets and that premature babies might have a disarray of the nitric oxide metabolic pathway. These findings show that CZE-LIFD is becoming a useful tool that could lead to a better understanding of the physiological and pathological roles of bioamine and to the development of therapeutic resources for several conditions.

Keywords: Biogenic amines, biomarker diagnostic, capillary zone electrophoresis, translational technologies

CE AND LIFD DEVELOPMENT

Stella Hjertén (1967) realized Capillary zone electrophoresis 48 years ago as another electrophoresis separation technique but it was not feasible at that moment. Mikkers et al. (1979) attempted to separate chemicals of a mixture by applying an electric field at the two ends of a 200 micrometer inside diameter glass tube filled with a conducting buffer. However the amounts of heat generated by Joule effect caused a band distortion (broadening) that rendered useless the technique. Nevertheless, these seminal articles presaged the upcoming era of micro-separation in fused silica tubing. The key factor to reduce band broadening was diminishing the current, which has a second power impact on the Joule heat, by increasing the resistance, which has a linear impact on the Joule heat, using capillaries of less than 75 micrometer inside diameter. This step was taken by Jorgenson and Lukacs (1981) who successfully separated the components of urine samples and detected them by a homemade UV detector. Once the band broadening due to the Joule effect was controlled the research focused in the detection techniques. The small amounts of analyte injected into the capillaries made hard to detect sub-micromolar concentrations. The detection of sub-micromolar concentrations was reached by Gassmann, Kuo and Zare (1985) who introduced a laser induced fluorescence detector. They detected high nanomolar concentrations of dansylated amino acids by focusing the 325 nm line of a Helium-cadmiun (HeCd) 5 mW laser on the window of a 75 micrometer inside diameter capillary. The volume at the window was 0.5 nanoliters initiating an unfinished race towards single molecule detectors for CZE-LIFD. Three years later Cheng and Dovichi (1988) pushed the limit of concentration detection by means of a Sheath Flow Cuvette, the 488 nm line of an argon-ion laser and Fluorescein Isothiocyanate Isomer 1 (FITC) towards the picomolar range. This article induced the adoption of the prefix Zepto and Yocto for 10^{-21} moles and 10^{-24} moles limits of mass detection and lowered by four orders of magnitude the limits of concentration detection of amino acids. A further improvement of the CZE-LIFD detector was introduced almost simultaneously by Mathies et al. (1992) and Hernandez et al. (1991) who set an epi-illumination or confocal laser induced fluorescence detector and detected low picomolar concentration of FITC derivatized arginine and 3.75 zeptomoles equivalent to 2,250 molecules in a less than 1 nanoliter volume.

With this powerful analytical technique we started to analyze amino acids in biomedical situations and proved that CZE-LIFD became a useful tool in translational medicine.

MENINGITIS

Cerebral spinal fluid of meningitis sick children was compared with CSF of age and gender matched controls. The amino acid glutamine was significantly less concentrated in the CSF of children suffering meningitis caused by *Haemophylus influenzae*, *Streptococcus pneumoniae* or *Neisseria meningitidis* (*meningococcus*). The concentration of glutamine was significantly lower in children infected with *H. influenzae* than in children infected with *S. pneumoniae* or *meningococcus*. The concentration of glutamine in children suffering viral meningitis was similar to the control group. The second day of treatment the children infected with *Streptococcus pneumoniae* or *meningococus* had normal levels of glutamine, but the children infected with *H. influenzae* still had significantly lower level of glutamine. By the tenth day the children had recovered and the glutamine levels were back to normal in all the groups. Glutamate concentrations were significantly higher in all the meningitis groups. In the bacterial meningitis group the glutamate levels were significantly higher than in the viral group. In general, these levels increased by the second day of treatment and they were still high at the 10th day of treatment. Again, the greatest change was observed in the CSF of children infected by *H. influenzae* (Tucci, 1997). On the whole these results suggested that there should be a chemical code for the CSF of meningitis sick children and that CZE-LIFD might be instrumental to determine the etiological agent of meningitis in particular patients. This piece of information might help to start antibiotherapy in a more specific way and help to prevent sequelae of this serious condition.

PREECLAMPSIA

Preeclampsia is a multisystem disease that affects 5-10% of pregnant women worldwide; the central nervous system, being one of the systems most disturbed. CZE-LIFD was used in an attempt to a) find biomarkers that will help an early and precise diagnosis of the disease and, b) explain underlying mechanisms of the disease. Levels of amino acids (arginine, GABA, glutamate, glutamine) and a polyamine (agmatine) were monitored in plasma and cerebrospinal fluid (CSF) of mild and severe preeclampsia compared to control patients. In order to measure agmatine we developed a method based on CZE-LIFD. Agmatine was derivatized with FITC in an alkaline buffer (20 mM Carbonate buffer). We prepared 1 ml of a 0.1 mg/ml solution of agmatine and mixed it with 5 microliters of a 1.25 mM solution of FITC in a 1:1 (V/V) acetone: 20 mM carbonate buffer. In this way we obtained a 6.25 micromolar solution of thiocarbamate of agmatine. Then we diluted this solution and obtained 6 concentrations ranging from 0 nanomolar to 60 nanomolar and 5 concentrations ranging from 0 nanomolar to 12 nanomolar. The points of the concentration vs. signal amplitude curve fitted a line with regression coefficients R = 0.998 and R = 0.981 respectively. The Limit of Detection (LOD) for this method was 2 nanomolar, the Limit of Mass Detection was 500 zeptomolar and the Limit of Quantitation (LOQ) was 6 nanomolar (Betancourt, 2012).

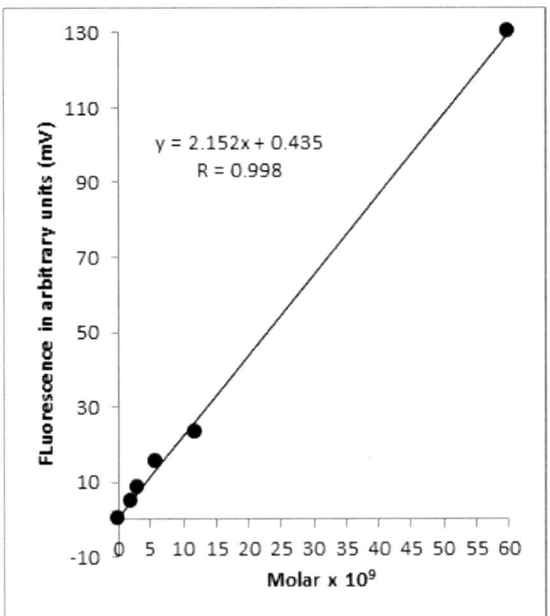

Figure 1. Agmatine dilutions ranging from 0 nanoM to 60 nanoM. The points of the concentration vs signal amplitude curve fitted a line with a regression coefficient R = 0.998.

Glutamate plasma levels were significantly and progressively increased in preeclampsia as the disease worsened, while CSF levels only increased in mild preeclampsia (Figure 3). On the contrary, arginine levels in plasma and CSF significantly decreased in mild and even more in severe preeclampsia (Figure 4). GABA levels also decreased in plasma and CSF of preeclampsia patients (Figure 5). Agmatine levels were increased only in plasma of preeclampsia patients (Figure 6). These results suggest that glutamate levels in plasma, as well as arginine levels, could be used as biomarkers on the severity of the disease. It could also explain the hyperexcitability of the nervous system observed in preeclampsia patients, probably due to an increase of the excitatory amino acid glutamate levels and a decrease of the main inhibitory amino acid GABA.

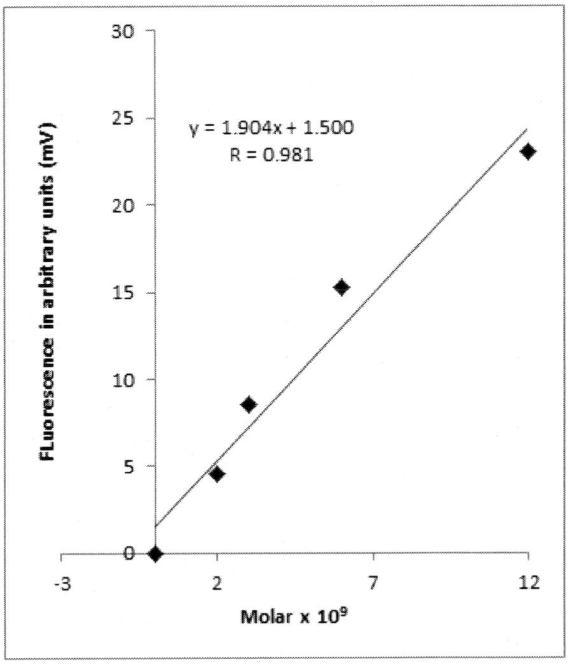

Figure 2. The Limit of Detection for this method was 2 nanoM and the Limit ofMass Detection was 500 Zeptomolar and the Limit of Quantization was 6 nanoM.

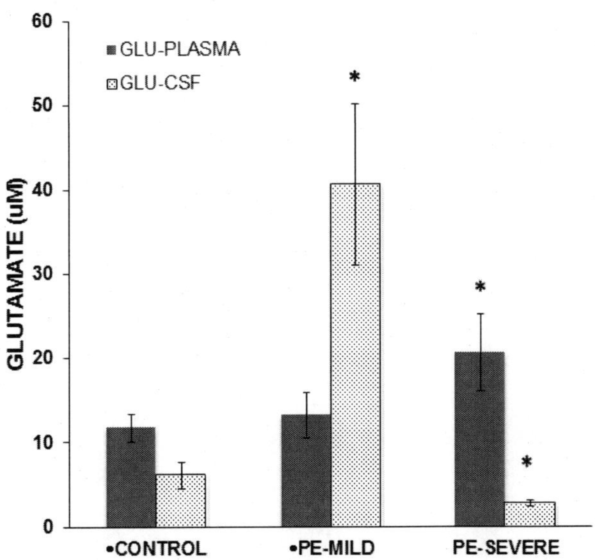

Figure 3. Plasma levels of glutamate progressively increased in mild and severe preeclampsia while a biphasic response was observed in CSF with a significant increase in mild and decrease in severe preeclampsia. Asterisk indicates $p < 0.05$ compared controls.

Nitric oxide (NO) has been found diminished in preeclampsia and such decrease is probably involved in the vasoconstriction and arterial hypertension observed in preeclampsia. Our results suggest that NO levels are low because of a decrease of its precursor, arginine. Moreover, the decarboxylation of arginine to agmatine instead of NO-citrulline could explain the significant increase in agmatine plasma levels observed in our study (Teran, 2012) as well as the low levels of NO detected in preeclampsia by others (López-Jaramillo, 2008).

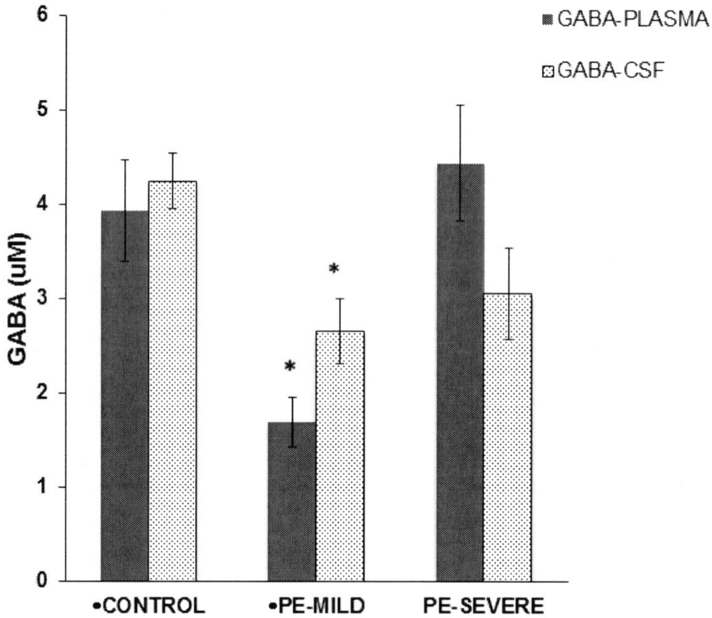

Figure 4. Plasma and CSF GABA levels significantly decreased in mild preeclampsia returning to control levels in severe patients. Asterisk indicates $p < 0.5$ compared to controls.

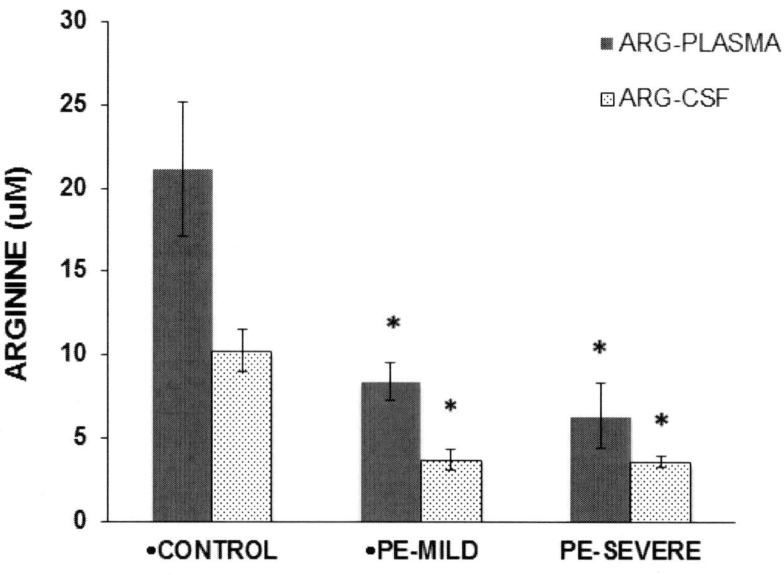

Figure 5. Arginine levels significantly decreased in mild and even more in severe preeclampsia patients both in plasma and CSF. Asterisks indicate $p < 0.05$ compared to controls.

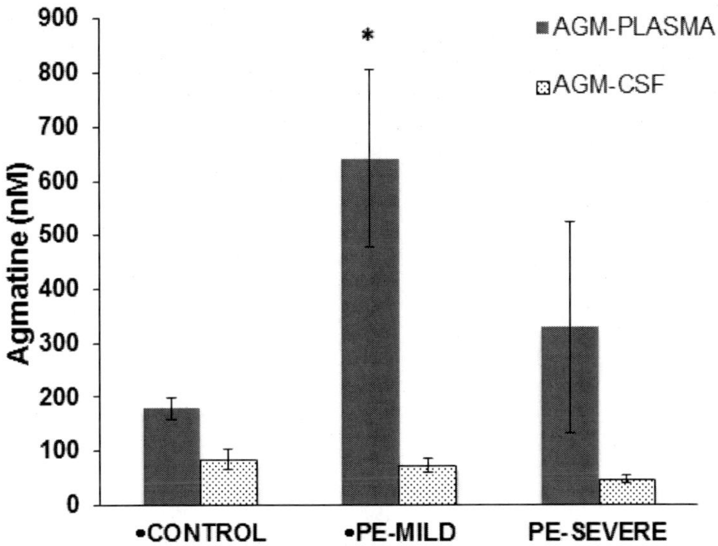

Figure 6. Agmatine levels in plasma significantly increased in mild preeclampsia with normal concentration in severe preeclampsia patients. Levels in CSF were undisturbed in mild and severe preeclampsia. Asterisk indicates $p < 0.05$ compared to controls.

Schizophrenia

The role of agmatine in central nervous system illness has been gaining support. Uzbay et al. (2010) have demonstrated that systemic injections of agmatine abolish pre-pulse inhibition (PPI) which is one of the hallmarks of schizophrenia-like behavior in experimental models of schizophrenia. Nevertheless, it is still unknown whether the actions of agmatine in the brain are physiological or pharmacological. Obviously, our method to monitor agmatine in the extracellular compartment of the brain in freely-moving animals might help to clear this issue. Brain microdialysis is an *in vivo* technique appropriate for monitoring changes of agmatine in the extracellular fluid (Hernández, 1986 and Hernández 1993). However, to the best of our knowledge, this technique has not been used to study physiological changes of polyamines in the brain. One of the reasons is that brain microdialysis requires analytical techniques for small volume samples with small masses of analytes. Therefore we decided to combine our analytical technique for agmatine determination and brain microdialysis.

In addition to agmatine, strong pharmacological evidence suggests an association between dopaminergic system disfunction and schizophrenia (Carlsson, 2004). The main antipsychotic drugs are dopamine (DA) receptor blockers and there is a clear correlation between affinity of a DA receptor blocker and its clinical efficiency to suppress schizophrenia symptoms (Seeman, 2005). Other neurotransmitter systems have been associated to schizophrenia. Specifically, it has been found that the blockade of glutamate NMDA receptors induces hallucinations and other symptoms of schizophrenia (Javitt, 1991). In addition, an association between mutations of the gene coding for neuregulin are correlated with schizophrenia too (Li, 2006). In an attempt to examine the relationship between these three chemicals (dopamine, glutamate and neuregulin) we have done experiments to figure out the way these chemicals are linked. Based upon the fact that neuregulin is a powerful stimulant of the dopamine system in the hippocampus (Kwon, 2008) and that agmatine is an NMDA blocker (Yang, 1999) we tested the effects of agmatine on DA system activity.

The perfusión of agmatine by reverse microdialysis increased extracelular dopamine by 237% in the striatum in rats ($p < .02$) (Figure 7). The three main metabolites of DA, i.e., dihydroxyphenylacetic acid (DOPAC), homovanillic acid (HVA) and 3-methoxytyramine (3-MT) increased too ($p < 0.04$, $p < 0.04$ and $p < 0.004$ respectively) (Figure 7)(Betancourt, 2013).

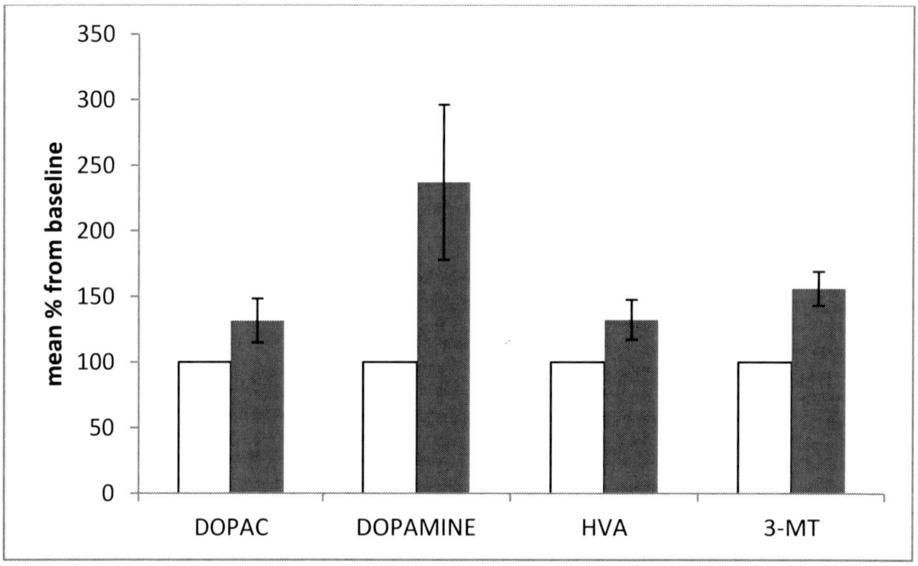

Figure 7. Agmatine injection significantly increased DA: 237% and this increase were statistically significant ($p < 0.02$).

Moreover, we found that agmatine is released exocytotically in the hippocampus, the same region where neuregulin enhances dopamine release (Betancourt, 2012). Therefore neuregulin might enhance dopamine release by releasing agmatine in local circuits of the hippocampus.

Parkinson's Disease

Polyamines in general are important modulators of cell functions, and are associated with neurodegenerative disease. Polyamines play an essential role in cell proliferation and differentiation and they are involved in many pathological conditions. Alterations in the expression and activity of the enzymes involved in polyamine metabolism as well as the actual levels of these enzymes have been reported in schizophrenia, affective disorders, anxiety, suicidal behavior and Parkinson´s disease (PD) (Fiori, 2008 and Gomes-Trolin, 2002). Although there are several techniques to measure putrescine, we thought that a technique based on CZE-LIFD might be useful. The reason is that it might be very sensitive and it will require very small sample volumes. To develop this technique we took a 2.8 micromolar solution of thiocarbamyl–putrescine and diluted it ten times in 20 mM carbonate buffer to obtain 1.4 μM, 0.7 μM, 350 nM, 175 nM, 87.5 nM, 43.7 nM, 21 nM, 10 nM and 5 nM concentrations. The standards and the samples were run in a 40 mM Sodium Dodecyl Sulphate and 20 mM Sodium Tetraborate buffer. The peak heights were measured and fit to a concentration vs. arbitrary units of fluorescence (mV) by means of regression analysis. There was a linear relation between concentration and signal amplitude in the whole range of concentrations (R = 0.998) and in the 5 lower concentrations (R = 0.995). The lowest concentration (5 nanomolar) produced a signal to noise ratio of 10:1. It means that 2 nanomolar concentrations were detectable.

Blood samples of PD and control patients were obtained with an automatic lancing device. Samples were collected into hematocrit tubes and immediately centrifuged for 5 min at 3000 RPM to separate plasma and red blood cells (RBC). The RBC portion of the sample was mixed with equal volume of water during 10 min, centrifuged for 5 min and the supernatant was deproteinized by combining with equal amounts of acetonitrile and again centrifuged for 5 minutes.

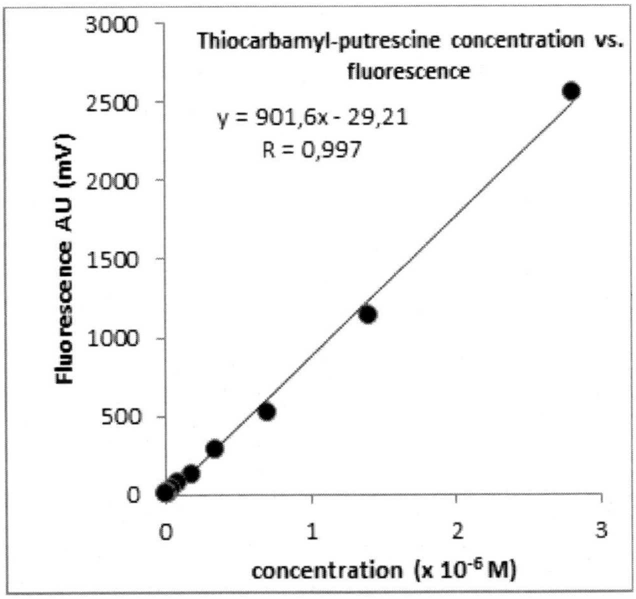

Figure 8. Putrescine concentration vs. arbitrary units of fluorescence (mV). There was a linear relation between concentration and signal amplitude in the whole range of concentrations (R = 0.998).

Plasma was directly deproteinized with isovolumes of acetonitrile and centrifuged for 5 minutes afterwards. Supernatants of deproteinized plasma and RBC were derivatized with FITC and thiocarbamyl-putrescine was monitored with CZE-LIFD. We found a significant increase of putrescine in the RBC and a non-significant increase of putrescine in plasma of PD patients vs controls (Figures 10 and 11).

These findings support a pathophysiological mechanism for PD and have great potential to provide a marker of PD. It has been found that aggregates of alpha-synuclein are present in DA cells of the brain stem (Lewy bodies) and it has been suggested that such aggregations are a step towards DA neurons degeneration (Forloni, 2000). Polyamines have a strong positive charge and conjugate with alpha-synuclein inducing protein aggregation and cellular degeneration (Fernandez, 2004 and Antony, 2002). By this mechanism a DA depletion can be caused in the terminal fields of the DA neurons. Therefore the increase of putrescine in RBC might be expression of a metabolic disorder that might lead to PD symptoms.

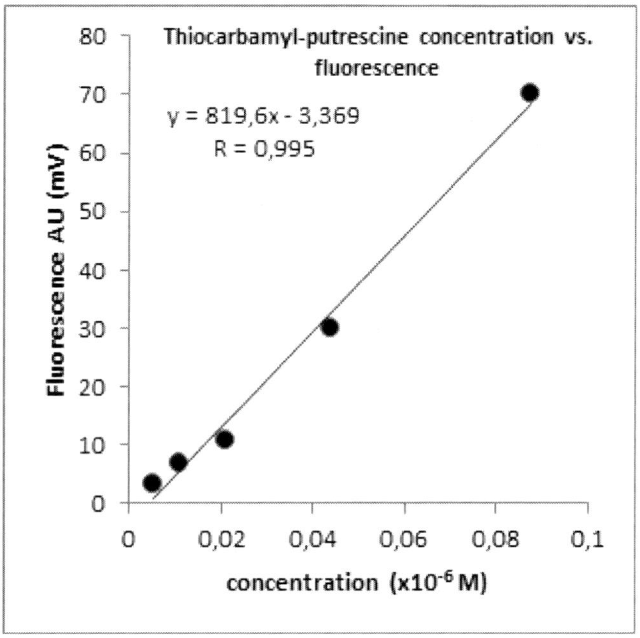

Figure 9. Linear relation offive low thiocarbamyl-putrescina concentrations (R = 0.995).

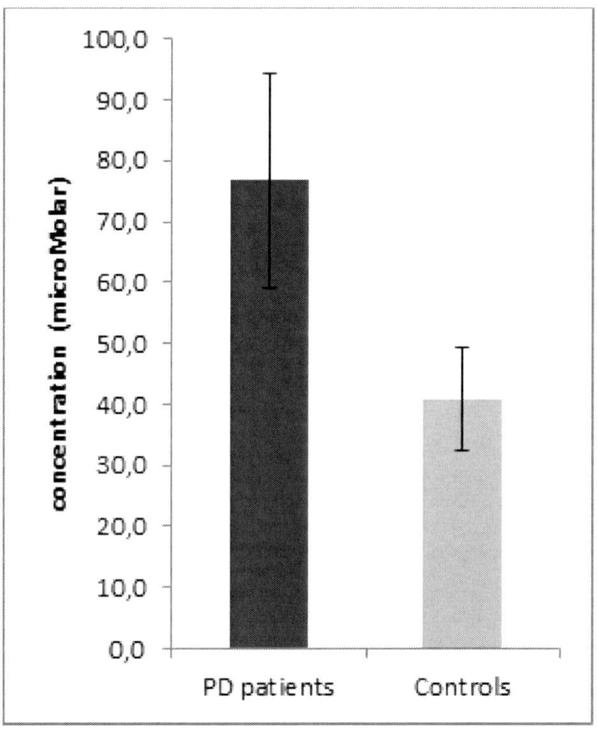

Figure 10. Concentration of putrescine in RBC of PD and control patients. Putrescine levels were significantly higher in PD patients.

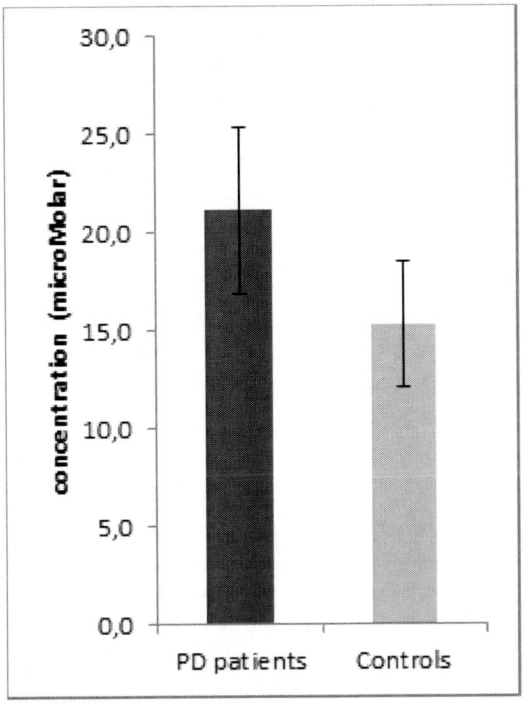

Figure 11. Concentration of putrescine in plasma of PD and control patients. There was no statistically significant difference.

PREMATURE BABIES

Although low levels of arginine and citrulline have been reported (Celik, 2013) and deficiency of the enzymes argininosuccinate synthetase (ASS) and argininosuccinatelyase (ASL) have been proposed (Wu, 2004), little attention has been paid to the relationship between arginine and citrulline in preterm babies.

Therefore, the correlation between the plasma level of arginine and citrulline in preterm babies as compared with full term neonates was investigated by means of CZE-LIFD. Blood samples were collected from the central via in the premature babies and from the umbilical cord in the mature babies.

The samples were derivatized with FITC and run in 25 micrometers inside diameter capillary. The running buffer was 20 mM Carbonate buffer at pH 10. The concentrations of arginine and citrulline were significantly lower in preterm babies than in normal neonates. There was a significant correlation between arginine and citrulline in normal neonates but there was no significant correlation between the two amino acids in the preterm babies. The low levels of arginine and citrulline in the premature babies suggest that they are not capable of synthesizing the required amount of arginine. The lack of correlation between arginine and citrulline in the preterm babies indicates that these babies do not convert adequately arginine in citrulline to produce Nitric Oxide.

CONCLUSION

The present review shows that CZE-LIFD is becoming a useful tool in clinical chemistry studies of Meningitis, Preeclampsia, Schizophrenia, Parkinson Disease and the degree of maturation of preterm babies. The thiocarbamyl-amine derivatives that FITC yields are products of high quantum efficiency. Therefore, picomolar concentration and zeptomolar mass detection of monoamines and polyamines are easy to reach. This allows measure them in different organic fluids including cerebral spinal fluid, blood and urine and in microdialyzate in experimental situations. In meningitis diagnosis CZE-LIFD is a valuable technique for etiological agent determination and for evolution and prognosis assessment. The metabolic pattern of amino acids in cerebral spinal fluid seems to depend on the microorganism that causes the meningitis. For instance, *Haemophylus influenzae* consumes large amounts of glutamine. A finding of low amounts of glutamine in the cerebral spinal fluid might indicate that *Haemophylus influenzae* is the etiological agent. Such hypothesis is testable in experimental model of meningitis. In Preeclampsia an increase of agmatine might indicate that arginine is being metabolized toward agmatine synthesis and very little towards NO synthesis. This deviation with a decrease of arginine strongly suggests that arginine might have therapeutic value in the treatment of preeclampsia. The administration of arginine might help to build up the synthesis of NO and curtail the trend to vasoconstriction and hypertension. The role of agmatine in schizophrenia is also interesting. Since agmatine is an NMDA receptor blocker, agmatine is a potential psychotogenic substance. The increase of dopamine induced by agmatine might cause schizophrenia symptoms due to overactivity of the dopaminergic system. This result suggest that a blockade of the enzyme that converts arginine in agmatine i.e., arginine decarboxylase might have some antipsychotic property and might be beneficial for the treatment of schizophrenia. The increase of putrescine in the RBC of PD patients suggest that decreasing the transformation of ornithine in putrescine might help to decrease the abnormally high concentration of polyamine in the cells of PD patients. This effect might be caused by inhibitors of the enzyme ornithine decarboxylase which might be tested as drugs retarding the degeneration of the dopaminergic neurons. Previously it has been reported that a supplement of arginine to premature babies helps to avert Necrotizing Enterocolitis (NEC). This serious condition consists in spontaneous necrosis of the intestinal mucosa. The lack of correlation between arginine and citrulline in preterm babies strongly suggest that they do not metabolize arginine to citrulline and nitric oxide. The lack of nitric oxide might contribute to severe vasoconstriction in the intestines of preterm babies. The finding here reported, i.e., significantly low levels of arginine and citrulline and lack of correlation between arginine and citrulline sheds some light on the mechanism of NEC in preterm babies.

The present findings show that CZE-LIFD is sensitive enough to study the physiological and pathological roles of different biogenic amines and provides a powerful tool for the study of the evolution of some neurodegenerative diseases and metabolic conditions.

REFERENCES

Antony T, Hoyer W, Cherny D, Heim G, Jovin TM, Subramaniam V. Cellular polyamines promote the aggregation of alpha-synuclein. *The Journal of biologicalchemistry*, 2002, 278(5):3235-40.

Betancourt, L; Rada, P; Paredes, D; Hernandez, L. In vivo monitoring of cerebral agmatine by microdialysis and capillary electrophoresis. *Journal of Chromatography* B, 2012, 880(1):58-65.

Betancourt, LR; Skirzewski, M; Catlow, B; Paredes, D; Rada, P; Hernandez,L. Local agmatine injection increases striatum extracellular dopamine in rats. 232.01/G24. *2013 Neuroscience Meeting Planner*. San Diego, CA: Society for Neuroscience, 2013. Online.

Carlsson, ML; Carlsson, A; Nilsson, M. Schizophrenia: from dopamine to glutamate and back. *Current Medicinal Chemistry*, 2004,11(3):267-77.

Celik, IH; Demirel, G; Canpolat, FE; Dilmen, U. Reduced plasma citrulline levels in low birth weight infants with necrotizing enterocolitis. *Journal of Clinical Laboratory Analysis,* 2013, 27(4):328-32.

Cheng, YF; Dovichi, NJ. Subattomole Amino Acid Analysis by Capillary Zone Electrophoresis and Laser Induced Fluorescence. *Science,* 1988, 242: 562-563.

Fernández, CO; Hoyer, W; Zweckstetter, M; Jares-Erijman, EA; Subramaniam, V; Griesinger, C; Jovin, TM. NMR of alpha-synuclein-polyamine complexes elucidates the mechanism and kinetics of induced aggregation. *The EMBO Journal*, 2004, 23(10):2039-46.

Fiori, LM; Turecki, G. Implication of the polyamine system in mental disorders. *Journal of Psychiatry and Neuroscience*, 2008, 33(2):102-10.

Forloni, G; Bertani, I; Calella, AM; Thaler, F; Invernizzi, R. Alpha-synuclein and Parkinson's disease: selective neurodegenerative effect of alpha-synuclein fragment on dopaminergic neurons in vitro and in vivo. *Annals of Neurology,* 2000, 47(5):632-40.

Gassmann, E; Kuos, JE; Zare, RN. Electrokinetic separation of chiral compounds. *Science*, 1985, 230: 813-814.

Gomes-Trolin, C; Nygren, I; Aquilonius, SM; Askmark, H. Increased red blood cell polyamines in ALS and Parkinson's disease. *Experimental Neurology,* 2002, 177(2):515-20.

Hernandez, L; Stanley, BG; Hoebel, BG. A small, removable microdialysis probe. Life Sciences, 1986, 39(26):2629-37.

Hernandez, L; Escalona, J; Joshi, N; Guzman, N. Laser-induced fluorescence and fluorescence microscopy for capillary electrophoresis zone detection. *Journal of Chromatography*, 1991, 559: 183-196

Hernandez, L; Joshi, N; Murzi, E; Verdeguer, P; Mifsud, JC; Guzman, N. Colinearlaser-induced fluorescence detector for capillary electrophoresis. Analysis of glutamic acid in brain dialysates. *Journal of Chromatography A*, 1993, 652(2):399-405.

Hjertén, S. Free zone electrophoresis. *Chromatographic Reviews*, 1967, 9(2):122-219.

Huang, XC; Quesada, MA; Mathies, RA. Capillary Array Electrophoresis Using Laser-Excited Confocal Fluorescence Detection. *Analytical Chemistry.* 1992, 64:967−972

Javitt, DC; Zukin, SR. Recent advances in the phencyclidine model of schizophrenia. *The American Journal of Psychiatry*. 1991 Oct;148(10):1301-8.

Jorgenson, JW; Lukacs, KD. Zone Electrophoresis in Open-Tubular Glass Capillaries. *Analytical Chemistry*,1981, 53: 1298-1302.

Kwon, OB; Paredes, D; Gonzalez, CM; Neddens, J; Hernandez, L; Vullhorst, D; Buonanno, A. Neuregulin-1 regulates LTP at CA1 hippocampal synapses through activation of dopamine D4 receptors. *Proceedings of the National Academy of Sciences of the USA*, 2008, 105(40):15587-92.

Li, D; Collier, DA; He, L. Meta-analysis shows strong positive association of the neuregulin 1 (NRG1) gene with schizophrenia. *Human Molecular Genetics*, 2006, 15(12):1995-2002.

López-Jaramillo, P; Arenas, WD; García, RG; Rincon, MY; López, M. The role of the L-arginine-nitric oxide pathway in preeclampsia. *Therapeutic Advances in Cardiovascular Disease,* 2008, 2(4):261-75.

Mikkers, FE; Everaerts, Verheggen, TP. High-performance free zone electrophoresis. *Journal of Chromatography A*, 1979, 169: 11-20.

Seeman, P; Lasaga, M. Dopamine agonist action of phencyclidine. *Synapse,* 2005, 58(4): 275-7.

Teran, Y; Ponce, O; Betancourt, L; Hernandez, L; Rada, P. Amino acid profile of plasma and cerebrospinal fluid in preeclampsia. *Pregnancy Hypertension,* 2012, 2 (4): 416–422.

Tucci, S; Pinto, C; Goyo, J; Rada, P; Hernández, L. Measurement of glutamine and glutamate by capillary electrophoresis and laser induced fluorescence detection in cerebrospinal fluid of meningitis sick children. *Clinical Biochemistry,* 1998, 31(3):143-50.

Uzbay, T; Kayir, H; Goktalay, G; Yildirim, M. Agmatine disrupts prepulse inhibition of acoustic startle reflex in rats. *Journal of Psychopharmacology*, 2010, 24(6):923-9.

Wu, G; Jaeger, LA; Bazer, FW; Rhoads, JM. Arginine deficiency in preterm infants: biochemical mechanisms and nutritional implications. *The Journal of Nutritional Biochemistry*, 2004, 15(8):442-51.

Yang, XC; Reis, DJ. Agmatine selectively blocks the N-methyl-D-aspartate subclass of glutamate receptor channels in rat hippocampal neurons. *The Journal of Pharmacology and Experimental Therapeutics,* 1999, 288(2):544-9.

In: Capillary Electrophoresis (CE)
Editor: Christian Reed
ISBN: 978-1-63483-122-2

Chapter 9

CAPILLARY ELECTROPHORESIS IN DETERMINATION OF STEROID HORMONES IN ENVIRONMENTAL AND DRINKING WATERS

Heli Sirén[1,*], Samira El Fellah[1], Aura Puolakka[1], Mikael Tilli[1] and Heidi Turkia[1,2]

[1]University of Helsinki, Department of Chemistry, Laboratory of Analytical Chemistry, University of Helsinki, Finland
[2]Turku University Hospital, Tykslab, Turku, Finland

ABSTRACT

Capillary electrophoresis (CE) was used to study residues of steroid hormones in influent and effluent waters of drinking water treatment plants. Steroids were of special interest, because they are slightly water-soluble. In general, their concentrations are at ng/L level in environmental waters, but cannot be totally purified from drinking waters.

In this research, a partial-filling micellar electrokinetic chromatographic (PF-MEKC) method was developed and optimized for separation and determination of neutral steroids and their metabolites. The micelle solution contained 1.5 mM sodium taurocholate and 29.5 mM SDS in 20 mM ammonium acetate (pH 9.68). The CE separations were detected with an UV detector at the steroid specific wavelength 247 nm. The optimization was made with six steroid standards.

The samples from water treatment plants were concentrated to 6:1000 (v/v) with solid-phase extraction (SPE) in nonpolar sorbents. The PF-MEKC method was very repeatable (r^2 0.99), which was detected from the migration times of the studied compounds. The relative standard deviations of electroosmosis and the steroids were 0.01-0.04% and 0.01-0.07%, respectively. Concentration ranges for the steroids were linear at 0.5-10 ng/μL range. The influent waters contained 3.22-68.3 ng/L of 4-androsten-17β-ol-3-one glucosiduronate, androstenedione, and progesterone. On the contrary, the effluent waters after the treatment contained those analytes at 2.72-27.9 ng/L level.

* Corresponding author: Heli Sirén, University of Helsinki, Department of Chemistry, A.I. Virtasen aukio 1, PO Box 55, FI-00014 University of Helsinki, Finland. E-mail: heli.m.siren@helsinki.fi.

Keywords: Steroid hormones, influent water, effluent water, partial filling, micellar electrokinetic chromatography

1. INTRODUCTION

The environmental waters contain many organic compounds as solids, water-soluble particles, and soluble chemicals, ions, and species. In addition, municipal wastewaters have various kinds of liquid and solid wastes, garbage, and chemicals, which are origin from living environment, institutions, commercial operators, and industrial sources.

Furthermore, the contaminated waters in environment exist natural and synthetic hormones and pharmaceutical compounds [1-5], and industrial chemicals [6, 7]. In addition, the waters may be effluents from pulp and paper, mining, biorefinary, and textile industries. Many chemical released into environment mimic activity of endogenous hormones such as estradiol. Especially, the environmental waters are contaminated from agricultural fluids and waste. Quality monitoring of inland, surface, transitional, coastal, and ground waters is obligatory for environmental waters. The Framework Directive (2000/60/EC) in European Union has a specific category for endocrine disrupting compounds (EDCs), which include among other compounds also steroid hormones and their metabolites (Figure 1).

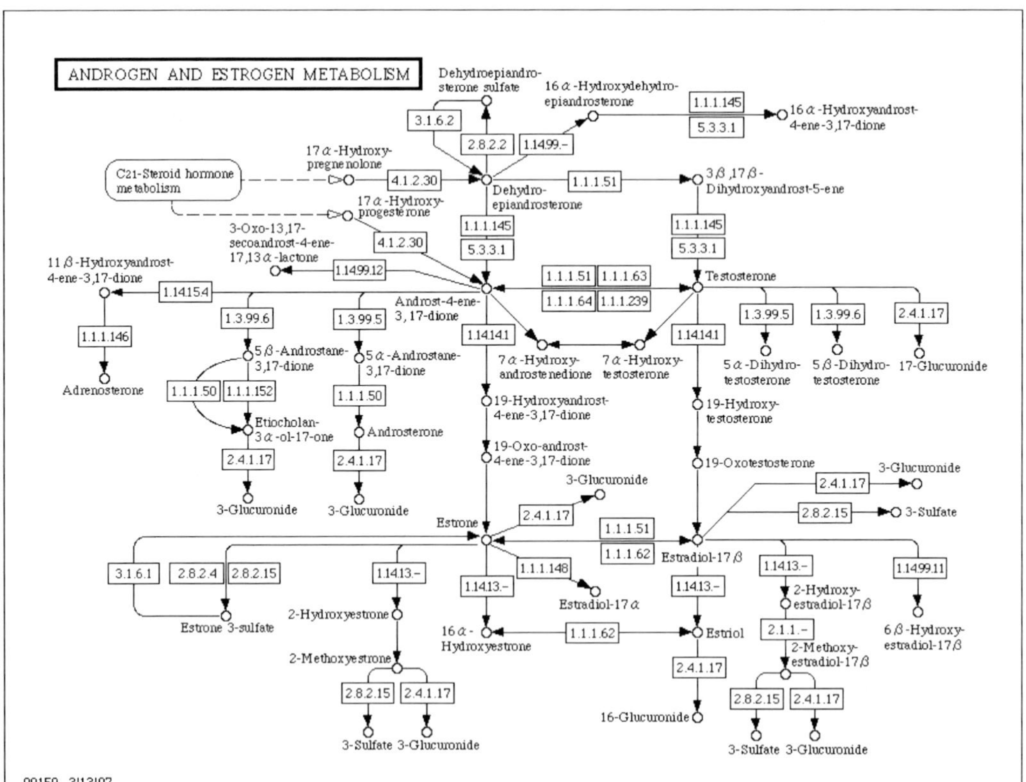

Ref. http://dwb4.unl.edu/Chem/CHEM869K/CHEM869KLinks/www.genome.ad.jp/kegg/metabolism_links/map/map00150.html.

Figure 1. The metabolic pathway of androgen and estrogen metabolism.

Natural steroids are listed into EDCs because of their biological disadvantages [6, 7]. Steroids are used in treatment of infertility, cancers, menstrual and menopausal hormonal disorders, and birth control [8]. They cause feminization of animal species, development of physical abnormalities and birth defects [9]. Steroid hormones were studied in human and environmental samples due to their toxicity to environment at low concentrations.

A report about the effluent waters from 29 municipal corporations to Grand River watershed in Ontario Canada describes about wastewaters mixed together. They were mixed with household's effluents and industrial waters. The waters of the river walleyes were also studied indirectly by analysing the wild fish. Steroids were noticed to enrich to the wild population [10] and had an effect on their sex.

Detection of organic compounds is challenging without mass spectrometric reference data with model compounds. In addition due to their low concentrations in water, only a few hundreds of the harmful compounds are included among regulations. Organic pollutants, primarily moved by diffusion in farming [11-14] contaminate waters and soils in the neighborhood. According to literature, depending on excretion in animals or humans, 10-90 % of drugs or steroid hormones administered were excreted into urine or faces as non-metabolized forms [11-13, 15].

There is no standardization, although widely accepted methods exist for determination of hormones in waters. The disadvantage is that mostly steroids are detected as parent compounds although they are released also as glucuronate and sulphate conjugates [16]. At present, there are several advanced analytical methods for detecting and quantifying emerging contaminats. Mostly, for steroids the methods used are gas chromatography (GC) or liquid chromatography (LC) with mass spectrometric detection (MS). To fulfil also the demands of low detection limits, recently LC MS/MS has been the most used method for the determination of all classes of pharmaceuticals in aqueous samples. Mainly electrospray ionization (ESI) or atmospheric pressure chemical ionization (APCI) were used for fragmentation of the compounds and to obtain reliable and selective identification for the steroid structures [16]. To obtain the low detection limits (LOD), sample preparation was shown to have an important role in developing the overall methodology for steroid hormones. The LODs of steroids reached with LC-MS/MS methods were higher than those obtained with GC-MS [17].

Usually, the steroid hormones are determined at ng/L level. Determination of individual steroids is more important than the total quantity of all the steroids detected. Due to that, capillary electrophoresis (CE) may also be utilized in steroid determination, although it is less sensitive than LC. It is well-known that CE gives better efficiency for separation of structurally similar compounds than LC. However, the main reason to prefer LC is the larger sample volume than in CE separation (2-100 μL versus 1-10 nL, respectively) [18]. Lately, trace concentrations of steroids were studied by liquid chromatography-electrospray ionization tandem mass spectrometry (LC-EI-MS/MS) in surface water, wastewater, and sludge samples [19]. It was a study developed by Chinese scientists, who made a sensitive and fast LC method. It was an on-line coupled system containing two mass spectrometers, which allowed reliable detection for 28 steroids.

The analytes were four estrogens (estrone, 17β-estradiol, 17α-ethynyl estradiol, and diethyl stilbestrol), 14 androgens (androsten-1,4-diene-3,17-dione, 17α-trembolone, 17β-trembolone, 4-androstene-3,17-dione, 19-nortesterone, 17β-boldenone, 17α-boldenone, testosterone, epi-androsterone, methyltestosterone, 4-hydroxy-androst-4-ene-17-dione, 5α-dihydrotestostrone, androsterone, and stanozolol), five progestrens (progesterone, ethynyl testosterone, 19-norethindrone, norgestrel, medroxyprogesterone), and five glucocortico-steroids (cortisol, cortisone, prednisone, dexamethasone).

Their recoveries from water were 90.6-119 % with the method limit of detection (MDL) between 0.01-0.24 ng/L.

According to many publications, the low ng/L detection concentrations in environmental samples could only be achieved by enrichment techniques. The need of 1000 – 10000 -fold preconcentration by SPE and LLE has been suggested for steroids detection from environmental waters [15]. Pre-concentration by solid-phase extraction conditions has turned out to be repeatable by using C18 cartridges for steroid hormones. When the analyses are made with CE and without pre-enrichment of the samples with SPE, steroids may be concentrated in-line during the analysis with partial-filling micellar electrokinetic chromatography (PF-MEKC). PF-MEKC is a modification of the conventional micellar method. MEKC is a useful mode of capillary electrophoresis because it can separate both neutral and charged analytes. It involves the addition of ionic surfactants to the separation solution. Surfactants form micelles, which have roughly a spherical structure with a hydrophobic interior and a hydrophilic exterior. The separation of the analytes is based on different partition of the steroids between the hydrophobic interior and the hydrophilic exterior. In comparison with conventional MEKC, partial-filling MEKC involves a small portion of the micellar solution placed after the main electrolyte solution. In that case, first the capillary is filled with the electrolyte followed by a small plug of micellar solution (most often sodium dodecyl sulfate, SDS), and finally a sample. Analytes will first migrate into the micellar plug, where they interact with the micelle and start the separation moving into the electrolyte solution and finally to the detector.

The PT-MEKC can also be modified and even improved by field amplified (FASI) or by other on-line sample concentration methods in order to introduce the sample as enriched zone to detection. The combined use of pressure (PA) and electric field amplification (FASI) was shown to improve the sensitivities of the analytes [20-24]. Recently, also a new on-line capillary pre-concentration electrokinetic injection with field amplified electrokinetic supercharging (EKS) was developed. It is a FASI method, which combines transient isotachophoresis (tITP) with sample concentration [25]. Sampling and sample preparation techniques and detection methods have shown crucial for sensitive detection in GC-MS, LC-MS and CE [15]. Sample preparation is the most demanding procedure before analysis and it should be reproducible and definite. The steroid analytics, which can utilize separation technique and continuous identification, need always a new development when the matrix is changed. In addition, identification of steroid hormones and their metabolites without sample preparation does not give the wanted information, because the organics in water samples disturb separation of steroids in CE, LC, and GC.

It has been noticed that also purified waters need cleaning, since industrial water removals form organic materials. Even the biorefinery industry produces mixture of sterols. Simultaneously, the need to have limits of detection lower than 0.5 ng/L resulted to improved purification with silica [26]. In present research, the aim was to examine, how much the drinking water contains human, animal, and plant based steroid hormones, from the influent and effluent waters of a purification plant which cannot be extracted with membranes or other sample purification methods.

The concentrations of steroid hormones, the natural estrogen, 17α-estradiol and the main components of the contraceptive pills (17α-ethynylestadiol, mestranol and levonorgestrel) and the metabolites, estrone and estriol could not be detected. The results showed that androgenic hormones were detected at higher concentrations than estrogenic hormones, which may be due to the higher excretion rates of androgens compared with oestrogens in humans [27]. Testosterone and its metabolised products, androsterone, etiocholanolone and dihydrotestosterone can all be detected [26-31].

The concentrations of androstenedione, androsterone, etiocholanolone, testosterone, 17β-estradiol, estriol, and estrone were 100, 1200, 6000, 180, 120, 75, 1100, and 1300 ng/L, respectively. The presence of trace organic chemical contaminants such as steroidal hormones in municipal wastewater has been the subject of increasing concern throughout recent decades [32]. Some of these trace organic chemical contaminants are known to have endocrine disrupting effects on aquatic organisms at low concentrations and others have been linked to ecological impacts due to acute and chronic toxicity mechanisms [33].

Pharmaceuticals in waters from lake and river systems, but also effluents and influents of some water purification works have been studied especially in Canada, North America. However, in many countries there are not known the exact drifts of steroid hormones into the debits of water sources [34]. Emerging wastewater treatment processes such as membrane bioreactors (MBRs) have attracted a significant amount of interest internationally due to their ability to produce high quality effluent suitable for water recycling. It is therefore important that their efficiency in removing hazardous trace organic contaminants is assessed [35].

Many hormone chemicals released into environment have been shown because endogeneous effects on wildlife and humans such as feminization of animal species. Methods of LC EI-MS/MS were used for analyses. Satisfactory detection limits and analyte recoveries were between 0.5-6 ng/L and 60% - 108%, respectively [36].

There is a lack of research for steroid metabolites and their transformation products in respect of characterization, occurrence and fate in all water types and especially in drinking water. The analytical techniques improvement allowed detecting traces of substances in any type of water [37]. In addition, new analysis techniques for on-line monitoring agricultural and industrial water removals need more attention.

Because the steroids are hormones, which are detected free or conjugated from animal and human body fluids, the flow of steroids should be continuously measured. Their determination requires concentration, extraction and clean-up prior to detection. Microextraction techniques have also been used for the determination of steroid hormones in biological (e.g., human urine, human serum, fish, shrimp and prawn tissue and milk) and environmental (e.g., wastewaters, surface waters, tap waters, river waters, sewage sludge, marine sediments and river sediments) samples [38].

The most recent applications are made in sorptive-microextraction modes, such as solid phase microextraction (SPME) with molecularly imprinted polymers (MIPs), in-tube solid-phase microextraction (IT-SPME), stir-bar sorptive extraction (SBSE), and microextraction in packed sorbent (MEPS). Researchers have modified old methods to incorporate procedures that use less-hazardous chemicals or that use smaller amounts of them.

Zarzycki et al. [39] have used temperature-dependent inclusion chromatography for fast screening of free steroids in various kind of environmental waters from Baltic Sea, and selected lakes and rivers of the Middle Pomerania in North Poland.

They used solid-phase extraction based on C18 sorbents and isocratic HPLC procedure for the quantification. The system established could be used for characterizing the compounds, which were from estradiol, which is a human steroid, produced by the fatal liver during pregnancy, to progesterone, which is an endogenous hormone in body. They all have different polarities and therefore their simultaneous extraction from waters need special validation for sensitive detection.

The aim of our research study was to validate and use a capillary electrophoresis (CE) method for determination of free steroid hormones and their metabolites in influent and effluent waters of drinking water treatment plants.

The waters were processed for drinking waters to households in city area in Finland. Special interest was focused for determination of those steroids that are slightly water-soluble and exist at low ng/L concentrations in urine and are transferred to environment.

2. EXPERIMENTAL

2.1. Chemicals

The steroids used for validation of the method are listed in Table 1. Their structures and the physical parameters of the steroids in the study and the chemicals are compiled in Tables 2 and 3, respectively.

Table 1. Steroid chemicals

Name of the steroid	Purity	CAS number	Manufacturer	Country
Androstenedione $C_{19}H_{26}O_2$	Assay ≥ 98%	63-05-8	Sigma-Aldrich Co.	Germany
Androsterone $C_{19}H_{30}O_2$	Assay (HPLC) 97.6%	53-41-8	Sigma-Aldrich Co.	Germany
4-androsten-9α-fluoro-17α-methyl-11β, 17β-diol-3-one $C_{20}H_{29}FO_3$	TLC: 1• purity	76-43-7	STERALOIDS, INC.	US
4-androsten-17β-ol-3-one glucosiduronate $C_{25}H_{36}O_8$	TLC: 1• purity	1180-25-2	STERALOIDS, INC.	US
17α-hydroxyprogesterone $C_{21}H_{30}O_3$	Assay ≥ 95%	68-96-2	Sigma-Aldrich Co.	Germany
17α-methyltestosterone $C_{20}H_{30}O_2$	(HPLC) ≥ 98%	58-18-4		
Progesterone $C_{21}H_{30}O_3$	Assay ≥ 98%	58-18-4	Sigma-Aldrich Co.	Germany
Testosterone $C_{19}H_{28}O_2$	Assay ≥ 98%	58-22-0	Sigma-Aldrich Co.	Germany

Table 2. Names, structures, CAS numbers, molar masses, water solubility, logP and pKa values of steroids in migration order

Name and structure (in migration order)	Molar mass [g/mol]	Predicted water solubility [mg/L]	Predicted logP	Predicted pKa value Strongest Acidic / Strongest Basic
1. 4-androsten-17β-ol-3-one glucosiduronate 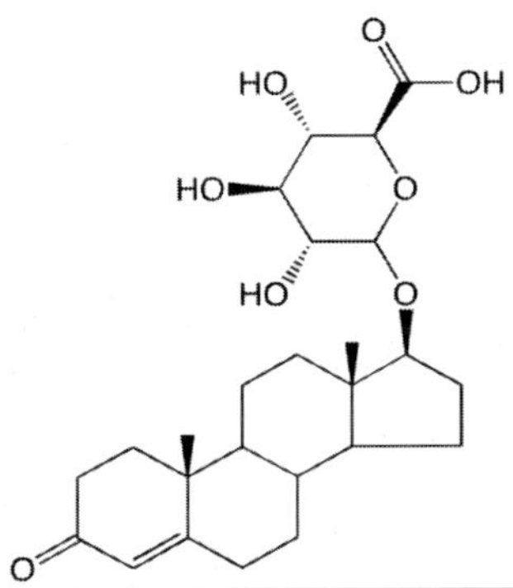	464.55	0.26	1.91	3.63 / -3.7
2. 4-androsten-9α-fluoro-17α-methyl-11β, 17β-diol-3-one 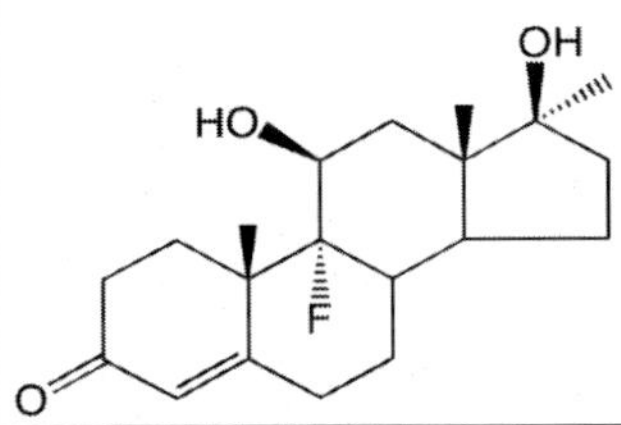	336.44	0.0452	2.38	13.6 / -3
3. Androstenedione 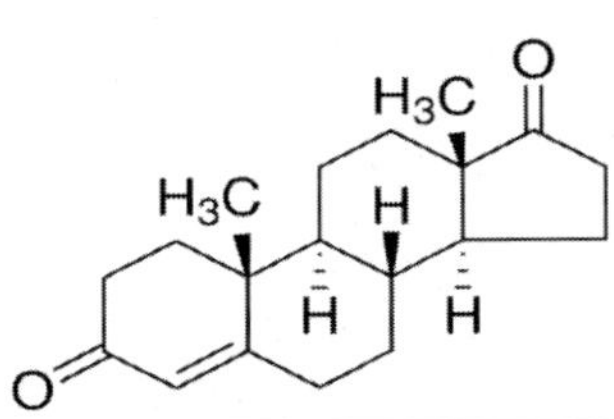	286.41	0.027	3.93	19.03 / -4.8
4. Testosterone 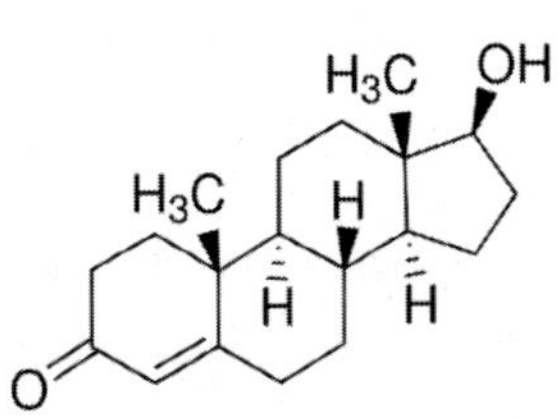	288.42	0.033	3.37	19.09 / -0.88

Table 2. (Continued)

Name and structure (in migration order)	Molar mass [g/mol]	Predicted water solubility [mg/L]	Predicted logP	Predicted pKa value Strongest Acidic / Strongest Basic
5. 17α-hydroxyprogesterone 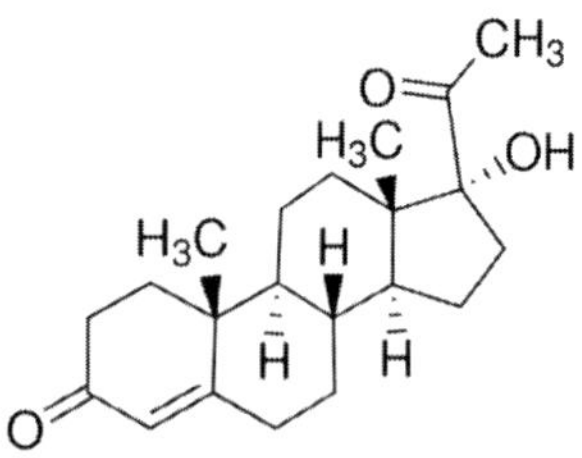	330.46	0.0219	3.4	12.7 / -3.8
6. 17α-methyltestosterone 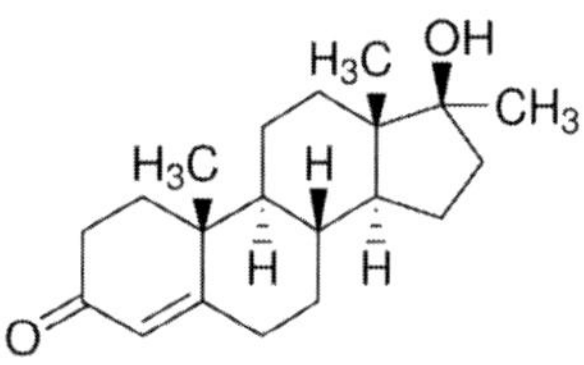	302.45	0.014	3.65	19.09 / -0.53
7. Progesterone 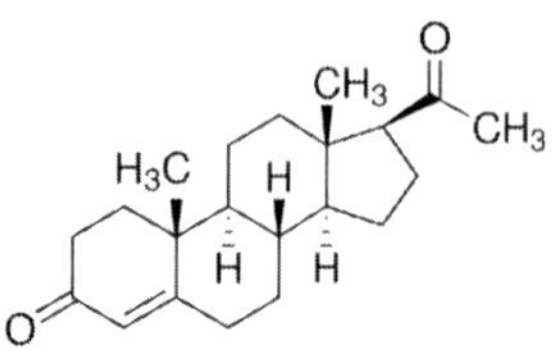	314.46	0.00546	4.15	18.92 / -4.8

2.2. Instruments

Micellar EKC separations were performed with a Hewlett-Packard 3^D CE system (Agilent, Waldbronn, Germany) equipped with a diode array detector, λ 190-600 nm). Bare fused silica capillaries (i.d. 50 μm, o.d. 375 μm) were purchased from Polymicro Technologies (TSP050375 3, 363-10, Phoenix, AZ, US).

The capillaries were cut to a total length of 80 cm and the detector window was burned to 71.5 cm. New capillaries were conditioned by flushing sequentially with 0.1 M NaOH in water, water and electrolyte solution for 20 min each.

The temperature during the analyses was 25°C. Positive polarity was used and voltage of + 25.00 kV was set as a constant. The current was detected during the analysis and it was typically remained at 16 − 18 μA. Detection with UV was simultaneously at 214, 220, 240, 247, and 260 nm, of which the pilot wavelength was 247 (± 2) nm.

Table 3. Other chemicals used in the study

Name	Purity	CAS number	Organization	Country
Ammonium acetate	Min. 98%	631-61-8	Sigma-Aldrich Co.	Germany
Ammonia solution	Min. 25%, Assay 31,5%	1336-21-6	VWR International S.A.S	France
Buffer solution pH 4 (phthalate), Stabilized	pH at 20°C 3.98pH pH at 25°C (calculated) 3.99pH	877-24-7	Fisher Scientific UK	UK
Buffer solution pH 7 (phosphate), Stabilized	pH at 20°C 7.02pH pH at 25°C (calculated) 7pH	7778-77-0	Fisher Scientific UK	UK
Buffer solution pH 10 (borate)	pH at 20°C 9.99pH	7732-18-5	Fisher Scientific UK	UK
Hydrochloric acid 1.0 mol/L (1.0 N) (A)	Analysis result 0.9995 mol/L, ± 0.0021 mol/L	7647-01-0	Oy FF-Chemicals Ab	Finland
Methanol	HPLC grade	67-56-1	Fisher Scientific UK	UK
Sodium dodecyl sulfate	Approx. 99%	151-21-3	Sigma-Aldrich Co.	Germany
Sodium hydroxide 1.0 mol/L (1N) (A)	Analysis result 1.0003 mol/L, ± 0.0021 mol/L	1310-73-2	Oy FF-Chemicals Ab	Finland
Taurocholic acid sodium salt hydrate	BioXtra, ≥ 95% (TLC)	345909-26-4	Sigma-Aldrich Co.	Germany

For separation of the steroids, the micellar solution was introduced at 0.5004 p.s.i (34.5 mbar) for 75 s (volume of the hydrodynamic injection 0.56 nL, CE Expert Lite, SCIEX). After the micellar solution, the sample was introduced at 0.725 p.s.i. (50 mbar) for 6 s (volume of the hydrodynamic injection 6.46 nL) from inlet of the capillary towards the detector. Before each analysis, the capillary was flushed with 0.1 M NaOH in water and with electrolyte solution for 2 min and 5 min, respectively. After every eighth run, the capillary was washed by flushing with 0.1 M NaOH in water, milli-Q water and electrolyte solution for 5 min each.

2.3. Other Instruments

The pH value of the electrolyte solution was adjusted using a MeterLab PHM 220 pH meter (Radiometer, Copenhagen, Denmark) and InoLab pH7110 (WTW) calibrated with 3-point-calibration pH 4.00, 7.00 and 10.00 commercial buffers (Fisher Scientific, Loughborough, UK).

The samples were centrifuged with MSE MISTRAL 1000 at 2000 rpm. All water used was purified with a Direct-Q UV Millipore water purification system (Millipore S.A., Molsheim, France).

2.4. Preparation of Standard Solutions

The stock solutions of 1000 mg/L of the steroids were prepared in methanol and stored at +4°C. The working solutions were prepared in methanol. The electrolyte solution used was 20 mM ammonium acetate (pH 9.5). It was stored at +4°C in a glass flask when not used. The pH was adjusted with 25% ammonia solution. The stock solution of 100 mM of SDS was prepared in the electrolyte solution and stored at room temperature in a volumetric flask. The stock solution of 100 mM sodium taurocholate was prepared in water and stored at +4°C in a glass flask. Neither electrolyte solution nor micelle solutions were filtered before use. The micelle solution was prepared from the stocks by pipetting the exact volumes into the glass vials of the CE instrument. The micellar solution was prepared by adding 1000 μL of 20 mM ammonium acetate buffer solution (pH 9.68), 440 μL of 100 mM sodium dodecyl sulphate in 20 mM ammonium acetate buffer solution (pH 9.68), and 50 μL of 100 mM sodium taurocholic acid sodium salt hydrate (in Milli-Q water) in this specific order.

2.5. Sampling and Sample Preparation of the Water Samples

The waters were collected in 2014 on March 18-19th and on March 24-25th from water treatment plants in Kajaani (abbreviation EF) and Turku (abbreviation WF) and from that in Porvoo (abbreviation SF), respectively. The plants are located in Eastern, Southern and Western Finland. The personnel of the water treatment plants sampled the influent and effluent waters in the research. The waters were sampled into 5 L-volume canisters, from which they were divided to three 1L water portions, which were used as the main samples (influent and effluent). In Kajaani (EF) one sample was taken before the biological filtration (biofilter). The influent and effluent waters (volume 1 L) were first filtrated through fiberglass and membrane (0.45 μm) filters. The process resulted in a) liquid fraction and b) solid fraction containing the particles. The liquid fractions were extracted with solid-phase polymer based reverse-phase material (Strata-X, Phenomenex, Copenhagen, Denmark). Before use, they (500 mg, 6 mL) were treated with methanol and water. The sample was introduced by pumping at 8 mL/min. After sampling, the materials were dried in vacuum for 30 min. Elution of the adsorbed compounds was made with 6 mL methanol.

2.6. Optimization of the PF-MEKC Separation

An existing PF-MEKC-UV method [22] was used as a starting point in optimization of the PF-MEKC. Testosterone (0.5 μg/mL – 20 μg/mL) was used as a reference compound, because its behavior was well studied in the earlier projects [21-23].

In total twelve different combinations of chemical and instrumental parameter were tested in order to find the best method for separation. The tested parameters were the injection pressure of micellar solution, sample volume in hydrodynamic injection, concentration of the electrolyte solution, capillary volume of micellar solution, separation voltage, temperature during separation, the temperature of the vial tray, the concentration of SDS and sodium taurocholate in micellar solution, dilution solution and ratio of samples, and the timing of injection. In addition, the impact of different preconditions of capillary were tested.

2.7. Identification of Steroids

Once the method was optimized, all compounds were first analyzed individually. Then their migration order, the separation efficiency of the method, and sensitivity was identified by adding one steroid at a time into their mixtures and analyzing with the validated CE method. In addition, sequential spiking with standards of 2-3 μg/mL identified steroids of samples from water treatment plants.

2.8. Calibration

For the concentration calibration, first the steroid stock solutions (1000 mg/L) were made to 500 mg/L or 100 mg/L solutions with methanol. Then, from the diluted solutions, five or six standard mixtures were prepared in methanol for working solutions. The calibration solutions were further diluted with methanol to get the final calibration concentrations.

Lastly, each of the calibration solutions were completed with 20 μl of 0.1 M NaOH. The concentration range used in calibration for 17α-hydroxyprogesterone was 0.5-6.0 μg/mL, that of 17α-methyltestosterone and progesterone were 0.5-8.0 μg/mL and that for all other steroids 0.5-10 μg/mL.

2.9. Data Handling

Averages of peak heights, areas and migration times were calculated by using equation (1)

$$\bar{x} = \frac{1}{n}\sum_{i=1}^{n} x_i, \tag{1}$$

where $\bar{x}$ is the average value and x_i an individual data point.

Variation of data points from the average value (standard deviation, STD) was calculated by using equation (2)

$$STD = \sqrt{\frac{\sum(x_i - \bar{x})^2}{N}} \tag{2}$$

where N is the number of measurements.

Variation relative to the average value (relative standard deviation, RSD) was calculated by using equation (3)

$$RSD = \frac{SD}{\bar{x}} \times 100\,\%. \tag{3}$$

3. RESULTS

3.1. Migration of the Steroids

The capillary electrophoresis method used for determination of steroids in effluent and influent waters is based on partial filling (PF) micellar electrokinetic capillary chromatography (MEKC). The micelle and the electrolyte solutions were sequentially introduced into the capillary. The micelle plug was short pseudostationary phase. It was placed between the main electrolyte solution in front and the sample zone behind of it. The purpose of the discontinuous solvent composition was to aid the nonionic steroids to move along the capillary and to separate from each other under the applied electric field. By optimization of the concentrations and chemical composition of both the solutions, the steroids were also concentrated during the movement to the UV detector.

In our research, the identification of the steroids was made with UV absorbance (the characteristic λ 247 nm, Figure 1). The analytes were separated from each other according as the first 4-androsten-17β-ol-3-one glucosiduronate followed by 4-androsten-9α-fluoro-17α-methyl-11,17β-diol-3-one, androstenedione, testosterone, 17α-hydroxyprogesterone, 17α-methyltestosterone, and progesterone (Figure 2). The PF-MEKC method is suitable for CE-ESI-MS/MS coupling, like earlier demonstrated with another project [22].

It can be an alternative for LC-MS/MS technique, like Carballa et al. [40] have shown. They identified 17α-ethynylestradiol, 17β-estradiol and two of its metabolites with LC-ESI-MS/MS. Detection limits of 17α-ethynylestradiol and 17β-estradiol and its metabolites were 0.5-6 ng/L. In the waters of wastewater treatment plant the concentrations of ECDs were from < 10 ng/L to nearly 1200 ng/L in the dissolved phase. Here, the PT-MEKC method was a new method in Agilent CE. It was noticed that steroid samples and solutions are in glass vials instead of plastics ones.

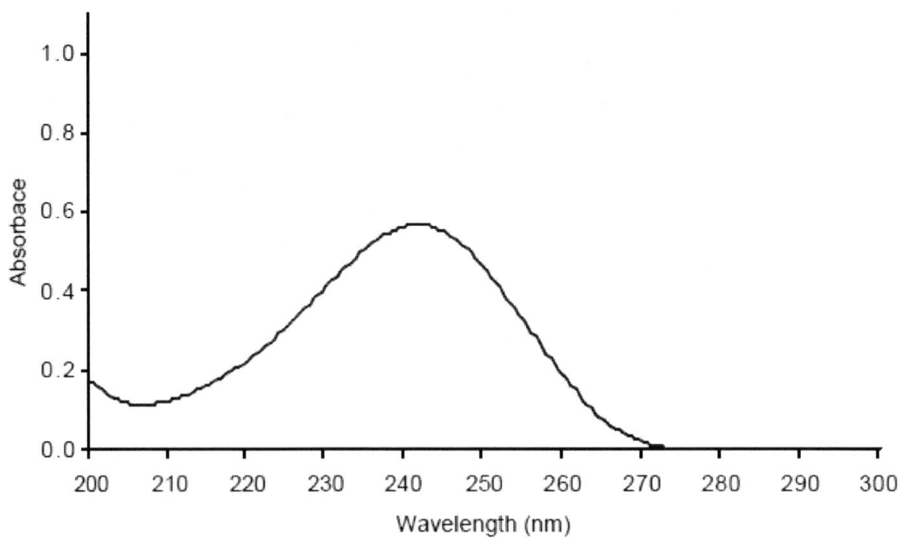

From www.sigma-aldrich.com.

Figure 2. UV spectrum of testosterone.

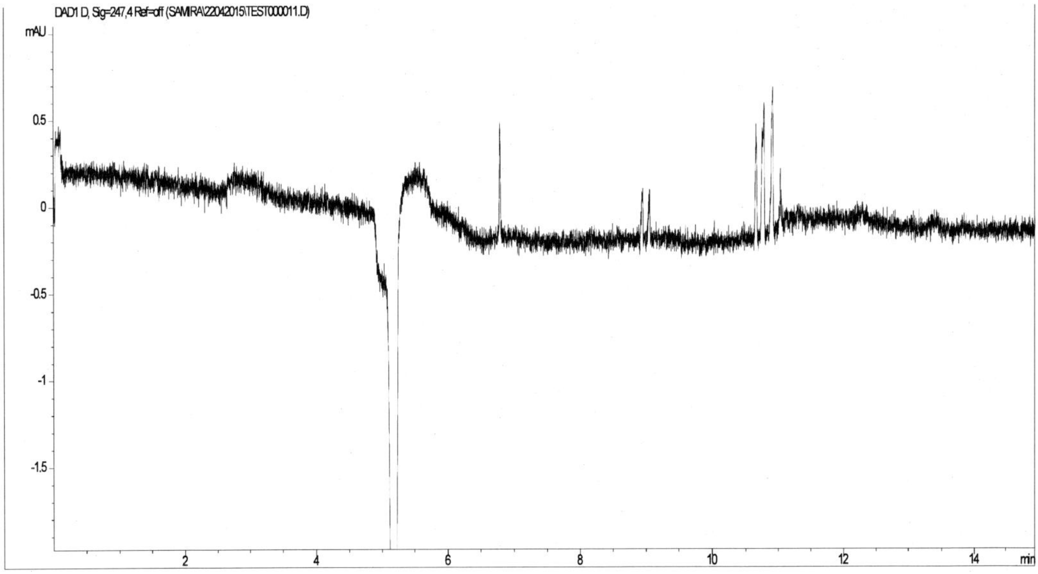

Figure 3. A PT-MEKC electropherogram of separation of the steroid hormones. Migration order of the steroids is 4-androsten-17β-ol-3-one glucosiduronate, 4-androsten-9α-fluoro-17α-methyl-11, 17β-diol-3-one, androstenedione, testosterone, 17α-hydroxyprogesterone, 17α-methyltestosterone, and progesterone.

The reason was that steroids and micelles adsorbed on the polymer surface and therefore contamination existed, when the vials were reused. The validation gave also information about periodical washing needs of the capillary and storing the electrolyte solutions in room temperature instead of in refrigerator (+4°C). The new information was that when to keep the micelle solution active it should be mixed from taurocholate, SDS and the electrolyte solution together with the sample preparation.

Experimental conditions: 20 mM ammonium acetate (prepared with milli-Q water purity) at pH 9.68 was used as a buffer electrolyte solution. The length of the capillary was 0.800 m and the detection window was at 0.715 m. The used capillary is silica based (Polymicro Technologies TSP050375 3, 363-10) with internal diameter of 50 μm and outer diameter of 375 μm. The monitored wavelength with UV detector was 247 (± 2) nm and the temperature was set at 25°C.

Positive polarity was used and voltage of +25.00 kV was set as a constant and current was approximately 17 μA during every run. The micelle was prepared by adding 1000 μL of 20 mM ammonium acetate buffer solution (pH 9.68), 440 μL of 100 mM sodium dodecyl sulfate in 20 mM ammonium acetate buffer solution (pH 9.68), and 50 μL of 100 mM sodium taurocholic acid sodium salt hydrate (in milli-Q water) in this specific order. Both the micelle and the sample were hydrostatically injected with a pressure of 34.5 mbar in 75.0 seconds and 50.0 mbar in 6.0 seconds, respectively.

The PT-MEKC method was very repeatable (Table 4), which is noticed from the absolute migration times of the steroids (RSD 0.029 - 0.046), electrophoretic mobilities (RSD 0.028 - 0.88) and the mobility of electroosmosis (RSD 0.015 - 0.038). The averages, standard deviations and relative standard deviations were calculated by using equation 1, 2, and 3. The method validation contained also using the method in other four Agilent CE instruments. The result was that the method was transferable from instrument to instrument.

Table 4. Electrophoretic mobility parameters of the steroids of the project

Name	Migration time [min]	Electrophoretic mobility $[m^2V^{-1}s^{-1}]$	Electroosmotic flow* $[m^2V^{-1}s^{-1}]$
testosterone	Average 15.78	Average 2.422E-08	Average 5.319E-08
	STD 0.733	STD 1.174E-09	STD 1.143E-09
	RSD 0.046	RSD 0.048	RSD 0.021
progesterone	Average 16.99	Average 2.260E-08	Average 6.128E-08
	STD 1.217	STD 1.984E-09	STD 2.323E-09
	RSD 0.072	RSD 0.088	RSD 0.038
17α-hydroxyprogesterone	Average 13.75	Average 2.781E-08	Average 6.566E-08
	STD 0.622	STD 1.500E-09	STD 1.803E-09
	RSD 0.045	RSD 0.054	RSD 0.027
androstenedione	Average 11.59	Average 3.294E-08	Average 6.251E-08
	STD 0.397	STD 1.299E-09	STD 1.575E-09
	RSD 0.034	RSD 0.039	RSD 0.025
4-androsten-9α-fluoro-17α-methyl-11,17 β -diol-3-one	Average 11.80	Average 3.260E-08	Average 6.091E-08
	STD 0.344	STD 1.083E-09	STD 1.423E-09
	RSD 0.029	RSD 0.033	RSD 0.023
4-androsten-17β-ol-3-one glucosiduronate	Average 8.16	Average 4.684E-08	Average 6.249E-08
	STD 0.630	STD 3.136E-09	STD 2.397E-09
	RSD 0.077	RSD 0.067	RSD 0.038
17α-methyltestosterone	Average 13.45	Average 2.422E-08	Average 6.450E-08
	STD 0.543	STD 1.174E-09	STD 1.369E-09
	RSD 0.040	RSD 0.048	RSD 0.021
androsterone	Average 13.60	Average 2.805E-08	Average 6.472E-08
	STD 0.392	STD 7.877E-10	STD 9.443E-10
	RSD 0.029	RSD 0.028	RSD 0.015

The measurements are done with 5-8 replicates.

*The mobity of electroosmosis is calculated from each of the analyses by using methanol as the neutral marker. Calculations made with the equation $\mu_{ep} = (L_{det} L_{tot}) / (U t_m)$ and $\mu_{eo} = (L_{det} L_{tot}) / (U t_{eo})$, where μ_{ep} and μ_{eo} are the electrophoretic mobilities of the analyte and electroosmosis, L_{det} is the length of the capillary to the detector, L_{tot} is the length of the total capillary, U is the applied voltage during the analysis, and $_{tm}$ and t_{eo} are the migration times of the analyte and electroosmosis (from the electropherogram), respectively.

3.2. Calibration

The concentration calibration for the steroids with the PT-MEKC method was made at concentrations 0.5-10 μg/mL. The steroid concentrations were correlated with the corresponding peak areas in the electropherograms by linear fitting (Table 5).

The LOD and LOQ values were 0.05-1.062 μg/mL and 0.501-10.62 μg/mL, respectively. Because the calculated volume of the sample in the capillary was only 6.50 nL, the steroid quantities were 0.325-6.903 pg and 3.257-69.03 pg, respectively.

3.3. Water Samples

The knowledge that the concentrations of estrogen and testosterone in urine samples are 1-40 μg/L and that the lowest detection needed is 100 ng/mL for androstenedione, testosterone, 17α-methyltestosterone, and progesterone, one liter of the environmental water was needed to be 100 -1000 times concentrated before the PT-MEKC analyses (Figure 4). The validation of the method informed that the SPE treatment was the fastest to prepare the samples for the analyses. The water purification with adsorption is fast and inexpensive. However, without molecular imprinting material the SPE is not analyte-specific. This fact has also noticed in other projects. Carballa et al. [41] studied three hormones, estrone, 17*β*-estradiol, and 17*α*-ethinylestradiol in waters of a municipal Sewage Treatment Plant in Galicia, Spain. They noticed significant concentrations of steroids only in the influent, which contained estrone and 17*β*-estradiol. In our earlier studies [42] we noticed with testosterone (T) and epitestosterone (E) in human urine, that after SPE the analytes were easily identified in the matrix. No derivatization of the analytes was required. Thus, the present PT-MEKC method is a tempting alternative for the conventional laborious GC–MS analysis of the steroids (Figure 5).

Table 5. Quality control values of the steroids studied with the validated PT-MEKC method

Compound	Linear equation	R^2 value	Concentration range [μg/mL]	LOD** [μg/mL]	LOQ*** [μg/mL]
4-androsten-17β-ol-3-one glucosiduronate	y = 1.2702x + 0.0054	0.9126	0.5-8.0	0.05	0.501
4-androsten-9α-fluoro-17α-methyl-11β, 17β-diol-3-one	y = 0.4688x + 0.022	0.9662	0.5-8.0	0.120	1.198
Androstenedione	y = 0.632x + 0.0294	0.9404	0.5-8.0	0.063	0.627
Testosterone	y = 0.779x + 0.2139	0.9624	0.5-8.0	0.942	9.420
17α-hydroxyprogesterone	y = 1.1506x – 0.354	0.9473	0.5-6.0	0.383	3.829
17α-methyltestosterone	y = 2.9447x – 3.0403	0.9688	0.5-10	1.062	10.62
progesterone	y = 4.3152x – 4.0771	0.9674	0.5-10	0.965	9.650

**LOD was measured from the electropherogram peak area of know steroid concentration (S, signal) divided with the average noise peak area (N, noise) with S/N = 3.

***LOQ was measured from the corresponding LOD of the steroid by multiplying with 10 (LOQ = 10 x LOD).

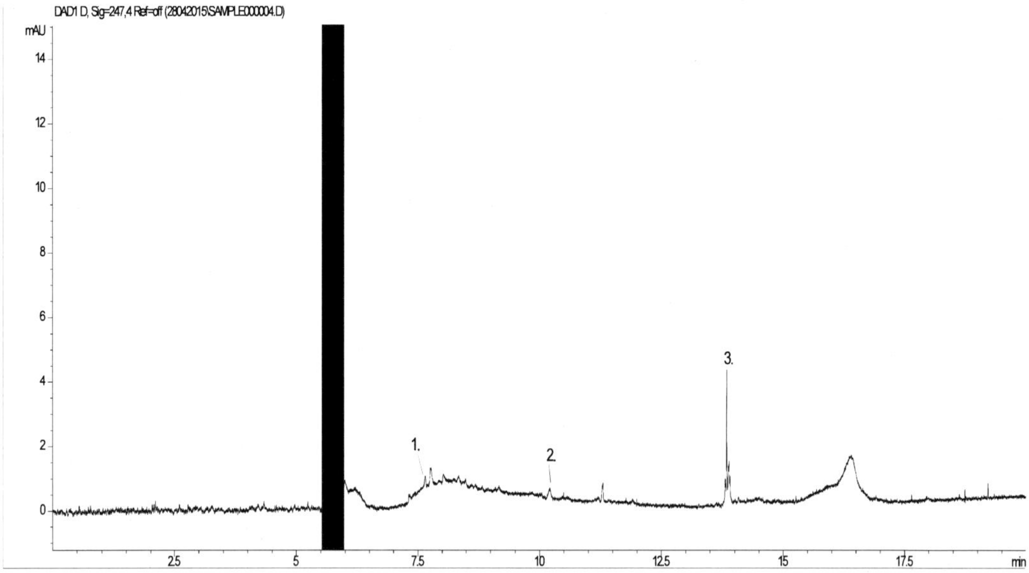

Figure 4. Chromatogram of the effluent from SF plant. The negative peak after the first intensive peak around 5 min is the marker for electroosmosis. Compounds identified with spiking of 3 µg/ mL standards. Detection at UV-247 nm. Compounds existing in the water: 1) 4-androsten-17β-ol-3-one glucoside, 2) androstenedione, and 3) progesterone. The details of sample concentration and clean-up are in Experimental. Before analysis, the sample was centrifuged at 2000 rpm for 10 min. Other details as in Figure 3.

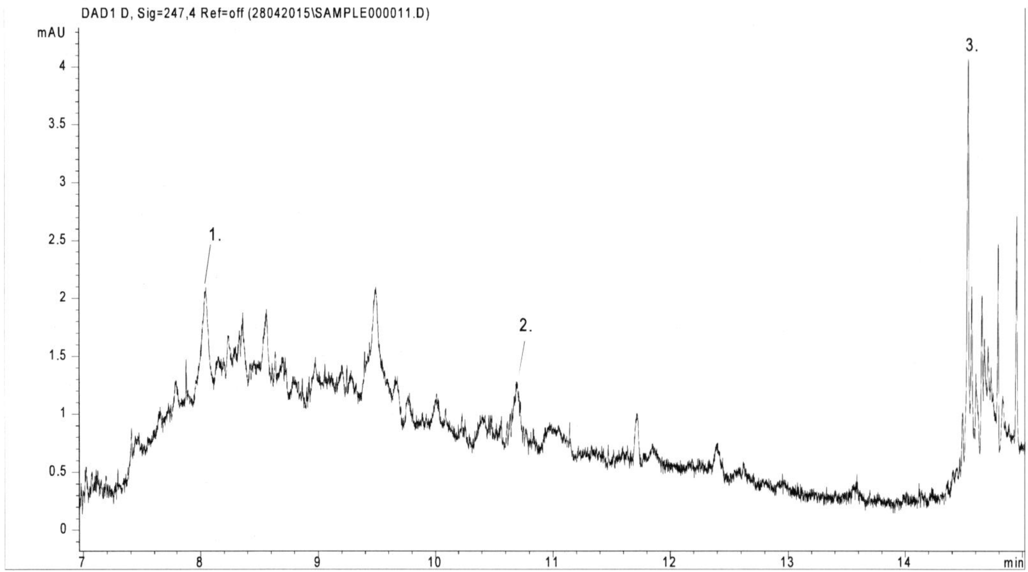

Figure 5. A PF-MEKC-UV electropherogram of effluent water from SF plant. Water concentrated with SPE. Peaks identified after enrichment contained 1) 4-androsten-17β-ol-3-one glucosiduronate, 2) androstenedione, and 3) progesterone. Detection at UV-247 nm. The injection volume to the 80-cm capillary was 6.5 nL. Before analysis, the sample was centrifuged at 2000 rpm for 10 min. The experimental conditions are as in Figure 3.

Table 6. Steroids in influent and effluent waters of water treatment plants

SF	Influent				Effluent			
	Migration time [min]	Area [min*mAU]	Height [mAU]	Calculated concentration [ng/L]	Migration time [min]	Area [min*mAU]	Height [mAU]	Calculated concentration [ng/L]
4-androsten-17β-ol-3-one glucosiduronate	8.046	4.696	0.920	56.5	7.646	0.779	0.400	16.7
Androstenedione	10.692	3.197	0.540	45.7	10.209	1.658	0.400	20.9
progesterone	14.536	3.520	3.700	3.22	13.837	2.974	4.200	2.72
4-androsten-17β-ol-3-one glucosiduronate	-	-	-	-	7.866	2.383	0.920	5.20
Androstenedione	11.902	2.329	0.890	29.4	11.489	1.817	0.800	22.9
progesterone	14.810	4.992	4.900	12.4	14.153	3.261	4.300	8.11
4-androsten-17β-ol-3-one glucosiduronate	8.173	10.078	1.800	19.8	7.981	3.221	0.970	6.31
Androstenedione	12.076	5.413	1.000	68.3	11.717	2.210	0.780	27.9
progesterone	15.177	8.237	5.400	28.5	14.539	5.398	4.800	18.7

Five repetitions with five injections. All the values are averages from the repeated measurements.

The low nanogram per liter range existing steroids in the effluent and the influent waters could be concentrated with solid-phase extraction enough for the PT-MEKC studies. The results showed that the studied influents and effluents of drinking water treatment plants contained notable amounts of 4-androsten-17β-ol-3-one glucosiduronate, androstenedione, and progesterone (Figure 5, Table 6). According to the results of the present study, both the influent and the effluent waters contained the steroids. In the plants after water purification, the effluent waters were remarkable cleaner than the influent water. The intake concentrations of 3.22-68.3 ng/L were decreased to 2.72-27.9 ng/L level.

CONCLUSION

Capillary electrophoresis could be used for comprehensive profiling of steroids in influent and effluent waters. Seven steroids were analyzed after solid phase extraction and enrichment of the waters. CE method was separating the free steroids from the conjugated ones. Thus, the present method was an alternative for both GC–MS/MS and LC-MS/MS methods.

ACKNOWLEDGMENTS

The authors like to thank the Foundation Maa- ja Vesitekniikan Tuki ry for financing the project during the years 2014-2015.

REFERENCES

[1] D. T. Bennier, Review of the environmental occurrence of alkylphenols and alkylphenol ethoxylates, *Water Qual. Res. J. Can.* 34 (1999) 79-122.

[2] S. A. Snyder, Occurrence, Treatment, and Toxicological Relevance of EDCs and Pharmaceuticals in Water. *Ozone: Science and Engineering* 30 (2008) 65-69.

[3] C. Ort, M. G. Lawrence, J. Rieckermann, A. Joss, Sampling for pharmaceuticals and personal care products (PPCPs) and illicit drugs in wastewater systems: Are your conclusions valid? A critical review. *Environ. Sci. Technol.* 44 (2010) 6024-6035.

[4] M. R. Servos, M. Smith, R. McInnis, K. Burnison, B. H. Lee, P. Seto, S. Backus, The presence of selected pharmaceuticals and the antimicrobial triclosan in drinking water in Ontario Canada, *Water Qual. Res. J. Canada* 42 (2007) 65-69.

[5] S. Jobling, R. Williams, A. Johnson, A. Taylor, M. Cross-Sorokin, M. Nolan, C. R. Tyler, R. van Aerle, E. Santos, G. Brighty, Predicted exposures to steroid estrogens in UK rivers correlate with widespread sexual disruption in wild fish populations, *Environmental Health Perspective* 114 (2006) 32-39.

[6] A. J. Whelton, L. K. McMillan, M. Connell, K. M. Kelley, J. P. Gill, K. D. White, R. Gupta, R. Dey, C. Novy. Residential Tap Water Contamination Following the Freedom Industries Chemical Spill: Perceptions, Water Quality, and Health Impacts, *Environmental Science and Technology* 49 (2015) 813-823.

[7] M. F. Wilson. Agriculture and Industry as Potential Origins for Chemical Contamination in the Environment. A Review of the Potential Sources of Organic Contamination, *Current Organic Chemistry* 17 (2013) 2972-2975.

[8] A. Arditsoglou, D. Voutsa. Partitioning of EDCs in inland waters and wastewaters in the coastal area of Thessaloniki, Northern Greece. *ESPR-Environ. Sci. Pollut. Res.* 17 (2010) 529-538.

[9] T. Vega-Morales, Z. Sosa-Ferrera, J. J. Santana-Rodriquez. Determination of alkylphenol polyethoxylates, bisphenol-A, 17α-ethynylestradiol and 17β-estradiol and its metabolites in sewage samples by SPE and LC/MS/MS. *Journal of Hazardous Materials* 183 (2010) 701-711.

[10] C. Stavrakakis, R. Colin, V. Hequet, C. Faur, P. Le Cloirec. Analysis of endocrine disturbing compounds in wastewater and drinking water treatment plants at the nanogram per litre level. *Environmental Technology* 29 (2008) 279-286.

[11] K. Kumar, S. C. Gupta, Y. Chander, A. K. Singh. Antibiotic use in agriculture and its impact on the terrestrial environment. *Advanced in Agronomy* 87 (2005) 1-54.

[12] A. Johnson, M. Jurgens. Endocrine active industrial chemicals: Release and occurrence in the environment. *Pure Appl. Chem.* 75 (2003) 1895-1904.

[13] L. S. Shore, M. Shemesh, Naturally produced steroid hormones and their release into the environment. *Pure Appl. Chem.* 75 (2003) 1859-1871.

[14] P. Diercxsens, J. Tarradellas. Soil contamination by some organic micropollutants related to sewage sludge spreading. *Int. Environ. Anal. Chem.* 28 (1987) 143-159.

[15] W. W. Buchberger. Current approaches to trace analysis of pharmaceuticals and personal care products in the environment. *J. Chromatogr. A* 1218 (2011) 603-618.

[16] B. J. Vanderford, J. E. Drewers, A. Eaton, Y. C. Guo, A. Haghani, C. Hoppe-Jones, M. P. Schluesener, S. A. Snyder, T. Ternes, C. J. Wood. Results of an interlaboratory

comparison of analytical methods for contaminants of emerging concern in water. *Anal. Chem.* 86 (2014) 774-782.

[17] M. Petrović, M. D. Hernando, M. S. Díaz-Cruz, D. Barceló, Liquid chromatography–tandem mass spectrometry for the analysis of pharmaceutical residues in environmental samples: A review. *Journal of Chromatography A* 1067 (2005) 1-14.

[18] S. Görög, Recent Advances in the Analysis of Steroid Hormones and Related Drugs, *Analytical Sciences* 20 (2004) 767-782.

[19] S. Liu, G. G. Ying, J. L. Zhao, F. Chen, B. Yang, L. J. Zhou, H. J. Lai, Trace analysis of 28 steroids in surface water, wastewater and sludge samples by rapid resolution liquid chromatography-electrospray ionization tandem mass spectrometry. *J. Chromatogr. A* 1218 (2011) 1367-1378.

[20] T. Sihvonen, A. Aaltonen, J. Leppinen, S. Hiltunen, H. Sirén, A novel capillary electrophoresis method with pressure assisted field amplified sample injection in determination of thiol collectors in flotation process waters. *Journal of Chromatography A* 1325 (2014) 234-240.

[21] L. K. Amundsen, J. T. Kokkonen, S. Rovio, H. Sirén. Analysis of anabolic steroids by partial filling micellar electrokinetic capillary chromatography and electrospray mass spectrometry *Journal of Chromatography A* 1040 (2004) 123-131.

[22] H. Sirén, T. Seppänen-Laakso, M. Oresic, Capillary electrophoresis with UV detection and mass spectrometry in method development for profiling metabolites of steroid hormone metabolism. *J. Chromatogr. B: Anal. Technol. Biomed. Life Sci.* 871 (2008) 375-382.

[23] H. Sirén, S. Vesanen, J. Suomi, Separation of steroids using vegetable oils in microemulsion electrokinetic capillary chromatography. *Journal of Chromatography B: Analytical Technologies in the Biomedical and Life Sciences* 945-946 (2014)199-206.

[24] L. K. Amundsen, J. T. Kokkonen, H. Sirén, Comparison of partial filling MEKC analyses of steroids with use of ESI-MS and UV spectrophotometry. *Journal of Separation Science* 31(5) (2008) 803-813.

[25] M. Matczuk, L. S. Foteeva, M. Jarosz, M. Galanski, B. K. Keppler, T. Hirokawa, A. R. Timerbaev, Can neutral analytes be concentrated by transient isotachophoresis in micellar electrokinetic chromatography and how much? *Journal of Chromatography A* 1345 (2014) 212-218.

[26] C. Stavrakakis, R. Colin, V. Hequet, C. Faur, P. Le Cloirec, Analysis of endocrine disturbing compounds in wastewater and drinking water treatment plants at the nanogram per litre level, *Environmental Technology* 29 (2008) 279-286.

[27] F. D. L. Leusch, M. R. van den Heuvel, H. F. Chapman, S. R. Gooneratne, A. M. E. Eriksson, L. A. Tremblay, Development of methods for extraction and in vitro quantification of estrogenic and androgenic activity of wastewater samples. *Comparative Biochemistry and Physiology, Part C: Toxicology and Pharmacology* 143C (2006) 117-126.

[28] F. D. L. Leusch, H. F. Chapman, M. R. van den Heuvel, B. L. L. Tan, S. R. Gooneratne, L. A. Tremblay, Bioassay-derived androgenic and estrogenic activity in municipal sewage in Australia and New Zealand, *Ecotoxicology and Environmental Safety* 65 (2006) 403-411.

[29] B. L. L. Tan, D. W. Hawker, J. F. Mueller, F. D. L. Leusch, L. A. Tremblay, H. F. Chapman, Comprehensive study of endocrine disrupting compounds using grab and

passive sampling at selected wastewater treatment plants in South East Queensland, Australia. *Environment International* 33(5) (2007) 654-669.

[30] L. N. Nguyen, F. I. Hai, S. Yang, J. Kang, F. D. L. Leusch, F. Roddick, W. E. Price, L. D. Nghiem, Removal of pharmaceuticals, steroid hormones, phytoestrogens, UV-filters, industrial chemicals and pesticides by Trametes versicolor: Role of biosorption and biodegradation, *International Biodeterioration and Biodegradation* 88 (2014) 169-175.

[31] Z. H. Liu, Y. Kanjo, S. Mizutani, Removal mechanisms for endocrine disrupting compounds (EDCs) in wastewater treatment — physical means, biodegradation, and chemical advanced oxidation: A review. *Science of the Total Environment* 407 (2009) 731-748.

[32] L. N. Nguyen, F. I. Hai, S. Yang, J. Kang, F. D. L. Leusch, F. Roddick, W. E. Price, L. D. Nghiem, Removal of pharmaceuticals, steroid hormones, phytoestrogens, UV-filters, industrial chemicals and pesticides by Trametes versicolor: Role of biosorption and biodegradation, *International Biodeterioration and Biodegradation* 88 (2014) 169-175.

[33] L. E. Panin, P. V. Mokrushnikov, V. G. Kunitsyn, B. N. Zaitsev, Interaction mechanism of anabolic steroid hormones with structural components of erythrocyte membranes, *Journal of Physical Chemistry B* 115(50) (2011) 14969-14979.

[34] G. R. Tetreault, C. J. Bennett, K. Shires, B. Knight, M. R. Servos, M. E. McMaster, Intersex and resproductive impairment of wild fish exposed to multiple municipal wastewater discharges, *Aquatic Toxicology* 104 (2011) 278-290.

[35] T. Trinh, B. van den Akker, R. M. Stuetz, H. M. Coleman, P. Le-Clech, S. J. Khan, Removal of trace organic chemical contaminants by a membrane bioreactor, *Water Science and Technology* 66 (2012) 1856-1863.

[36] T. Vega-Moealez, Z. Sosa-Ferrera, J. J. Santa-Rodriquez, Determination of alkylphenol polyethoxylates, bisphenol-A, 17α-ethylestradiol and 17β-estradiol, and its metabolites in sewage samples by SPE and LC-MS/MS. *Journal of Hazardous Materials* 183 (2010) 701-711.

[37] S. Mompelat, B. Le Bot, O. Thomas, Occurrence and fate of pharmaceutical products and by-products, from resource to drinking water, *Environmental International* 35 (2009) 803-814.

[38] J. Aufartová, C. Mahugo-Santana, Z. Sosa-Ferrera, J. J. Santana-Rodríguez, L. Nováková, P. Solich, Determination of steroid hormones in biological and environmental samples using green microextraction techniques: An overview. *Anal. Chim. Acta* 704(1-2) (2011) 33-46.

[39] P. K. Zarzycki, E. Wlodarzyk, M. J. Baran, Determination of endocrinine disturbing compounds using temperature-dependent inclusion chromatography II: Fast screening of free steroids and related low-molecular-mass compounds fraction in the environmental samples derived from surface waters, treated and untreated sewage waters as well as activated sludge material. *J. Chromatogr. A* 1216 (2009) 7612-7622.

[40] T. Vega-Morales, Z. Sosa-Ferrera, J. J. Santana-Rodriquez, Determination of alkylphenol poly-ethoxylates, bisphenol-A, 17a-ethynylestradiol and 17b-estradiol and its metabolites in sewage samples by SPE and LC/MS/MS. *Journal of Hazardous Materials* 183 (2010) 701-711.

[41] M. Carballa, F. Omil, J. M. Lema, M. Llompart, C. García-Jares, I. Rodríquez, M. Gómez, T. Ternes, Behaviour of pharmaceuticals, cosmetics, and hormones in a sewage treatment plant, *Water Research* 38 (2004) 2918-2926.

[42] L. K. Amundsen, T. K. Nevanen, K. Takkinen, S. Rovio, H. Sirén, Microscale immunoaffinity SPE and MEKC in fast determination of testosterone in male urine, *Electrophoresis* 28(18) (2007) 3232-3241.

In: Capillary Electrophoresis (CE)
Editor: Christian Reed

ISBN: 978-1-63483-122-2

Chapter 10

CAPILLARY ELECTROPHORESIS WITH LASER-INDUCED FLUORESCENCE DETECTION: CHALLENGES IN DETECTOR DESIGN, LABELING AND APPLICATIONS

Marketa Vaculovicova, Vojtech Adam and Rene Kizek *

Department of Chemistry and Biochemistry, Mendel University in Brno, Brno, Czech Republic

Central European Institute of Technology, Brno University of Technology, Technicka, Brno, Czech Republic

ABSTRACT

In CE, the synchronization of three major elements - injection, separation, and detection – is responsible for successful analyte determination. All these parts are indispensable and failure of either of them spoils the whole analysis.

The current goal of determination of extremely low concentrations in extremely low sample amounts leads to developments especially in detection part of the setup. It is most commonly realized by the UV/Vis photometric detection; however, its drawback is in a relatively low sensitivity. Nevertheless, by utilization of fluorescence detection even picomolar levels can be reached. Currently, a variety of both covalent and non-covalent labeling probes from the area of either small organic molecules or nanomaterial-based labels with high quantum yields is available.

Besides the development of fluorescent labels, also instrumental advances in the field of detector design enhance the sensitivity and applicability of this detection mode. In hard competition with other techniques, especially mass spectrometry, the fluorescence detection remains important player with significant advantages.

In this chapter is summarized not only the state of the art of the instrumental developments but also labeling strategies utilizing well-established and modern

* E-mail: kizek@sci.muni.cz, tel. : +420 545 133 350, fax.: +420 545 212 044. Department of Chemistry and Biochemistry, Mendel University in Brno, Zemedelska 1, CZ-613 00 Brno, Czech Republic, European Union

fluorescent tags. Finally, selected applications of capillary electrophoresis with laser-induced fluorescence detection are highlighted.

Keywords: Laser-induced fluorescence detection, labeling, nanomaterial

INTRODUCTION

Capillary electrophoresis (CE) can be variably coupled with numerous detection techniques. Each of them has its own advantages as well as disadvantages in terms of selectivity, sensitivity and/or versatility. Both, on-capillary and off-capillary detection modes are available. On-capillary detection is a nondestructive approach minimizing the band broadening and enabling the employment of several detectors simultaneously (either consecutively or at the same detection point, however the off-capillary methods may provide additional information such as molecular mass. Photometric (or absorbance) detection is outstanding due to its versatility and therefore it is the most commonly used technique. However, the internal diameter of the capillary (generally 10-100 μm) defines the optical path length. Therefore, relatively high analyte concentrations, extended path length flow cells (bubble cell, Z-cell), or preconcentration techniques are required to reach satisfactory results.

Currently, mass spectrometric detection is attracting enormous attention in both in-line (electrospray ionization, ionization by inductively coupled plasma) and off-line (matrix assisted laser desorption/ionization) mode [1]. The biggest advantage of this detection is that the analyte identification by its molecular mass information is provided. The unique resolution of current instruments allows the identification of thousands of analytes within a single separation run.

Besides the above mentioned detection modes, a number of others including electrochemical [2], chemiluminescence [3] and/or electrochemiluminescence [4] methods are employed, however one outstanding technique - laser-induced fluorescence detection (LIF) - has to be highlighted especially due to its extremely low limits of detection (~ 10^{-9} – 10^{-12} mol L^{-1}).

Even though the range of detection techniques used in CE is very wide, none of them covers all the aspects of universal, sensitive, selective, miniaturized and/or easy to use detection. Therefore, combined detection techniques have been developed, integrating two or three detection modes in a single device [5].

INSTRUMENTATION

Detector Design

Excitation light source, detection cell and fluorescence detection device are the three key parts of the laser-induced fluorescence detector.

Generally, xenon and deuterium lamps belong to the light sources providing the biggest variability in the excitation wavelength selection and by using appropriate excitation filter, the wavelength from deep ultra violet to near infrared can be selected. However, due to the divergence of the light, additional collimating optics has to be employed to focus the light to the center of the capillary and also the optical output per wavelength is relatively low. To overcome these limitations, utilization of lasers as excitation light sources, providing the coherent light with sufficient optical power, is an option. Nevertheless, the single wavelength light sources compromise the flexibility of the application. Furthermore, the spectral coverage is lower in the case of lasers compared to the lamp-based light sources, and especially in the deep ultra violet region, the selection is limited. An alternative is currently represented by the light emitting diodes (LEDs) especially due to their small dimensions, stable output, long lifetimes and low costs. The availability of UV LEDs is steadily increasing as well as their optical power. In general, LEDs provide a broader spectral bandwidth than lasers (spectral half-width typically 20–30 nm).

The excitation light passes through number of optical components (filters, slits, lenses) and capillary walls before exciting the analytes and subsequently the resulting fluorescence is emitted into all directions and therefore to detect the maximum of the emitted fluorescence is a challenge addressed in number of detection cell designs (Figure 1).

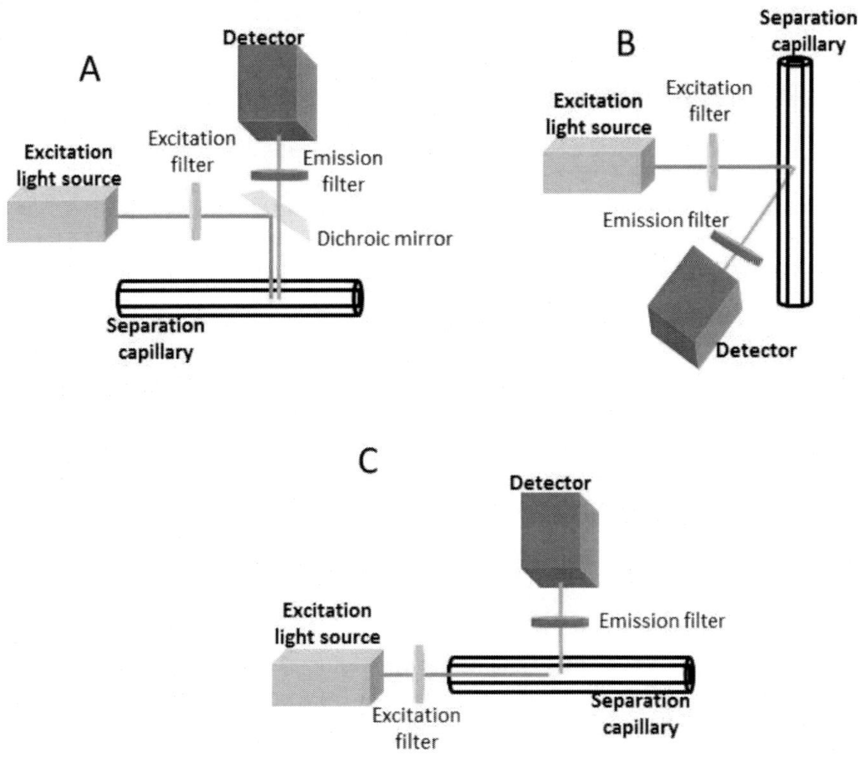

Figure 1. Schemes of different cell designs for CE-LIF. A) collinear B) right-angle C) in-column excitation.

To widen the flexibility of the CE detection, modular detectors enabling either absorbance, LIF or both detection modes are beneficial. Moreover, utilization of the modular system suitable for easy exchange of excitation sources (LEDs) and particular optical filters (Figure 2) opens the options for use of just one detection system for analysis of a variety of fluorophores in the number of applications. In comparison with bulky, expensive laser light source, such system is more versatile and cost effective.

To detect the emitted light, photomultipliers (PMT) are utilized in most of the cases, due to their high sensitivity and low noise. A semiconductor analogue of the PMT is the avalanche photodiode (APD). The typical quantum efficiency ranges from 75 to 85%, the active area diameter may vary from mm to tens of micrometers and the spectral response ranges from UV to IR [6]. The compact dimensions of APDs and the lower price compared with those of PMTs are advantageous and therefore they are excellent alternative for lab-on-chip applications.

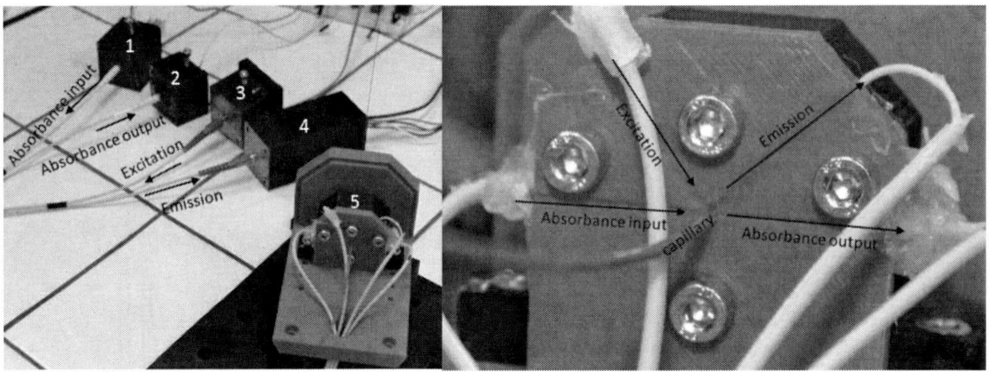

Figure 2. A) Components of dual detector (absorbance, fluorescence) - 1 – exchangeable absorbance LED light source, 2 – absorbance detector – photodiode, 3 – exchangeable fluorescence LED light source (excitation) with replaceable filter holder, 4 – fluorescence detector – PMT (emission) with replaceable filter holder, 5 – detection point; B) Detail of the detection point – optical fibers focused into the capillary.

In applications requiring the spatial signal distribution, charged-coupled devices (CCDs) are employed. The initial costs of sensitive CCDs are significantly higher compared with APDs or PMTs. For these reasons, CCDs are usually used in imaging applications or in wavelength-resolved measurements in bench-top systems [6].

Miniaturization in CE-LIF

General trend of analytical instrumentation miniaturization is pronounced in CE more than in any other method especially due to its relatively simple instrumental setup. Development of miniaturized microfluidic electrophoretic chips is closely connected to the development in the area of miniaturized light sources – laser diodes and LEDs. Laser diodes are significantly smaller compared to lasers, which is advantageous in miniaturization and portability point of view. Since 1960s, LEDs became commercially available emitting in the red range of spectra. Subsequently, shorter wavelengths appeared and currently it is possible to obtain LEDs emitting the light in deep-UV (approx. 240 nm). From the application point of view, UV-LEDs are more applicable especially in the bioanalytical area; however, the optical power of these LEDs is still significantly lower compared to the longer-wavelength ones because more complicated formation of semiconductor junctions with higher bandgaps is required [7]. Compared to laser-based detectors, simplified instrumentation can be obtained if a LED and a silicon photodiode are used instead of a laser source and photomultiplier tube, respectively; however, the reduced sensitivity is a trade-off [8].

In general, LIF detection is the most widely applied in miniaturized microfluidic electrophoretic devices due to the fact that the amount of analyte in the chip channel is even lower than in the conventional CE and therefore extremely sensitive detection is needed [9-11].

In general, two approaches can be distinguished in the field of coupling LIF detection with microchip CE: 1) off-chip approach is joining the macro-scale detection devices with micro-scale detection points by waveguides, optical fibers, pinholes and slits, 2) on-chip approach is integrating the components of the detection device straight to the chip creating a compact, portable platform with minimum external connections.

The off-chip arrangement is instrumentally simpler and therefore more often used in the research studies. On the other hand, commercially available instruments are also in the off-chip mode taking advantage of the disposability of individual chips.

Even though the commercial chip CE platforms are usually dedicated to routine analyses and therefore protected from any changes within the instrument setup, some minor modifications can be done including incorporation of the isolation step (Figure 3) [12].

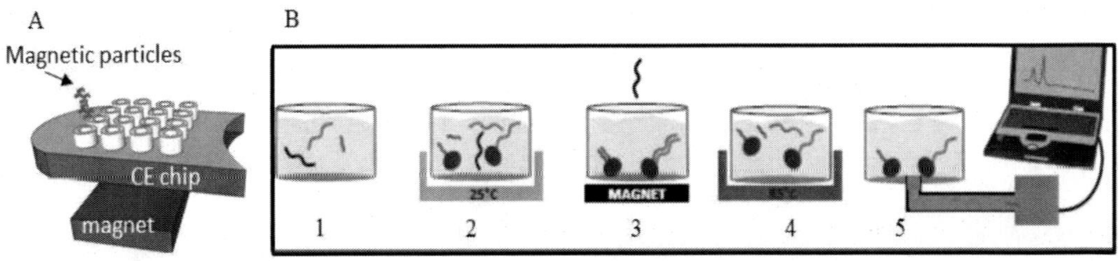

Figure 3. Scheme of the application of the commercial CE chip platform for specific isolation of molecules by magnetic particles. A) Commercial CE chip with magnetic particles placed inside the sample well and magnet underneath to manipulate them, B) Scheme of the isolation process for extraction of target oligonucleotides (1 – sample solution containing interferent (black), 2 – hybridization of target analytes with specific probes on the nanoparticle surface, 3 – immobilization by magnet and removal of interferent, 4 – release analytes from the nanopraticles by elevated temperature, separation of the released analytes by chip CE),

However, to reach truly miniaturized lab-on-chip systems, the integration of all components (excitation light source, filters, lenses, mirrors and detection device) into the chip is required. Therefore, the development of so-called organic electronics is advancing.

Organic light emitting diode (OLED) is a special type of diode formed by emissive electroluminescent layer of organic compound emitting light as a response to electric current. Currently, OLEDs are finding their applications in digital displays (television, computer monitors, and mobile phones); however, also their utilization in detection systems is also convenient. The big advantage of OLEDs compared to the conventional LEDs is their flat shape, which makes them easy to apply into microfluidic devices and to bring them into close proximity of the channel. Moreover, the fabrication in any size and shape by photolithography techniques is possible. The disadvantage can be seen in their relatively broad emission spectra requiring the additional application of excitation filters [13] to eliminate as much as possible the part of the excitation light, which overlaps with the emission spectrum of analytes in order to suppress the background interference [14]. Also quite low light intensity may be a limitation. The summary of selected properties of light sources is given in Figure 4.

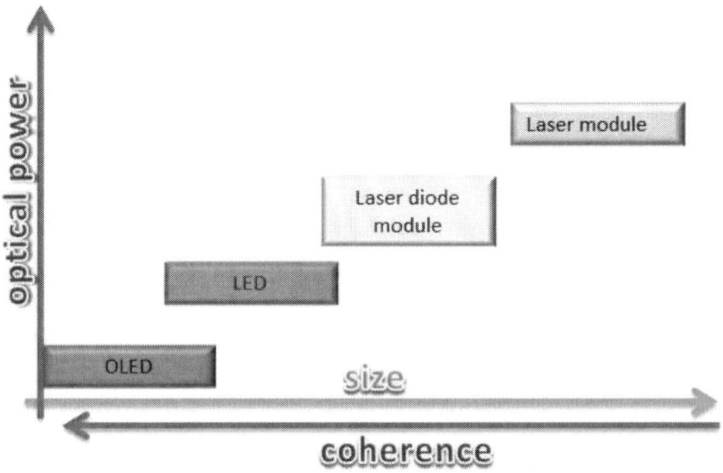

Figure 4. Excitation light sources and their selected properties (optical power, light coherence and size).

Not only organic light sources but also organic detectors are an option for compact, inexpensive, biodegradable and disposable microfluidic applications. Organic photodiodes, for instance, offer the best potential for future lab-on-chip technology, as they are inexpensive, are easily fabricated, have a large dynamic range, and are highly sensitive.

Label-Free LIF Detection

The advantage represented by the selectivity of the LIF detection given by the required match between absorption properties of the analyte and excitation properties of the light source is at the same time the factor limiting the applicability for native forms of analytes. Even though there are some exceptions (e.g., riboflavin, doxorubicin, ellipticine), the intrinsic fluorescence of most of the analytes can be excited only by UV light sources (200-400 nm), because the optimal excitation wavelength depends on the energy gap between ground state and excited state. Fluorescence of most organic molecules occurs via a radiative $S1 \rightarrow S0$ transition. This transition is located around or above 300 nm for most aromatic moieties [15].

The advances of native fluorescence LIF detection is closely connected to the development in the area of UV light sources (solid state lasers, laser diodes and LEDs), UV transparent optical components (lenses, filters, mirrors, optical fibers, etc.) The optical transparency of the used materials has to be adapted for this type of detection and therefore fused silica and/or borosilicate glass are the optimal materials. For microfluidic applications mostly polydimethylsiloxane and poly(methyl methacrylate) are employed.

In the area of peptides/protein research, tryptophan, tyrosine and phenylalanine are the key amino acids of interest. Similarly to other fluorophores, tryptophan fluorescence is strongly dependent on pH. The maximum values are obtained between pH 9 and 11.

Contrary, at acidic and neutral pH (3–8) the emission intensity is reduced for 30% [6]. The optical properties of three fluorescent amino acids at neutral pH are given in Table 1 [16].

Besides protein and peptide applications, native CE-LIF has been used for a number of other applications such as in the fields of cell [17, 18] neurotransmitter [19], vitamins [20, 21], and drug [22] analysis.

Table 1. Optical properties of fluorescent amino acids

	Absorption maximum (nm)	Emission maximum (nm)	Quantum yield
Tryptophan	280	348	0.2
Tyrosine	274	303	0.14
Phenylalanine	257	282	0.04

FLUORESCENT LABELING FOR LIF

In case of non-fluorescent analytes, the derivatization by fluorescent probes providing the appropriate optical properties, is an option. Such labeling may be carried out in various arrangements such as the pre-capillary, in-capillary and post-capillary mode.

Pre-column labeling procedures are still the most frequently used ones for labeling purposes. The advantages of this arrangement include the applicability for reagents requiring conditions such as elevated temperatures or long reaction times. In addition, purification of the reaction product to remove the excess reagent is possible [23]. However, this may increase the total analysis time and often, the time required for fluorescent labeling exceeds the time required to conduct the CE separation [24]. A special type of pre-column derivatization is the on-line mode, in which the derivatization takes place by mixing the analytes with the reagent just before the capillary using an on-line coupled reactor (i.e., T-junction). This mode requires a reaction with short interaction time and low reactant volumes. The next step is the on-capillary derivatization technique, where the reaction is taking place inside the capillary during electrophoresis. The reaction components (analyte and reagent) are brought together either in the tandem or in the sandwich mode. Due to the different migration velocities, the reactants mix during migration through the capillary. Generally, the on-capillary derivatization mode is suitable for very small sample volumes, since dilution is reduced to the minimum. Finally, post-capillary derivatization is another technique for labeling the analyte prior the detection. The labeling is taking place after the CE separation and therefore among the advantages belongs that the analytes are separated in the native form, thus avoiding interferences from side products as well as band broadening, caused by multiple derivatization reactions. However, the negative effects on peak efficiency, loss of analyte, incomplete reactions and higher baseline noise are obvious [23]. Moreover, the instrumental requirements such as low dead-volume and band distortion effects have to be taken into account.

In general, the optimal derivatization reagent should provide high quantum yield, high Stokes' shift and resistance to quenchers [25]. The obstacle brought by most of the derivatization procedures is the problem of multiple labeling. Especially, in the case of proteins and other biomolecules, the labeling reaction may lead to molecules differing in number of attached fluorophores and resulting in the formation of multiple products. This problem is particularly pronounced in case of the proteins labeled by amino-reactive labels due to their reactivity not only with the terminal amino group of the protein but also with lysine side chain amino group. As a result, a wide envelope of products differing in electrophoretic mobilities is formed.

Fluorescent labeling is applicable to the number of analytes such as carbohydrates [26, 27], lipids [28, 29] and nucleic acids [30]; however, the biggest group undergoing this procedure is covering proteins, peptides and amino acids [31-33].

Non-Covalent Labeling

In non-covalent labeling, a number of mechanisms including hydrophobic interactions, electrostatic interactions and/or hydrogen bonding are involved. The exact nature of these interactions is often difficult to determine, but the evidence of interaction is provided by a change in the emission of the fluorophore–analyte complex relative to that of the free, unreacted reagent [24].

This type of derivatization is advantageous especially due to its minimal sample preparation (usually the analyte and label are just mixed together), the reaction is taking place at biological conditions (i.e., neutral temperatures and pHs) and provides short analysis times (therefore applicable for on-column derivatization). Moreover, the application for post-column labeling purposes is enabled providing benefits especially in the analyses of analyte-fluorophore conjugates with short half-life.

On the other hand, this type of labeling is often less sensitive than the covalent derivatization, the interactions tend to be less selective than covalent ones and non-covalently labeled protein complexes are typically less robust [24].

Even though the cyanine dyes (Cy) belong to a group of the oldest synthetic labeling probes, they find applications in a number of areas. They are able to interact with the biomolecule either through covalent or non-covalent bonding. Commonly known as Cy3 and Cy5 became the fluorophores of choice primarily due to their remarkable photostability, large absorption cross sections and fluorescence efficiencies and compatibility with common lasers. Cy dyes are used for labeling proteins as well as nucleic acids. Depending on the structure, they cover the spectrum from IR to UV. Cy dyes are cationic molecules in which two heterocyclic units are joined by a polyene chain [25, 34] One particularly important Cy dye labeling mechanism is the intercalation into the double helical DNA exhibiting large fluorescence enhancements upon binding [35].

Covalent Labeling by Organic Dyes

The covalent interaction between the derivatization dye and the targeted protein/peptide may be carried out employing various functional groups including -NH_2, -COOH, -SH and/or –OH depending on the aim of the analysis. The derivatization principles are common to all fluorescence-based visualization techniques (HPLC, GC, microscopy, imaging). In most cases, N-terminus amino group of the protein or lysine side-chain is used for labeling and the

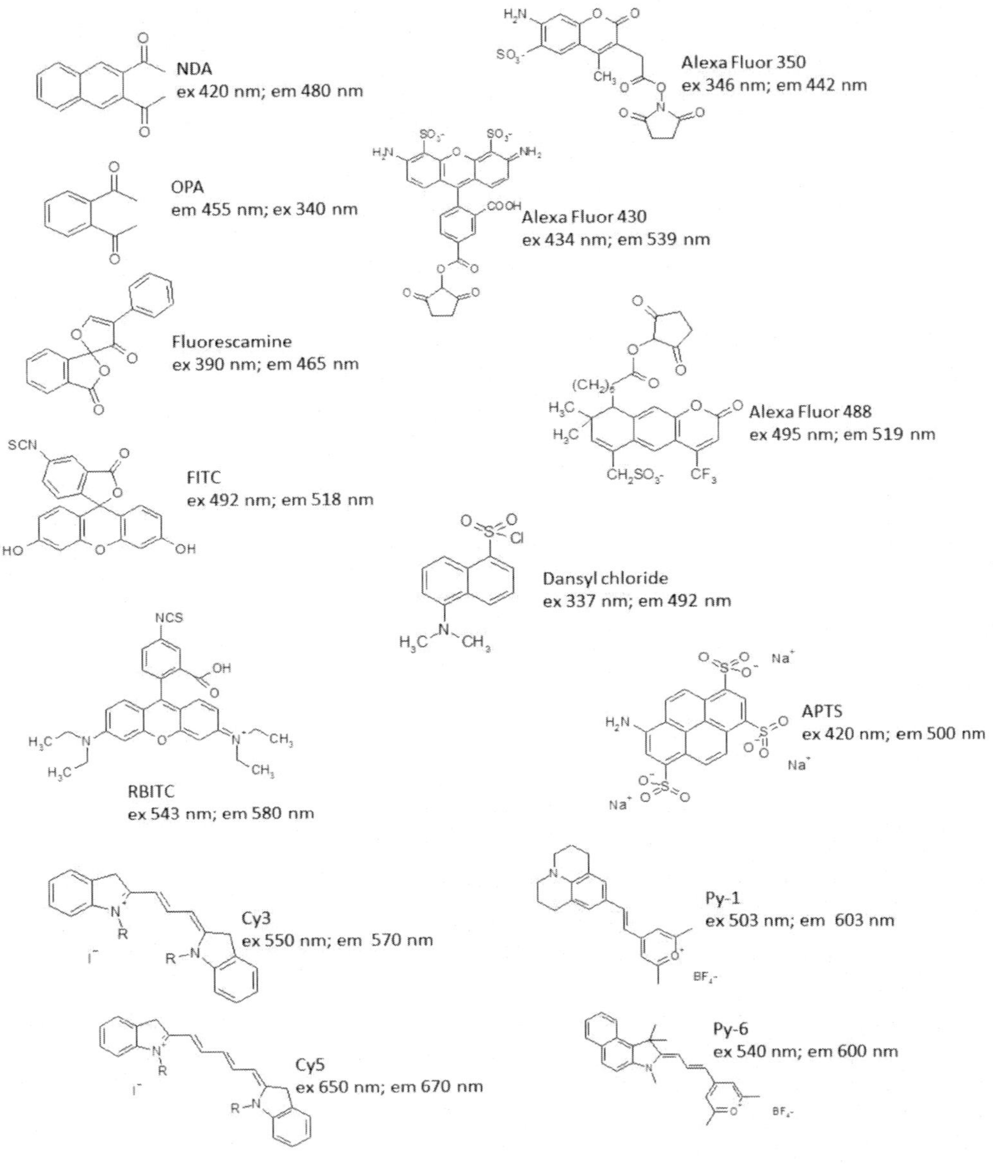

Figure 5. Structures of selected commonly used derivatization reagents and their excitation and emission wavelengths. NDA- naphthalene-2,3-dicarboxaldehyde, OPA – ortho-phthalaldehyde, FITC - Fluorescein isothiocyanate, RBITC - Rhodamine B isothiocyanate, APTS - 8-Aminopyrene-1,3,6-Trisulfonic Acid, Cy3 and Cy5 – cyanine dyes, Py-1 and Py-6 – chameleon dyes.

number of derivatization dyes for this type of functional group is inexhaustible (structures of selected examples is shown in Figure 5). The appropriate label can be chosen based on the excitation source wavelength. Conventional dyes used in CE, suitable for UV excitation among others include ortho-phthalaldehyde (OPA), naphthalene-2,3-dicarboxaldehyde (NDA) and/or dansyl chloride. Fluorescamine and NDA are moreover beneficial because they are nonfluorescent until reacted with a primary amine. This simplifies the derivatization procedure because the purification from the unreacted dye is eliminated. Alternatively, the so-called chameleon dyes undergoing a remarkable color change upon conjugation to a protein, (typically changing from blue to red) have been introduced [36]. Moreover, the reaction with a primary amine results in a product where the positive charge of the amine is retained, and quantum yields are strongly enhanced.

Fluorescein isothiocyanate (FITC), Rhodamine B isothiocyanate (RBITC) are traditional dyes used for visualization in visible range of spectra [23]. In addition, Cy dyes belong to the covalent labels and furthermore modern commercial sets of dyes covering the whole spectral range are available (e.g., Alexa Fluor®).

FLUORESCENT LABELING BY NANOMATERIALS

Currently, fluorescent nanomaterials have become an important group of labels employed for a variety of applications including derivatization for CE purposes. In general, nanomaterial-based fluorescent tags are based mostly on quantum dots (QDs) - nanocrystals made mostly from semiconductor nanomaterials such as elements from groups II and VI or groups III and V of the periodic table. They are known for their size (1-10 nm) and size-dependent optical and electronic properties caused by quantum confinement. QDs as new generation of fluorophores have several advantages over conventional ones. In addition, QDs have one unique characteristic incomparable with organic fluorophores; the ability of tuning the emission range as a result of the core size regulation during synthesis follows quantum confinement. QDs broad excitation spectra and narrow defined emission peak allow multicolor QDs to be excited from one source without the emission signal overlap [37, 38], also 10-100 times lager molar extinction coefficient than fluorophores results in brighter probes, compared to the conventional fluorophores [39, 40]. This induce large Stokes shift (difference between peak absorption and peak emission wavelengths) of QDs in a range of 300-400 nm as well valuable for multiplexing [41]. These advantages enable imaging and/or tracking multiple molecular targets at the same time as well as elimination of background autofluorescence, which can emerge in biological samples causing detection of mixed signals from autofluorescence and fluorescence of the administered fluorophores. Therefore, fluorescence lifetime plays an important role and QDs, with their lifetime of 20-50 ns, have superiority over fluorophores with their few-nanosecond fluorescence lifetime, as well as size-tunable absorption and emission spectra [42]. Further notable advantage is the high quantum yield ranging from 40% to 90% and due to their inorganic core; they are highly resistant to the photobleaching and/or chemical degradation [43, 44].

QDs are an order of magnitude bigger than organic dyes, which represent a problem if the probe size is important [42]. Further as shortcomings, their synthesis costs and high toxicity of the used precursors are usually stated. Their overall toxicity remains a subject of discussions although possible solutions are given by the development of alternative ways of synthesis such as "green synthesis" [45-47] or biosynthesis [48-50] of QDs.

QDs application in biology considers their linking to the biomolecules of the interest. Ideally, bioconjugation chemistry should fulfill several conditions such as control over the number (ratiometric conjugation) of biomolecules attached per one QD; control over the orientation of the attached biomolecule; provide desirable distance between QDs and biomolecules; without altering the function of biomolecules or QDs. Generally speaking, the bioconjugation chemistries could be divided to covalent and non-covalent conjugation [51].

Non-covalent coupling refers to electrostatic interaction and direct adsorption between QDs and targeted biomolecule [52-55]. Most of the currently used covalent conjugation methods are borrowed from conventional protein labeling chemistry and use carboxyl, amine and thiol groups for coupling via crosslinkers (Figure 6).

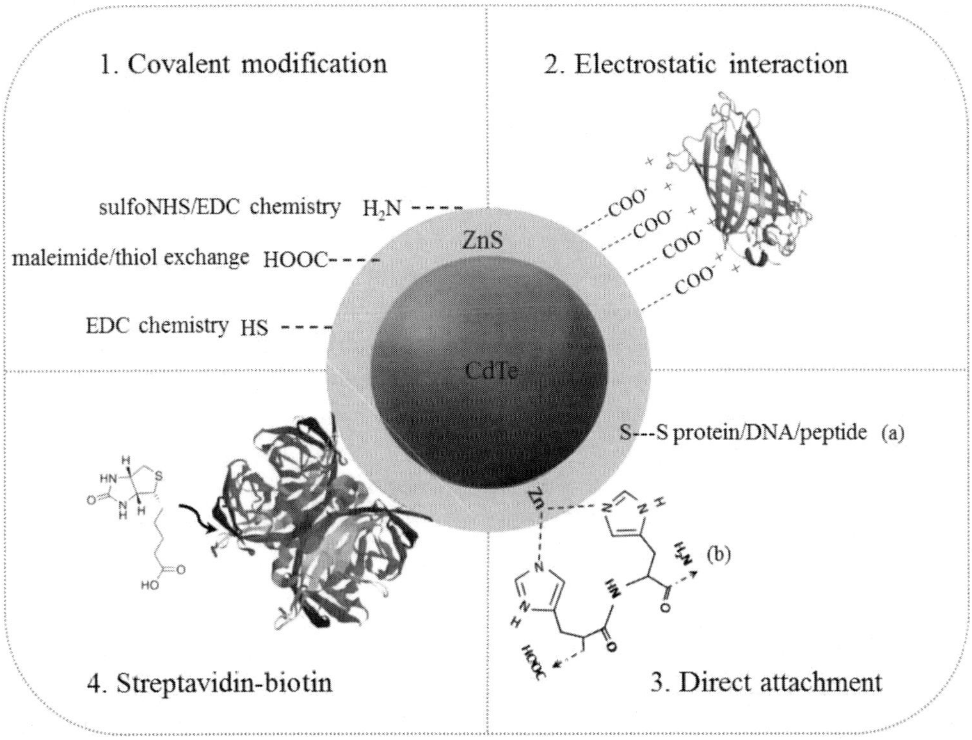

Figure 6. Methods of conjugation biomolecules to QDs. Schematic diagram with the most common bioconjugation method. Method 1 shows covalent modification via crosslinking chemistry regarding the functional groups present on the QDs surface. Method 2 uses electrostatic interactions between opposite charged QDs and biomolecules, in the scheme protein. Method 3 shows direct attachment of biomolecules to the metal atoms on the QDs surface via dative thiol bond (a) or metal affinity coordination (b). Method 4 uses the non-covalent streptavidin-biotin interaction. Adapted from [56].

Carbodiimide chemistry or EDC (N-(3-dimethylaminopropyl)-N'-ethylcarbodiimide) mediated condensation of the carboxyls and amines to an amide bond is the most popular method of the bioconjugation. In practice, bonding is performed in the presence of N-hydroxysulfosuccinimide (sulfo-NHS) for improving the solubility of reagents and increasing the coupling efficiency. EDC crosslinking is most efficient under the acidic condition and requires a large excess of the EDC to prevent competing hydrolysis.

However, it is very easy to obtain and cheap, EDC coupling usually require a lot of empirical optimization, purification steps followed by a common lack of the orientation control, possible crosslinking between proteins or QDs and proteins. Further sulfosuccinimidyl 4-(N-maleimidomethyl) cyclohexane-1-carboxylate sulfo-SMCC as heterobifunctional crosslinker was used. It has an amine-reactive sulfo-NHS-ester group and sulfhydryl reactive maleimide group and it can be used for two-step conjugation under near-physiological conditions [57].

Streptavidin-biotin linking is very commonly employed conjugation chemistry. Streptavidin is a tetramer isolated from *Streptomyces avidinii* with very strong affinity towards biotin. Streptavidin-biotin interaction is one of the strongest non-covalent interactions known in nature. Complex is formed very fast; four biotins could be bound with one streptavidin and the resulting complex exhibits strong stability at extreme pH, temperature and even to denaturing agents. Due to the small size of biotin any molecule can be biotinylated which is why this conjugation is widely applied. Streptavidin modified QDs as well as biotinylated QDs have been used in practice. Streptavidin-biotin complex is followed by shortcomings such as unwanted crosslinking and lack of possibility to control orientation of the molecules on the QDs surface [58]

Very good alternative to covalent or streptavidin-biotin interaction is polyhistidine-metal-affinity conjugation. Conjugation is based on the ability of the histidine's imidazole side chain to chelate transitional metals such as zinc(II), one of the dominant elements on the QDs surface. However, it remains unclear whether polyhistidine (His_n) tags bind directly to the QDs surface due to the possible defects in the surface coverage or His_n tags displace ligands. Further histidine metal affinity coordination in combination with the chemoselective ligation reaction provide an excellent control over the QDs bioconjugation regarding the ratio of attached biomolecules, orientation and distance between QDs and conjugates [59]. A great advantage and challenge hidden in this method is the ability to engineer the desired proteins/peptides with appended polyhistidine sequence for assembly onto the QDs surface through the metal affinity coordination. In addition, no reactive chemistry is involved so the purification steps have been avoided, simplifying the procedure.

Pioneers of the non-covalent interaction between QDs and biomolecules are Mattoussi et al. They have used engineered maltose binding protein (MBP) consisting a positively charged domain which electrostatically interact with negatively charged QD. However, this approach has limited application in the biological environment, since the electrostatic interactions are not specific or stable enough [60, 61]. A variety of conjugation methods have been proposed and investigated in order to facilitate the biolabeling and improve QDs application. The choice of the proper bioconjugation method is strongly dependent on the biomolecule of interest. What should be kept on mind is that alongside bioconjugations improvement, surface modification advancement is very important aspect as well, almost inseparable.

CONCLUSION

CE is a powerful separation technique ideal for attractive miniaturized applications, however the sensitive detection of extremely low amount of analyte is challenging and therefore intensively investigated. This aim can be reached either by the development of light source and optical components with superior properties, design of the detection cell with maximal effectivity, and/or synthesis of derivatization probes with high quantum yields, wide spectral coverage and easy conjugation with a variety of functional groups. Combining all these parts together will enable to detect analytes present in extremely low concentrations by simple routine procedures using portable and universal devices. Moreover, current boom in nanomaterial research is opening the way to new, more efficient dyes and labels.

ACKNOWLEDGMENTS

Financial support by CZ.1.07/2.3.00/30.0039 is highly acknowledged.

REFERENCES

[1] Kleparnik, K (2015). Recent advances in combination of capillary electrophoresis with mass spectrometry: Methodology and theory. *Electrophoresis, 36*, 159-178.

[2] Yu, H, Xu, XY, Sun, JY, and You, TY (2012). Recent progress for capillary electrophoresis with electrochemical detection. *Cent. Eur. J. Chem, 10*, 639-651.

[3] Garcia-Campana, AM, Lara, FJ, Gamiz-Gracia, L, and Huertas-Perez, JF (2009). Chemiluminescence detection coupled to capillary electrophoresis. *Trac-Trends Anal. Chem., 28*, 973-986.

[4] Guo, LH, Fu, F, and Chen, GN (2011). Capillary electrophoresis with electrochemiluminescence detection: fundamental theory, apparatus, and applications. *Anal. Bioanal. Chem., 399*, 3323-3343.

[5] Opekar, F, and Stulik, K (2011). Some important combinations of detection techniques for electrophoresis in capillaries and on chips with emphasis on electrochemical principles. *Electrophoresis, 32*, 795-810.

[6] Schulze, P, and Belder, D (2009). Label-free fluorescence detection in capillary and microchip electrophoresis. *Anal. Bioanal. Chem., 393*, 515-525.

[7] Bui, DA, and Hauser, PC (2015). Analytical devices based on light-emitting diodes - a review of the state-of-the-art. *Anal. Chim. Acta, 853*, 46-58.

[8] Schwarz, MA, and Hauser, PC (2001). Recent developments in detection methods for microfabricated analytical devices. *Lab Chip, 1*, 1-6.

[9] Krizkova, S, Adam, V, and Kizek, R (2009). Study of metallothionein oxidation by using of chip CE. *30*, 4029-4033.

[10] Krizkova, S, Hrdinova, V, Adam, V, Burgess, EPJ, Kramer, KJ, Masarik, M, and Kizek, R (2008). Chip-based CE for avidin determination in transgenic tobacco and its comparison with square-wave voltammetry and standard gel electrophoresis. *67*, S75-S81.

[11] Zitka, O, Krizkova, S, Huska, D, Adam, V, Hubalek, J, Eckschlager, T, and Kizek, R (2011). Chip gel electrophoresis as a tool for study of matrix metalloproteinase 9 interaction with metallothionein. *32*, 857-860.

[12] Vaculovicova, M, Smerkova, K, Sedlacek, J, Vyslouzil, J, Hubalek, J, Kizek, R, and Adam, V (2013). Integrated chip electrophoresis and magnetic particle isolation used for detection of hepatitis B virus oligonucleotides. *Electrophoresis, 34*, 1548-1554.

[13] Kuswandi, B, Nuriman, Huskens, J, and Verboom, W (2007). Optical sensing systems for microfluidic devices: A review. *Anal. Chim. Acta, 601*, 141-155.

[14] Yao, B, Luo, G, Wang, LD, Gao, YD, Lei, GT, Ren, KN, Chen, LX, Wang, YM, Hu, Y, and Qiu, Y (2005). A microfluidic device using a green organic light emitting diode as an integrated excitation source. *Lab Chip, 5*, 1041-1047.

[15] Shear, JB (1999). Multiphoton-excited fluorescence. *Anal. Chem., 71*, 598A-605A.

[16] Ghisaidoobe, ABT, and Chung, SJ (2014). Intrinsic Tryptophan Fluorescence in the Detection and Analysis of Proteins: A Focus on Forster Resonance Energy Transfer Techniques. *Int. J. Mol. Sci., 15*, 22518-22538.

[17] Cecala, C, Rubakhin, SS, Mitchell, JW, Gillette, MU, and Sweedler, JV (2012). A hyphenated optical trap capillary electrophoresis laser induced native fluorescence system for single-cell chemical analysis. *Analyst, 137*, 2965-2972.

[18] Yeung, ES (1999). Study of single cells by using capillary electrophoresis and native fluorescence detection. *J. Chromatogr. A, 830*, 243-262.

[19] Lapainis, T, Scanlan, C, Rubakhin, SS, and Sweedler, JV (2007). A multichannel native fluorescence detection system for capillary electrophoretic analysis of neurotransmitters in single neurons. *Anal. Bioanal. Chem., 387*, 97-105.

[20] Hu, L, Yang, X, Wang, C, Yuan, H, and Xiao, D (2007). Determination of riboflavin in urine and beverages by capillary electrophoresis with in-column optical fiber laser-induced fluorescence detection. *J. Chromatogr. B, 856*, 245-251.

[21] Chen, J, Sun, JZ, and Liu, SH (2013). Determination of Riboflavin in Cereal Grains by Capillary Electrophoresis with Laser-Induced Fluorescence Detection with On-Line Concentration. *Anal. Lett., 46*, 887-899.

[22] Kishikawa, N, and Kuroda, N (2014). Analytical techniques for the determination of biologically active quinones in biological and environmental samples. *J. Pharm. Biomed. Anal., 87*, 261-270.

[23] Underberg, WJM, and Waterval, JCM (2002). Derivatization trends in capillary electrophoresis: An update. *Electrophoresis, 23*, 3922-3933.

[24] Colyer, C (2000). Noncovalent labeling of proteins in capillary electrophoresis with laser-induced fluorescence detection. *Cell Biochem. Biophys., 33*, 323-337.

[25] Patonay, G, Salon, J, Sowell, J, and Strekowski, L (2004). Noncovalent labeling of biomolecules with red and near-infrared dyes. *Molecules, 9*, 40-49.

[26] Lamari, FN, Kuhn, R, and Karamanos, NK (2003). Derivatization of carbohydrates for chromatographic, electrophoretic and mass spectrometric structure analysis. *J. Chromatogr. B, 793*, 15-36.

[27] Anumula, KR (2014). Single tag for total carbohydrate analysis. *Anal. Biochem., 457*, 31-37.

[28] de Oliveira, MAL, Porto, BLS, Faria, IDL, de Oliveira, PL, Barra, PMD, Castro, RDC, and Sato, RT (2014). 20 Years of Fatty Acid Analysis by Capillary Electrophoresis. *Molecules, 19*, 14094-14113.

[29] Montealegre, C, Verardo, V, Marina, ML, and Caboni, MF (2014). Analysis of glycerophospho-and sphingolipids by CE. *Electrophoresis, 35*, 779-792.

[30] Lin, YW, Chiu, TC, and Chang, HT (2003). Laser-induced fluorescence technique for DNA and proteins separated by capillary electrophoresis. *J. Chromatogr. B, 793*, 37-48.

[31] Ban, E, and Song, EJ (2013). Recent developments and applications of capillary electrophoresis with laser-induced fluorescence detection in biological samples. *J. Chromatogr. B, 929*, 180-186.

[32] Kasicka, V (2014). Recent developments in capillary and microchip electroseparations of peptides (2011-2013). *Electrophoresis, 35*, 69-95.

[33] Poinsot, V, Ong-Meang, V, Gavard, P, and Couderc, F (2014). Recent advances in amino acid analysis by capillary electromigration methods, 2011-2013. *Electrophoresis, 35*, 50-68.

[34] Levitus, M, and Ranjit, S (2011). Cyanine dyes in biophysical research: the photophysics of polymethine fluorescent dyes in biomolecular environments. *Q. Rev. Biophys., 44*, 123-151.

[35] Armitage, BA (2005). Cyanine dye-DNA interactions: Intercalation, groove binding, and aggregation. In DNA Binders and Related Subjects, Volume 253, M.J. Waring and J.B. Chaires, eds. (Berlin: Springer-Verlag Berlin), pp. 55-76.

[36] Craig, DB, Wetzl, BK, Duerkop, A, and Wolfbeis, OS (2005). Determination of picomolar concentrations of proteins using novel amino reactive chameleon labels and capillary electrophoresis laser-induced fluorescence detection. *Electrophoresis, 26*, 2208-2213.

[37] Alivisatos, AP, Gu, WW, and Larabell, C (2005). Quantum dots as cellular probes. In Annual Review of Biomedical Engineering, Volume 7. (Palo Alto: Annual Reviews), pp. 55-76.

[38] Probst, CE, Zrazhevskiy, P, Bagalkot, V, and Gao, XH (2013). Quantum dots as a platform for nanoparticle drug delivery vehicle design. *Adv. Drug Deliv. Rev., 65*, 703-718.

[39] Yu, WW, Qu, LH, Guo, WZ, and Peng, XG (2003). Experimental determination of the extinction coefficient of CdTe, CdSe, and CdS nanocrystals. *Chem. Mater., 15*, 2854-2860.

[40] Sun, J, and Goldys, EM (2008). Linear absorption and molar extinction coefficients in direct semiconductor quantum dots. *J. Phys. Chem. C, 112*, 9261-9266.

[41] Fu, AH, Gu, WW, Larabell, C, and Alivisatos, AP (2005). Semiconductor nanocrystals for biological imaging. *Curr. Opin. Neurobiol., 15*, 568-575.

[42] Walling, MA, Novak, JA, and Shepard, JRE (2009). Quantum Dots for Live Cell and In Vivo Imaging. *Int. J. Mol. Sci., 10*, 441-491.

[43] Wolfgang, JP, Teresa, P, and Christian, P (2005). Labelling of cells with quantum dots. *Nanotechnology, 16*, R9-R25.

[44] Zrazhevskiy, P, Sena, M, and Gao, XH (2010). Designing multifunctional quantum dots for bioimaging, detection, and drug delivery. *Chem. Soc. Rev., 39*, 4326-4354.

[45] Huang, PC, Jiang, Q, Yu, P, Yang, LF, and Mao, LQ (2013). Alkaline Post-Treatment of Cd(II)-Glutathione Coordination Polymers: Toward Green Synthesis of Water-Soluble and Cytocompatible CdS Quantum Dots with Tunable Optical Properties. *ACS Appl. Mater. Interfaces, 5*, 5239-5246.

[46] Beri, RK, and Khanna, PK (2011). "Green" Synthesis of Cadmium Selenide Nanocrystals: The Scope of 1,2,3-Selendiazoles in the Synthesis of Magic-Size Nanocrystals and Quantum Dots. *J. Nanosci. Nanotechnol., 11*, 5137-5142.

[47] Ahmed, M, Guleria, A, Rath, MC, Singh, AK, Adhikari, S, and Sarkar, SK (2014). Facile and Green Synthesis of CdSe Quantum Dots in Protein Matrix: Tuning of Morphology and Optical Properties. *J. Nanosci. Nanotechnol., 14*, 5730-5742.

[48] Bao, HF, Hao, N, Yang, YX, and Zhao, DY (2010). Biosynthesis of biocompatible cadmium telluride quantum dots using yeast cells. *Nano Res., 3*, 481-489.

[49] Huang, HQ, He, MX, Wang, WX, Liu, JL, Mi, CC, and Xu, SK (2012). Biosynthesis of CdS Quantum Dots in Saccharomyces Cerevisiae and Spectroscopic Characterization. *Spectrosc. Spectr. Anal., 32*, 1090-1093.

[50] Sturzenbaum, SR, Hockner, M, Panneerselvam, A, Levitt, J, Bouillard, JS, Taniguchi, S, Dailey, LA, Khanbeigi, RA, Rosca, EV, Thanou, M, et al. (2013). Biosynthesis of luminescent quantum dots in an earthworm. *Nat. Nanotechnol., 8*, 57-60.

[51] Algar, WR, Prasuhn, DE, Stewart, MH, Jennings, TL, Blanco-Canosa, JB, Dawson, PE, and Medintz, IL (2011). The Controlled Display of Biomolecules on Nanoparticles: A Challenge Suited to Bioorthogonal Chemistry. *Bioconjugate Chem., 22*, 825-858.

[52] Stanisavljevic, M, Chomoucka, J, Dostalova, S, Krizkova, S, Vaculovicova, M, Adam, V, and Kizek, R (2014). Interactions between CdTe quantum dots and DNA revealed by capillary electrophoresis with laser-induced fluorescence detection. *35*, 2587-2592.

[53] Stanisavljevic, M, Vaculovicova, M, Kizek, R, and Adam, V (2014). Capillary electrophoresis of quantum dots: Minireview. *Electrophoresis, 35*, 1929-1937.

[54] Tmejova, K, Hynek, D, Kopel, P, Krizkova, S, Blazkova, I, Trnkova, L, Adam, V, and Kizek, R (2014). Study of metallothionein-quantum dots interactions. *117*, 534-537.

[55] Krejcova, L, Nejdl, L, Merlos, MAR, Zurek, M, Matousek, M, Hynek, D, Zitka, O, Kopel, P, Adam, V, and Kizek, R (2014). 3D printed chip for electrochemical detection of influenza virus labeled with CdS quantum dots. *54*, 421-427.

[56] Sapsford, KE, Pons, T, Medintz, IL, and Mattoussi, H (2006). Biosensing with luminescent semiconductor quantum dots. *6*, 925-953.

[57] Sapsford, KE, Algar, WR, Berti, L, Gemmill, KB, Casey, BJ, Oh, E, Stewart, MH, and Medintz, IL (2013). Functionalizing Nanoparticles with Biological Molecules: Developing Chemistries that Facilitate Nanotechnology. *113*, 1904-2074.

[58] Blanco-Canosa, JB, Wu, M, Susumu, K, Petryayeva, E, Jennings, TL, Dawson, PE, Algar, WR, and Medintz, IL (2014). Recent progress in the bioconjugation of quantum dots. *263*, 101-137.

[59] Prasuhn, DE, Blanco-Canosa, JB, Vora, GJ, Delehanty, JB, Susumu, K, Mei, BC, Dawson, PE, and Medintz, IL (2010). Combining Chemoselective Ligation with Polyhistidine-Driven Self-Assembly for the Modular Display of Biomolecules on Quantum Dots. *4*, 267-278.

[60] Mattoussi, H, Mauro, JM, Goldman, ER, Anderson, GP, Sundar, VC, Mikulec, FV, and Bawendi, MG (2000). Self-assembly of CdSe-ZnS quantum dot bioconjugates using an engineered recombinant protein. *122*, 12142-12150.

[61] Rosenthal, SJ, Chang, JC, Kovtun, O, McBride, JR, and Tomlinson, ID (2011). Biocompatible Quantum Dots for Biological Applications. *18*, 10-24.

In: Capillary Electrophoresis (CE)
Editor: Christian Reed

ISBN: 978-1-63483-122-2

Chapter 11

APPLICATION OF CAPILLARY ZONE ELECTROPHORESIS METHODS FOR POLYPHENOLS AND ORGANIC ACIDS SEPARATION IN DIFFERENT EXTRACTS

***Eugenia Dumitra Teodor*[1], *Florentina Gatea*[1], *Georgiana Ileana Badea*[1], *Alina Oana Matei*[1] *and Gabriel Lucian Radu*[2]**

[1]National Institute for Biological Sciences, Centre of Bioanalysis, Bucharest, Romania
[2]University "Politehnica" Bucharest, Faculty of Applied Chemistry and Materials Science, Bucharest, Romania

ABSTRACT

Capillary electrophoresis has proved to be a good alternative technique to high performance liquid chromatography for the investigation of various compounds due to its good resolution, versatility, simplicity, short analysis time and low consumption of chemicals and samples.

This chapter presents a synthesis of our work regarding applications of capillary electrophoretic methods (capillary zone electrophoresis with diode array detection): the separation of small-chain organic acids from plants extracts, wines, lactic bacteria fermentation products, and the separation of polyphenolic compounds from propolis extracts, plant extracts and wines.

Quantitative evaluation of organic acids in plants and foodstuff is important for flavour and nutritional studies, and also could be used as marker of bacterial activity. Organic acids occurring in foods are additives or end-products of carbohydrate metabolism of lactic acid bacteria. A good selection of lactic acid bacteria, in terms of content in organic acids, allows the control of mould growth and improves the shelf life of many fermented products and, therefore, reduces health risks due to exposure to mycotoxins.

On the other side, the largely studied group of phytochemicals is polyphenols, an assembly of secondary metabolites with various chemical structures and functions and biological activities, which are produced during the physiological plant growth process as a response to different forms of environmental conditions.

The methods for separation and quantification of organic acids and polyphenolic compounds were validated in terms of linearity of response, limit of detection, limit of quantification, precisions (i.e., intra-day, inter-day reproducibility) and recovery. The methods are simply, rapid, reliable and cost effective.

Keywords: Capillary zone electrophoresis, polyphenols, organic acids, plant extracts, propolis extracts, lactic acid bacteria extracts

INTRODUCTION

Capillary electrophoresis (CE) is a promising analytical technique for the separation of different compounds in complex matrices as plants due to its high separation efficiency. Capillary electrophoresis has proved to be a good alternative technique to high performance liquid chromatography (HPLC) for the study of various compounds owing to its good resolution, versatility and simplicity, short analysis time and low consumption of chemicals and samples. UV–Vis absorption is the detection technique most usually used [1, 2], even if nowadays, CE coupled to mass spectrometry (MS) is gaining increased attention. For obtaining a superior separation in CE it is necessary to optimize several parameters, such as buffer (background electrolyte-BGE) type, concentration and pH, type and dimensions of capillary, work temperature, voltage and injection mode, etc.

In the last years, a wide assortment of chromatographic and electrophoretic methods were developed and validated for quantification of different categories of natural compounds or additives in foods. García-Cañas et al. (2014) reviewed the CE methods for discrimination of amino acids, peptides, proteins, phenolic compounds, carbohydrates, DNA fragments, vitamins, small organic and inorganic compounds, toxins, pesticides, additives and other minor compounds in different products [3].

Generally, plants contain considerable amounts of organic substances with a diversity of metabolites which includes more than 200,000 compounds. The most important plant metabolites, present at concentrations ranging between 20-100 μmol g^{-1} in raw materials, are polysaccharides, polyols, amino acids, and organic acids [4, 5].

The largely studied group of phytochemicals is polyphenols, an assemblage of secondary metabolites with various chemical structures and functions, which are being produced during the physiological plant growth process as a response to different forms of environmental conditions [6]. Their biological activities have been extensively studied during the last decades, providing strong evidences of their health benefits potential. The latter are mainly endorsed by their antioxidant properties, since they can act as free-radical scavengers, electron or hydrogen donors and strong metal chelators, having neuroprotective effects and thus preventing the lipid peroxidation, DNA damage, etc. [7-10]. As a consequence, radical scavenger compounds are nowadays gaining increasing interest and the consumption of food rich in antioxidants is greater than ever.

Medicinal plants, vegetables and fruits are the major source of natural antioxidants [11]. Besides that, propolis contains predominantly polyphenolic compounds including flavonoids and cinnamic acid derivatives which appear to be the principal components responsible for its biological activities [12]. Propolis has a long history of being used in traditional medicine dating back to 300 BC [13] and has been reported to have a broad spectrum of biological activities, namely anticancer, antioxidant, antiinflammatory, antibiotic and antifungal activities [14].

Clinical experiments provided evidence that several polyphenolic compounds such as phenolic acids (both hydroxybenzoic and hydroxicinnamic acids), flavonoids (catechin, quercetin, myricetin, kaempferol) and other polyphenols (epigallocatechins, resveratrol), could induce apoptosis in cancer cells [15-18]. Another aspect is that polyphenolic compounds may contribute to Alzheimer's disease-modifying activity by reducing the generation of amyloid-β (Aβ) peptides that are critical for disease onset and progression [19]. Ho et al. (2013) study indicated that quercetin-3-*O*-glucuronide derivatives (from red wine and some plants) found accumulated in the brain are capable of interfering with the generation of Aβ peptides and may lower the relative risk for developing Alzheimer's disease (AD) dementia [20].

Organic acids are involved as intermediate or end products in different fundamental pathways in plant metabolism and catabolism; for example the citrate, succinate, malate, fumarate, and acetate in the acetyl coenzyme A form, play an important role in the Krebs cycle which is the central energy yielding cycle of the cell [21]. Some of these short-chain organic acids serve as precursor for a variety of products, such as acetate or formate, others, such as malate, are involved in respiration and photosynthesis processes or in detoxification (oxalate and citrate) [22, 23]. Organic acids are responsible for the taste, the flavour, the microbial stability, and the product consistence of plant derived beverages and are used in food preservation because of their effects on bacteria [24, 25].

The chemical study of organic acids, as part of metabolomics analysis, provides biochemical information on cellular functioning and on pathways affected by stress or disease. Qualitative and quantitative determination of a large number of organic acids metabolites provides an overall view of the biochemical status of the cell [4].

Considering the CE methods developed for organic acids quantification, all these methods are focused on aliphatic organic acids or on lactic acid and derivatives [26, 27, 3]. From our point of view, we considered that a method for the simultaneous quantification of small-chain aliphatic organic acids and aromatic derivatives of lactic acid could be useful in analysis of bacteria fermentation products or in other extracts. Although the indirect UV detection was the most common used detection mode in capillary zone electrophoresis (CZE) for the determination of organic acids in these samples, direct UV detection seems to be more suitable due to the stability of the baseline [24, 28-30].

Organic acids occurring in foods are additives or end-products of carbohydrate metabolism of LAB. Lactic and acetic acids are the main products of the carbohydrates fermentation by LAB. Among organic acids, lactic, acetic, phenyllactic (PLA) and p-OH-phenyllactic acids (H-PLA) produced by LAB play a role in inhibiting fungal and bacterial growth [31-34]. Caproic, propionic, butyric and benzoic acids were also evidenced for antifungal effect [35, 36]. New biological conservation of foods includes diverse strategies such as biopreservation by lactic acid bacteria (LAB) or by their antimicrobial metabolites. These microorganisms are widely used for the production of fermented foods and are also part of intestinal microflora. LAB has a considerable role in food fermentations due to its impact on flavor changes and as a preservative, thus helping to afford food safety by inhibiting pathogen growth [37].

Phenyllactic acid (PLA) has the ability to inhibit some pathogenic bacteria such as *Listeria monocytogenes*, *Staphylococcus aureus*, *Escherichia coli*, and *Aeromonas hydrophila* according to some studies [38]. Recently, Rodríguez-Pazo et al. (2013) made an evaluation of *Lactobacillus plantarum* CECT-221 capacity to produce antimicrobial compounds acid using HPLC technique (including the novel phenyllactic acid); extracts obtained after fed-batch fermentation were assayed as an antimicrobial against pathogens such as *Pseudomonas aeruginosa, Salmonella enterica, Listeria monocytogenes, and Staphylococcus aureus.* The bacteriocin activity was evaluated against *Carnobacterium piscicola* [39].

In this chapter, our results for separation of small-chain organic acids and lactic acid derivatives from plants extracts and bacteria fermentation products, and polyphenols from propolis and plants extracts are presented. 12 organic acids (formic, oxalic, succinic, malic, tartaric, acetic, citric, lactic, butyric, benzoic, phenyllactic and hydroxyphenyllactic) were quantified in 15 minutes and 20 polyphenolic compounds (resveratrol, pinostrobin, acacetin, chrysin, rutin, naringenin, isoquercitrin, umbelliferone, cinnamic acid, chlorogenic acid, galangin, sinapic acid, syringic acid, ferulic acid, kaempferol, luteolin, coumaric acid, quercetin, rosmarinic acid and caffeic acid) in less than 27 min. The methods are simple, reliable, were partially validated and successfully applied on different real samples.

PART I APPLICATION OF CAPILLARY ELECTROPHORESIS FOR POLYPHENOLS QUANTIFICATION IN DIFFERENT EXTRACTS

Equipment and Method

Electrophoretic separation was carried out using an Agilent CE instrument with DAD detector (software ChemStation) and CE standard bare fused-silica capillary (Agilent Technologies, Germany) with internal diameter of 50 μm and effective length of 72 cm. Prior to use, the capillary was washed successively with basic solutions: 10 min with 1N NaOH, 10 min with 0.1 N NaOH followed by ultra pure water for 10 min and buffer for 20 min. The capillary was flushed between runs with 0.1M NaOH for 1 min, H_2O for 1 min and background electrolyte for 2 min. The electrolyte was refreshed after 3 consecutive runs.

Sample injection was performed using the hydrodynamic mode (35 mbar/12 sec) while the capillary was maintained at constant temperature of 30°C.

The simultaneous separation of polyphenolic compounds was obtained using 45 mM tetraborate buffer with 0.9 mM SDS (pH = 9.35 adjusted with HCl 1 M) as background electrolyte. BGE was filtered on 0.2 μm membranes (Millipore, Bedford, MA, USA) and degassed before use. The applied voltage was 30 kV; direct UV absorption detection was carried out from 200 to 360 nm and the quantification of samples was performed at 280 nm.

Samples Preparation

The propolis sample was collected in 2012 from Dambovita County, Romania. The sample was homogenized and frozen at −18°C and an aliquot (100 g) was grounded to powder by hand into a porcelain mortar. 10 g frozen propolis was mixed with 100 mL of different solutions (distilled water; glycine buffer 0.1 M, pH = 2.5; acetate buffer 0.1 M, pH = 5; phosphate buffer 0.1 M, pH = 7.4 and carbonate buffer 0.1 M, pH = 9). The suspensions were maintained for 15 min under stirring at 70°C and then cooled at room temperature. The mixtures were left for maceration for 10 days at room temperature and then were filtered through Whatman no. 1 filter paper, adjusted to 100 mL with the same solutions, and then filtered on 0.2 μm Millipore filters before the analysis.

The purpose of using different media for propolis extraction was to increase the solubility of certain compounds (especially flavonoids) in aqueous media. It is known that the concentration ranges of flavonoids in extracts are limited due to their restricted solubility, but there are some parameters that can improve the solubility, such as temperature [40.41], nature of the solvents [42] and the pH [43]. Depending on pH, the hydroxyl groups of polyphenols are more or less ionized and this could influence the solubility of compounds in aqueous solutions.

The plants (*Mentha aquatica* and *Origanum* from Plafar Company) were dried for one week at room temperature (RT) and finely minced with a Grindomix GM200 grinder. The extraction was made at RT, during 7 days, with a mixture of ethanol: water (70% (v/v)) in 1:10 ratio (g/V). After that, the extracts were centrifuged 15 minutes at 5000 rpm and supernatants were collected, adjusted to 10 mL and filtered (0.2 μm Millipore Bedford, MA, USA). Samples were diluted (if necessary) in BGE.

The wine samples (two sorts of red wine, from the market) were filtered using 0.2 μm membranes and injected undiluted in the instrument.

Method Development

Several CE methods were considered for polyphenolic compounds separations [44-48]. Several BGEs were examined for polyphenolic compounds separations (e.g., phosphate, borate) merely or combined with different surfactants (SDS). Our CE method belongs to CZE category with direct UV detection. The anionic surfactant, SDS, improves the separation but was under critical concentration level for a micellar chromatography. The procedure based on tetraborate buffer at alkaline pH was optimized for the best separation of 20 compounds. Tetraborate concentration, SDS concentration and pH value were slightly varied for an optimum separation in shortest time possible.

Effect of Concentration of BGE and Anionic Surfactant

With increasing buffer concentration (electroosmotic flow-EOF- reduced) resolution increase but the migration time also increase [49]. In our attempt to improve the method we used three different migration buffers (40-50mM) at the same value of pH and SDS.

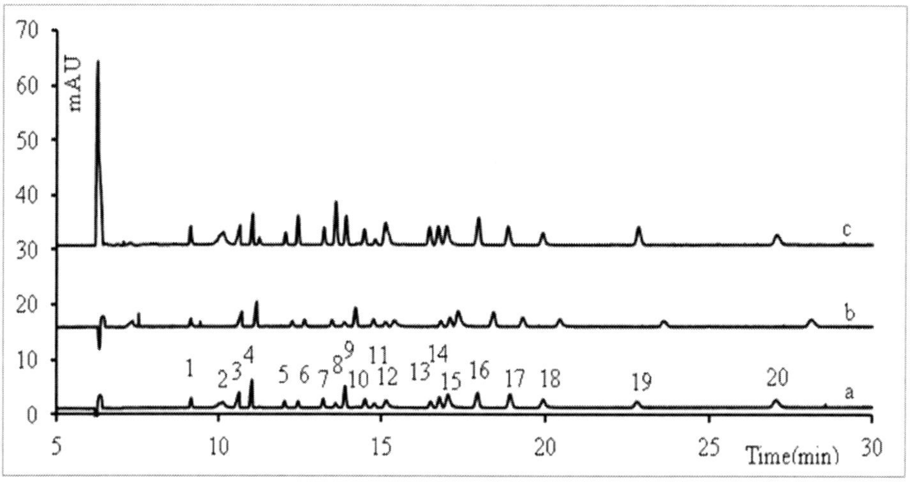

Figure 1. Effect of sodium borate concentrations for the separation of 20 polyphenolic compounds; (1) resveratrol,(2) pinostrobin,(3) acacetin, (4) chrysin, (5) rutin, (6) naringenin, (7) isoquercitrin, (8) umbelliferone, (9) cinnamic acid, (10) chlorogenic acid, (11) galangin, (12) sinapic acid, (13) syringic acid, (14) ferulic acid, (15) kaempferol, (16) luteolin, (17) coumaric acid, (18) quercetin, (19) rosmarinic acid and (20) caffeic acid.

a) 45 mM borate and 0.9 mM SDS pH 9.35 (working conditions).
b) 40 mM borate and 0.9 mM SDS pH 9.35.
c) 50 mM borate and 0.9 mM SDS pH 9.35.

However, it can be seen that at 50 mM concentration the migration time increased and the peaks resolution decreased compared with concentration of BGE considered optimal (45 mM, as could be seen in Figure 1a).

The effect of SDS concentrations was studied within the domain 0.45-1.35 mM. The buffer concentration of sodium tetraborate was maintained at 45 mM and pH 9.35. As shown in Figure 2a, at 0.45 mM SDS concentration the migration time increased and pinostrobin could not be separated, while at a higher concentration of SDS (Figure 2c) was observed that peaks resolution decreases in parallel with increasing migration times. Therefore, the optimum concentration of SDS was considered to be 0.9 mM (Figure 2a).

Effect of BGE pH

The pH of the electrolyte buffer has considerable influence on the separation of the analytes. The experiments were performed varying the pH from 9.25 to 9.45 preserving the others conditions. As presented in Figure 3 the best result for the separation of 20 polyphenolic compounds at different pH values was obtained at pH 9.35 (Figure 3c), respectively the elution of all compounds (including pinostrobin (see peak 2 in Figure 3c)) and the shortest runtime.

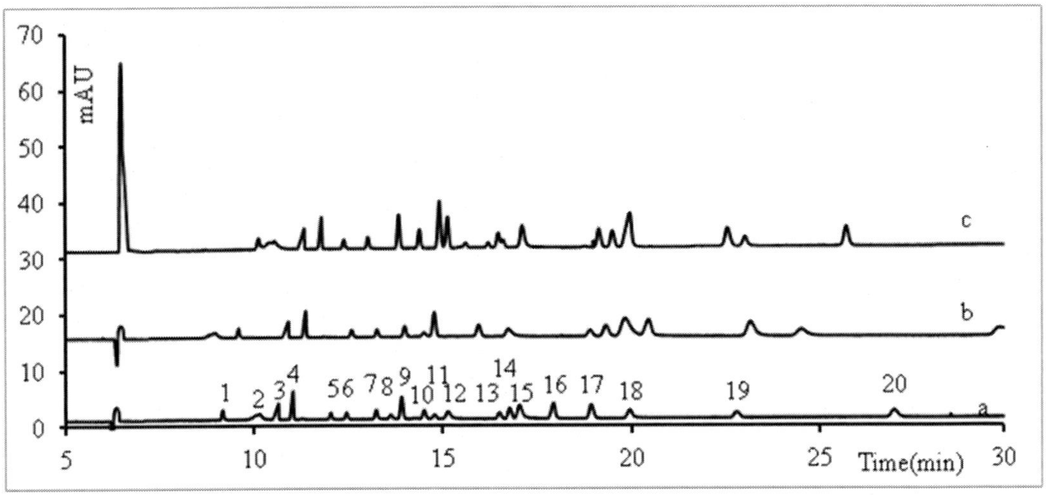

Figure 2. Effect of SDS concentrations for the separation of 20 polyphenolic compounds; (1) resveratrol,(2) pinostrobin,(3) acacetin, (4) chrysin, (5) rutin, (6) naringenin, (7) isoquercitrin, (8) umbelliferone, (9) cinnamic acid, (10) chlorogenic acid, (11) galangin, (12) sinapic acid, (13) syringic acid, (14) ferulic acid, (15) kaempferol, (16) luteolin, (17) coumaric acid, (18) quercetin, (19) rosmarinic acid and (20) caffeic acid.

a) 45 mM borate and 0.9 mM SDS pH 9.35 (working conditions).
b) 45 mM borate and 0.45 mM SDS pH 9.35.
c) 45 mM borate and 1.35 mM SDS pH 9.35.

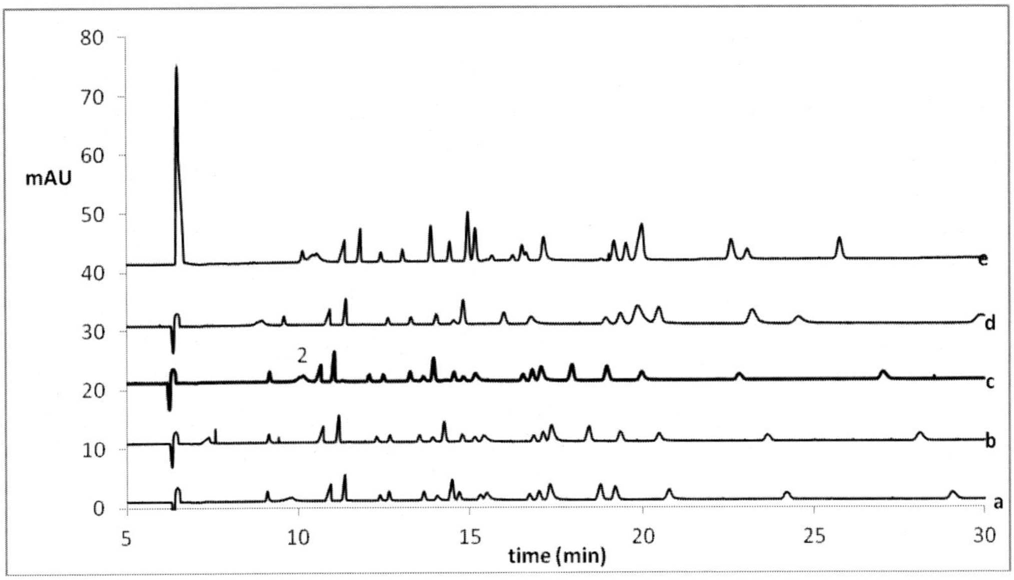

Figure 3. Comparison of electropherograms obtained at different pH of BGE (45 mM sodium borate with 0.9 mM SDS); a) pH = 9.25; b) pH = 9.3; c) pH = 9.35 (working condition); d) pH = 9.4; e) pH = 9.45.

(SpringerScience + Business Media New York, Food Analytical Methods, Capillary Electrophoresis Method Validation for 20 Polyphenols Separation in Propolis and Plant Extracts, 2015, 8, 1197-1206; DOI:10.1007/s12161-014-0006-5, F. Gatea, E. D. Teodor, A. O. Matei, G. I. Badea, G. L. Radu, original copyright notice) "With kind permission of Springer Science + Business Media."

Other Operating Conditions

The 72 cm length capillary was used for an optimum separation and the highest 30 kV voltage was applied for reducing the runtime to ~27 min.

The addition of organic solvent in the BGE was also tested, the small amount of organic solvent (methanol, acetonitrile) added leading to a decrease in resolution. Hydrodynamic injection time (5−15 s) was also studied to increase sensitivity. An injection time of 12 s (35 mbar) was selected as an optimal (35 mbar/12 sec) for a good resolution.

Table 1. Performance characteristics of the method for polyphenolic compounds separation

Compound	t_R (min)	The linear regression equations (μg mL^{-1})	R^2	Linearity range of response μg mL^{-1}	LoD μg mL^{-1}	LoQ μg mL^{-1}
Resveratrol	9.14 ± 0.12	y = 0.775x -0.202	0.998	2.5-50	0.06	0.19
Pinostrobin	10.10 ± 0.19	y = 0.489x + 2.680	0.997	5.0-50	1.75	5.77
Acacetin	10.68 ± 0.12	y = 1.512x - 0.366	0.999	2.5-50	0.02	0.08
Chrysin	11.01 ± 0.18	y = 1.771x + 1.089	0.999	2.5-50	0.12	0.41
Rutin	12.01 ± 0.15	y = 0.477x + 0.178	0.999	2.5-50	0.12	0.39
Naringenin	12.39 ± 0.18	y = 0.476x + 0.576	0.998	2.5-50	1.64	5.40
Isoquercitrin	13.16 ± 0.19	y = 0.879x - 0.459	0.999	2.5-50	0.09	0.31
Umbelliferone	13.56 ± 0.19	y = 0.322x + 0.904	0.998	2.5-50	0.98	3.25
Cinnamic Acid	13.86 ± 0.21	y = 1.857x + 1.653	0.999	2.5-50	0.09	0.29
Chlorogenic Acid	14.42 ± 0.25	y = 0.806x + 0.287	0.999	2.5-50	0.47	1.55
Galangin	14.72 ± 0.22	y = 0.377x + 0.514	0.999	2.5-50	0.40	1.32
Sinapic Acid	15.04 ± 0.25	y = 1.226x - 0.549	0.999	2.5-50	0.03	0.10
Syringic Acid	16.36 ± 0.36	y = 0.782x - 0.592	0.999	2.5-50	1.16	3.84
Ferulic Acid	16.66 ± 0.32	y = 1.114x + 0.295	0.998	2.5-50	0.08	0.27
Kaempferol	16.89 ± 0.33	y = 2.965x - 3.679	0.998	2.5-50	0.08	0.25
Luteolin	17.83 ± 0.34	y = 2.012x - 0.037	0.998	2.5-50	0.03	0.09
Coumaric Acid	18.75 ± 0.22	y = 2.034x + 0.253	0.998	2.5-50	0.08	0.27
Quercetol	19.82 ± 0.21	y = 1.541x - 2.855	0.998	2.5-50	0.02	0.07
Rosmarinic Acid	22.65 ± 0.50	y = 0.763x + 2.665	0.998	2.5-50	1.03	3.40
Caffeic Acid	26.80 ± 0.68	y = 1.287x + 3.051	0.999	2.5-50	0.37	1.24

(SpringerScience + Business Media New York, Food Analytical Methods, Capillary Electrophoresis Method Validation for 20 Polyphenols Separation in Propolis and Plant Extracts, 2015, 8, 1197-1206; DOI:10.1007/s12161-014-0006-5, F. Gatea, E. D. Teodor, A. O. Matei, G. I. Badea, G. L. Radu, original copyright notice) "With kind permission of Springer Science + Business Media."

Validation of the Electrophoretic Method

The main parameters used in the validation of the methodology are: the selectivity, linearity, precision, accuracy (recovery), limit of detection (LoD) and limit of quantification (LoQ) and the results are presented in Tables 1-3. LoD and LoQ used to assess sensitivity were estimated using a signal-to-noise ratio of 3 and 10, respectively. Detection limits for the samples resulted between 0.02 μg mL^{-1} for quercetin and acacetin, and 1.75 μg mL^{-1} for pinostrobin. Linearity ranges used for compound quantification were satisfactory, presenting correlation coefficients (r^2) between 0.997 and 0.999 for all 20 compounds.

The repeatability of the method was studied by repeated injections of the polyphenols mixtures (standards) 5 times in the same day (intra-day precision), whereas the reproducibility assimilated to inter-day precision was assessed by triplicate injections in 3 different days (Table 2). The results are reported in terms of relative standard deviation (RSD). The RSD values for repeatability did not exceed 4.86% for intra-day assays and 5.07 for inter-day assays. Quantification limits maintained between 0.07 μg mL^{-1} for quercitin and 5.77 μg mL^{-1} for pinostrobin.

In order to verify the applicability of the proposed method for various types of polyphenolic extracts the recovery tests were performed for an ethanolic sample of *Origanum vulgare* (diluted 20x) and an aqueous sample of propolis (diluted 50x) spiked with known concentrations of standard solutions (Table 3). The recovery assays presented results between 87.4% and 114. 2% for *Origanum vulgare* sample and between 85.0% and 111.0% for propolis sample. Regarding all validation parameters, the method complies with validation requirements and it is suitable for the analysis of selected samples.

Table 2. Precision results obtained for the CZE separation method

Compound	Intra-assay Precision[a] (%, n = 5)	Inter-assay Precision[a] (%, n = 2x5)	Inter-assay Precision[b] (%, n = 2x5)	Inter-assay Precision[c] (%, n = 2x5)
Resveratrol	3.51	4.14	2.91	2.51
Pinostrobin	2.36	2.07	2.82	3.04
Acacetin	3.35	4.35	3.72	2.43
Chrysin	1.15	2.25	1.32	1.59
Rutin	2.06	3.57	1.05	1.31
Naringenin	3.81	5.07	4.15	5.52
Isoquercitrin	4.82	3.17	2.29	2.08
Umbelliferone	4.02	5.64	4.62	4.61
Cinnamic Acid	2.55	3.48	3.19	2.06
Chlorogenic Acid	4.86	4.72	2.37	2.50
Galangin	4.35	4.27	2.37	3.34
Sinapic Acid	2.34	5.46	3.21	3.87
Syringic Acid	3.87	3.61	2.09	3.06
Ferulic Acid	3.60	3.69	3.34	2.52
Kaempferol	2.70	4.73	3.21	2.39
Luteolin	5.28	4.95	3.19	2.95
Coumaric Acid	4.59	4.91	3.82	2.51
Quercetol	4.40	4.05	4.35	2.74
Rosmarinic Acid	3.94	5.34	3.88	3.69
Caffeic Acid	4.45	4.81	5.12	3.74

[a] Standards concentration: 10 μg mL^{-1}; [b] standards concentration: 17 μg mL^{-1}; [c] standards concentration: 23 μg mL^{-1}.

Table 3. Recovery values (%) of polyphenols in tested samples: *Origanum vulgare* and propolis extracts

Compound	Spiked concentration			
	10 μg mL^{-1}	15 μg mL^{-1}	10 μg mL^{-1}	15 μg mL^{-1}
	Origanum vulgare[a]		Propolis[b]	
	Recovery (%)			
Resveratrol	99.8 ± 3.2	96.4 ± 2.2	102.5 ± 4.3	100.0 ± 3.9
Pinostrobin	108.6 ± 2.9	114.2 ± 2.3	106.0 ± 3.8	106.0 ± 4.8
Acacetin	112.5 ± 3.4	107.0 ± 4.7	97.7 ± 3.0	85.0 ± 5.7
Chrysin	98.6 ± 1.7	101.7 ± 2.7	89.0 ± 4.7	96.0 ± 3.7
Rutin	94.4 ± 2.7	100.5 ± 4.0	99.0 ± 3.6	90.0 ± 3.9
Naringenin	92.7 ± 3.0	95.4 ± 4.7	91.5 ± 2.9	94.0 ± 4.9
Isoquercitrin	94.6 ± 4.0	95.0 ± 2.5	97.4 ± 4.8	97.0 ± 2.5
Umbelliferone	100.9 ± 2.5	101.0 ± 3.3	98.0 ± 3.0	93.7 ± 3.7
Cinnamic Acid	96.8 ± 2.8	98.0 ± 2.0	90.7 ± 4.3	96.0 ± 3.8
Chlorogenic Acid	98.7 ± 2.6	102.0 ± 3.4	87.8 ± 3.5	92.8 ± 2.6
Galangin	96.3 ± 2.6	103.4 ± 3.6	105.7 ± 4.4	111.0 ± 5.0
Sinapic Acid	109.9 ± 1.9	94.4 ± 3.8	109.0 ± 4.6	97.0 ± 3.8
Syringic Acid	96.0 ± 2.6	105.0 ± 2.5	106.5 ± 2.4	93.0 ± 4.6
Ferulic Acid	87.4 ± 3.6	89.4 ± 2.5	92.0 ± 4.0	91.0 ± 3.8
Kaempferol	110.0 ± 3.5	95.6 ± 3.7	93.0 ± 4.6	96.0 ± 2.4
Luteolin	105.6 ± 2.5	111.0 ± 2.7	103.0 ± 4.0	98.0 ± 4.8
Coumaric Acid	115.4 ± 4.0	105.6 ± 3.6	100.0 ± 2.7	103.0 ± 5.8
Quercetol	88.2 ± 3.0	88.7 ± 3.4	91.0 ± 3.9	98.0 ± 2.9
Rosmarinic Acid	90.0 ± 2.5	89.0 ± 2.7	103.0 ± 2.9	98.0 ± 4.5
Caffeic Acid	92.7 ± 2.6	87.5 ± 3.4	87.0 ± 5.6	89.0 ± 5.6

Recovery values expressed as ((average observed concentration)/(nominal concentration)) x100.
[a] diluted 20x, [b] diluted 50x.

(SpringerScience + Business Media New York, Food Analytical Methods, Capillary Electrophoresis Method Validation for 20 Polyphenols Separation in Propolis and Plant Extracts, 2015, 8, 1197-1206; DOI:10.1007/s12161-014-0006-5, F. Gatea, E. D. Teodor, A. O. Matei, G. I. Badea, G. L. Radu, original copyright notice) "With kind permission of Springer Science + Business Media."

Samples Analysis

Different types of samples were analyzed, respectively propolis extracts, plant extracts and red wines. The content of polyphenolic compounds found in analyzed samples is shown in Tables 4 and 5 and electropherograms of three samples are presented in Figures 4-6.

The composition of propolis depends on the vegetation of the area from where is collected [12]. Propolis from temperate zones (Europe, Asia, North America, etc.) contains usually phenolic compounds, including some flavonoids, aromatic acids and their esters originated mainly from the poplar buds (*Populus spp.*) exudates, which appear to be the principal source of propolis [50, 51]. Looking at the results obtained in our study on aqueous Romanian propolis extracts (Table 5) the major components were flavonoids (chrysin, pinostrobin, quercetin, naringenin, galangine) and phenolic acids (caffeic, coumaric, ferulic, cinnamic). These results are in accordance with data of other authors, which found flavonoids and phenolic acid esters as main constituents in Bulgarian, respectively Anatolian, Greek and Romanian propolis samples [52-55].

Table 4. Concentrations of polyphenols in propolis samples

Compound	P1[a] µg mL^{-1}	P2[b] µg mL^{-1}	P3[c] µg mL^{-1}	P4[d] µg mL^{-1}	P5[e] µg mL^{-1}
Pinostrobin	4.2 ± 1.4	7.4 ± 0.8	7.8 ± 0.5	19.5 ± 1.5	23.6 ± 2.6
Acacetin	3.5 ± 0.2	1.8 ± 0.1	6.9 ± 1.4	11.4 ± 1.5	12.7 ± 2.5
Chrysin	371.3 ± 5.4	31.4 ± 1.5	95.9. ± 3.2	368.5 ± 18.4	719.5 ± 23.4
Rutin	15.7 ± 1.4	19.7 ± 2.5	15.9 ± 2.6	27.7 ± 2.6	35.7 ± 2.8
Naringenin	49.2 ± 3.0	41.6 ± 3.7	46,9 ± 3.3	123.7 ± 8.4	780.3 ± 37.3
Isoquercitrin	29.3 ± 2.2	71.5 ± 6.3	65.3 ± 4.4	66.3 ± 4.6	26.8 ± 2.3
Cinnamic Acid	129.2 ± 4.3	132.9 ± 4.4	306.5 ± 8.5	657.3 ± 17.4	1012.3 ± 24.4
Chlorogenic Acid	7.6 ± 0.4	8.6 ± 0.5	9.4 ± 0.6	10.5 ± 0.5	12.5 ± 1.4
Galangin	42.9 ± 2.2	113.4 ± 6.4	148.9 ± 8.6	120.6 ± 7.5	156.1 ± 11.5
Syringic Acid	9.7 ± 1.7	18.8 ± 2.3	32.6 ± 3.8	34.4 ± 4.8	45.9 ± 6.4
Ferulic Acid	36.4 ± 3.3	48.9 ± 5.4	149.7 ± 8.3	416.6 ± 24.4	357.8 ± 17.5
Kaempferol	6.8 ± 0.6	9.3 ± 1.5	29.0 ± 3.0	4.4 ± 0.6	5.0 ± 0.6
Luteolin	14.8 ± 2.3	11.9 ± 4.7	28.2 ± 5.3	16.2 ± 4.5	19.3 ± 4.3
Coumaric Acid	160.3 ± 9.5	249.7 ± 17.4	543.5 ± 28.4	939.3 ± 58.8	649.9 ± 35.5
Quercetin	1.9 ± 0.1	3.8 ± 0.4	9.6 ± 1.5	30.9 ± 5.2	36.4 ± 6.4
Caffeic Acid	455.6 ± 36.3	1475.3 ± 75.2	3401.3 ± 187.5	4456.5 ± 243.4	2208.2 ± 123.4

[a]Propolis aqueous extract; [b]Propolis extract at pH 2.5; [c]Propolis extract at pH 5; [d]Propolis extract at pH 7.4; [e]Propolis extract at pH 9. Samples were analyzed in triplicate (Mean ± SD).

Table 5. Concentrations of polyphenols in plant extracts and wines (µg mL^{-1})

Compound	M[a]	O[b]	Red wine 1	Red wine 2
Resveratrol	nd	nd	5.0 ± 0.1	6.5 ± 0.5
Rutin	19.4 ± 1.4	24.9 ± 2.7	2.1 ± 0.1	2.9 ± 0.4
Naringenin	16.7 ± 4.8	31.0 ± 2.0	nd	nd
Isoquercitrin	3.3 ± 0.5	nd	nd	nd
Umbelliferone	nd	nd	nd	nd
Cinnamic Acid	6.2 ± 0.4	1.5 ± 0.1	nd	2.3 ± 0.3
Chlorogenic Acid	5.3 ± 0.3	8.5 ± 0.5	nd	nd
Galangin	nd	nd	nd	nd
Sinapic Acid	nd	3.2 ± 0.2	nd	nd
Syringic Acid	nd	15.0 ± 2.2	14.5 ± 1.7	17.5 ± 1.5
Ferulic Acid	5.5 ± 0.6	nd	15.8	nd
Kaempferol	2.2 ± 0.2	nd	4.5 ± 0.5	6.1 ± 0.8
Luteolin	nd	1.9 ± 0.2	nd	nd
Coumaric Acid	nd	nd	nd	7.1 ± 0.7
Quercetin	6.4 ± 1.5	12.7 ± 2.2	nd	9.5 ± 1.1
Rosmarinic Acid	67.8 ± 4.2	1998.5 ± 92.5	nd	nd
Caffeic Acid	49.9 ± 4.3	73.7 ± 6.4	6.9 ± 1.3	10.1 ± 1.2

[a]*Mentha aquatica* ethanolic extract; [b]*Origanum vulgare* ethanolic extract; Samples were analyzed in triplicate (Mean ± SD) n.d., not detected.

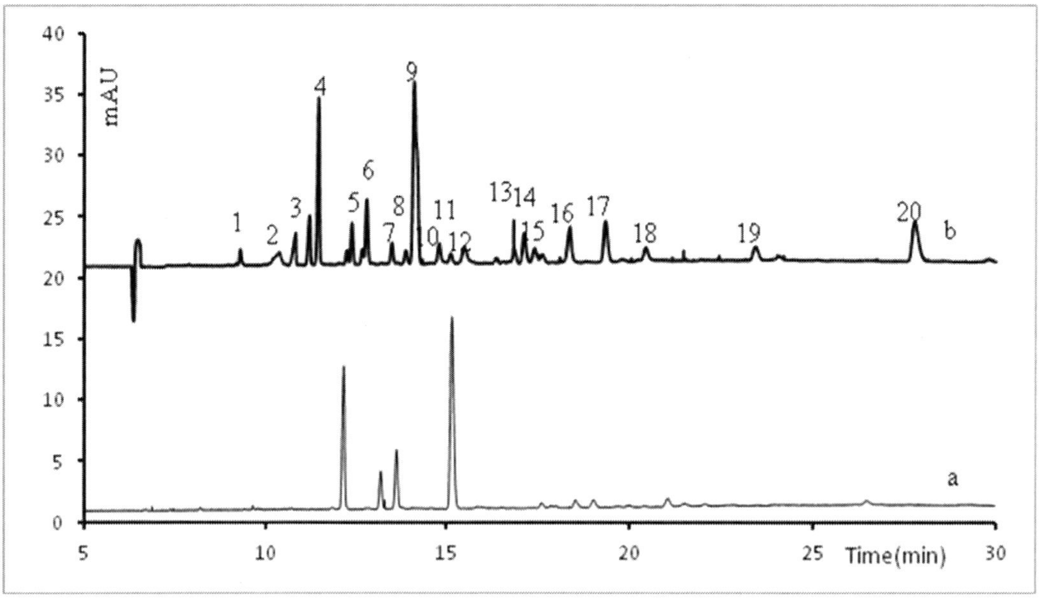

Figure 4. Electropherograms of b) standards and a) propolis sample (pH 7.4) diluted x 10 ; (2) pinostrobin, (3) acacetin, (4) chrysin, (5) rutin, (6) naringenin, (7) isoquercitrin, (9) cinnamic acid, (10) chlorogenic acid, (11) galangin, (12) sinapic acid, (13) syringic acid, (14) ferulic acid, (15) kaempferol, (16) luteolin, (17) coumaric acid, (18) quercetin,(19) rosmarinic acid and (20) caffeic acid.

The results obtained showed that the different environments of extraction resulted in various amounts of polyphenols in propolis extracts, the highest concentrations of flavonoids being found in propolis extract at pH 9, while the highest concentrations of caffeic, coumaric and ferulic acids was found in propolis extract at pH 7.6 (see Table 5). The carbonate buffer 0.1 M, pH = 9 was the most efficient medium for polyphenols extraction from propolis. Chrysin, considered the reference flavonoid in poplar propolis, which was reported in high amounts in Romanian samples previously reported (1.6 mg/g propolis [55], was found in our samples in higher amounts, respectively ~ 3.7 mg/g in neutral media and 7.2 mg/g in alkaline media (pH = 9).

Concerning the sample of *Origanum*, rosmarinic acid was representatively found, in concordance with other study from Romania and with other studies on *Origanum vulgare* [56, 57]. Our results regarding polyphenols composition of *Origanum vulgare* ethanolic extract are similar to those reported in Lithuania, India and Greece [58-60].

The sample of *Mentha aquatica* presented the same compounds previously reported in literature [61], namely rosmarinic acid, quercetin, naringenin, caffeic acid, chlorogenic acid, and other compounds such as rutin and ferulic acid found by HPLC-DAD-MS in our previous study (submitted manuscript, Teodor et al. 2015). In addition, cinnnamic acid, isoquercitrin and kaempherol were identified using CE method in the *Mentha aquatica* sample (Figure 6).

The samples of wines (two sorts of red wine) presented the same compounds previously reported in literature for different regions [46, 48], namely resveratrol, rutin, kaempferol, quercetin, ferulic acid, coumaric acid, syringic acid and caffeic acid.

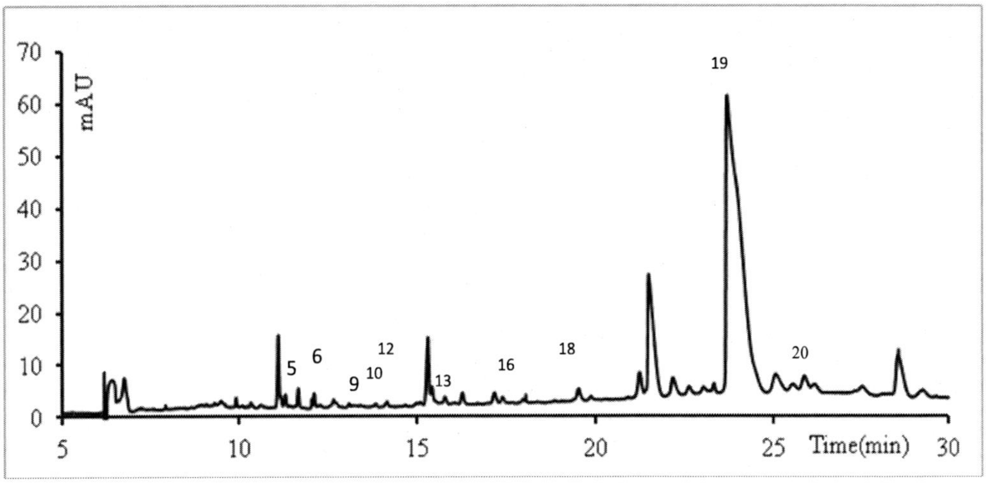

Figure 5. Electropherogram of *Origanum* extract; (5) rutin, (6) naringenin, (9) cinnamic acid, (10) chlorogenic acid, (12) sinapic acid, (13) syringic acid, (16) luteolin, (18) quercetin,(19) rosmarinic acid and (20) caffeic acid.

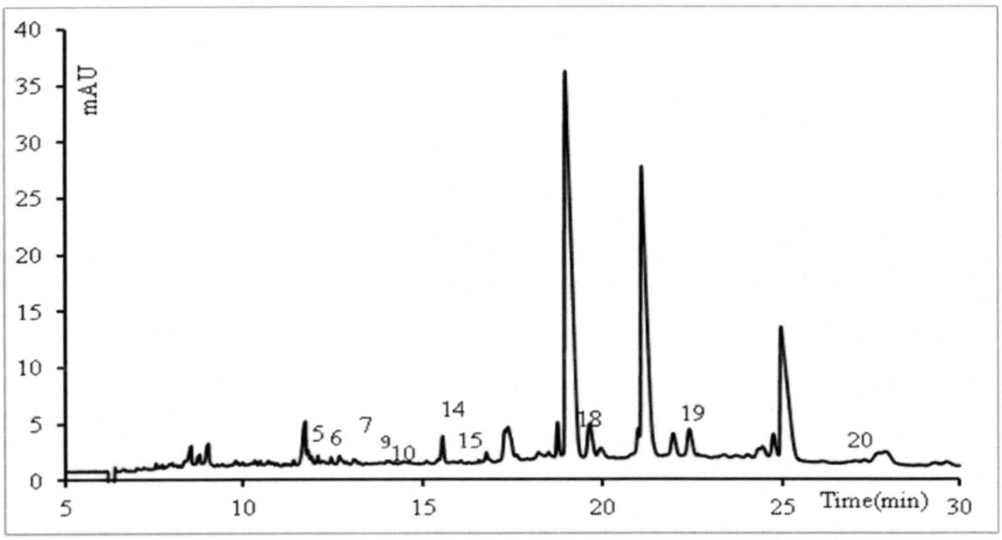

Figure 6. Electropherograms of mint sample; (5) rutin, (6) naringenin, (7) isoquercitrin, (9) cinnamic acid, (10) chlorogenic acid, (14) ferulic acid, (15) kaempferol, (18) quercetin,(19) rosmarinic acid and (20) caffeic acid.

We can conclude that the method is reliable and suitable for a large category of real samples.

Part II. Application of Capillary Electrophoresis for Organic Acids Quantification in Different Extracts

Chemicals and Reagents

All the reagents were of analytical grade (purity > 98%): DL-lactic acid and butyric acid from Fluka (Buchs, Switzerland), acetic acid from Riedel-de-Haën (Germany), L-(+)- tartaric acid, formic, citric acid, benzoic acid, succinic acid, malic acid, LD-p-hydroxyphenyllactic acid (HO-PLA) and phenyllactic acid (PLA) from Sigma-Aldrich (USA). Phosphoric acid 85% and oxalic acid were purchased from Merck (Germany), cetyltrimethylammonium bromide (CTAB) from Loba Chemie (Austria), HPLC-grade water, 0.1N and 1N sodium hydroxide solutions were purchased from Agilent Technologies (USA). Solvents (Merck, Germany) and solutions were filtered on 0.2 μm membranes (Millipore, Bedford, MA, USA) and degassed prior to use. Stock solutions for each standard were prepared at a concentration of 1 mg mL^{-1} in water and stored at +4^{0}C. Working solutions were prepared daily by diluting the stock solutions.

Samples

Three types of medicinal plants were analyzed, chamomile *(Matricaria recutita,* Asteraceae), linden (lime, *Tilia platyphyllos*, Tiliaceae) and mint (menthe, *Mentha piperita*, Lamiaceae). The samples of plants were obtained from different brands of medicinal teas available on the Romanian market. The samples were prepared in infusion and decoction form. For infusions, one tea bag (approx. 1g) of each plant category was minced in a mortar (homogenized), mixed with 200 mL hot distilled water (100° C) and let to infuse for 5 minutes; when the solution was cold it was filtered through a 0.2 μm Millipore filter and injected undiluted in the instrument. Samples preparation of tea decoction form: the same amount of sample, 1 g (1 tea bag) from each category of medicinal plants was added over boiling water and boiled for 5 minutes, and then was left to cool, adjusted to 200 mL, filtered and injected undiluted into the instrument.

The wine samples (one sort of white wine and one sort of red wine) were filtered using 0.2 μm membranes and injected undiluted in the instrument.

Lactic acid bacteria strains used for this study were isolated from infant faeces, dairy products and fermented vegetables and identified using standardized kit API 50CHL (bioMérieux, Marcy l'Etoile, France). The results were integrated using the API Web software.

The strains were: *Lactobacillus plantarum* GM3, *Lactobacillus rhamonsus* E4.2, *Lactobacillus plantarum* E2.4, *Lactobacillus fermentum* 428ST, *Weissella paramesenteroides* Fta, *Lactococcus lactis* FFb, *Lactococcus lactis* F2a3 and *Lactobacillus paracasei ssp. paracasei* 47.1a.

150 μl fresh culture bacterial strains were used to inoculate 15 mL de Man Rogosa Sharpe (MRS) broth (Oxoid, Basingstoke, UK). Bacterial strains were grown at 37^{0}C, 18-24 h, without stirring. Bacterial cultures were centrifuged at 5,000x*g* for 10 min. The supernatants resulted were collected, filtered on 0.2 μm membranes (Millipore, Bedford, MA,

USA), diluted (1:20) in the same mode with standard solutions and injected into the instrument.

Capillary Electrophoresis Method

Electrophoretic separation was carried out using an Agilent CE instrument with DAD detector and CE standard bare fused-silica capillary having 50 μm internal diameter and 72 cm total length (63 cm effective length). Prior to use, the capillary was washed successively with basic solutions: 10 min with 1N NaOH, 10 min with 0.1 N NaOH followed by Ultra Pure Water 10 min and buffer 20 min.

Based on our previous experience [62, 63], the CE method selected belongs to reversed polarity category, and the conditions are the following:

- The applied voltage was -20 kV and the best UV detection was performed at 200 nm (direct detection).
- Sample injection was performed using the hydrodynamic mode, 35 mbar/12 sec, while the capillary was maintained at constant temperature of 25^0C.
- The used background electrolyte contains 0.5M H_3PO_4, 0.5 mM CTAB as cationic surfactant (pH adjusted with NaOH to 6.24) and with 15% methanol as organic modifier, filtered on 0.2 μm membranes (Millipore, Bedford, MA, USA) and degassed before use.
- The organic acids order of elution was formic, oxalic, succinic, malic, tartaric, acetic, citric, lactic, butyric, benzoic, PLA and H-PLA acid, and the analysis time of 20 minutes.
- The capillary was flushed between runs with 0.1M NaOH for 1 min, H_2O for 1 min and the background electrolyte for 2 min.

Validation

After we established the optimal conditions for the separation, the selectivity, linearity, precision, accuracy (recovery), limit of detection and limit of quantification were calculated.

The method selected is based on Galli and Barbas (2004) with several variations to obtain better resolution between the aliphatic acids [30]. The addition of methanol in BGE improved the separation for malic, tartaric and acetic acids, and addition of BGE in standards and samples matrix accelerated the ionization of analytes and improved the signal for lactic acid. Also, the signal was slightly improved by increasing the time of injection (12 seconds, a small stacking effect). We obtained satisfactory results only with a capillary having 50 μm internal diameter and 72 cm total length.

Finally, the optimum results for the simultaneous separation and quantification of 9 small chain aliphatic organic acids and 3 aromatic organic acids were obtained using 0.5 M H_3PO_4 and 0.5 mM CTAB (pH = 6.24) with 15% (V/V) methanol as background electrolyte (BGE) and BGE:water (1:1, V:V) as the sample matrix. Due to the presence of other components in LAB strain extracts, the method of standard additions was used for the identification of organic acids, comparing their migration time with the migration times obtained for standard organic acids (see Figure 7).

The linearity of the response was established for each organic acid by building the calibration curves. The levels of concentrations varied according to the type of organic acid: one level (for benzoic acid, phenyllactic and HO phenyllactic acid), fourteen levels (for formic, succinic, malic, tartaric, acetic, citric and butyric acid) to twenty-one levels (for lactic acid). Each calibration point corresponded to three injections.

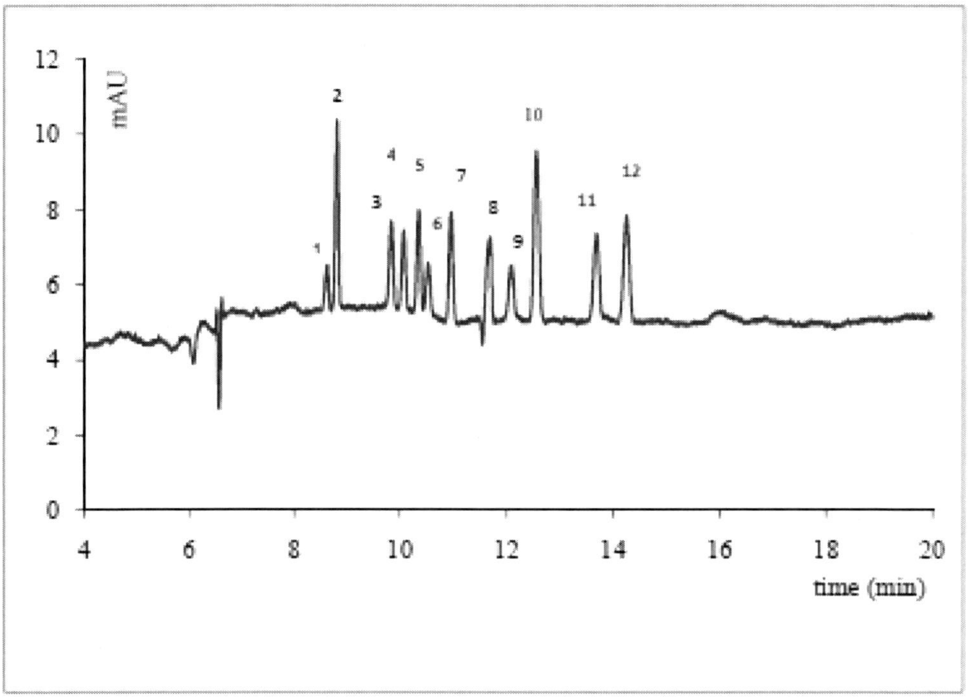

Figure 7. Electropherogram for standards: 1-formic acid, 2-oxalic acid, 3-succinic acid, 4-malic acid, 5-tartaric acid, 6-acetic acid, 7-citric acid, 8-lactic acid, 9-butyric acid, 10-benzoic acid, 11-PLA, and 12-H-PLA.

In Table 6 are presented the equations of regression lines which have good linearity in the range 5.00–140.00 μg mL^{-1} for formic, succinic, malic, tartaric, acetic, citric and butyric acid, 2.00-80.00 μg mL^{-1} for oxalic acid, 7.50-210.00 μg $mL^{-1}$for lactic acid and 0.25-10.00 μg mL^{-1} for benzoic, PLA and H-PLA.

The regression coefficients (R^2) ranged between 0.995 and 0.999. LoD and LoQ used to assess sensitivity were estimated using a signal-to-noise ratio of 3 and 10, respectively. The LoDs values ranged between 0.001 and 1.43 μg mL^{-1}, and LoQs ranged from 0.004 to 4.72 μg mL^{-1}.

The sensitivity of the method is good for a CZE method, for example the detection limit for benzoic acid 0.003 μg mL^{-1} is comparable with 0.0006 μg mL^{-1} obtained by Zhu et al. (2012) by field enhancement sample stacking (FESS) capillary electrophoresis [64].

Table 6. Performance parameters of method for organic acids separation

Organic acid	t_R (min)	The linear regression equations ($\mu g\ mL^{-1}$)	R^2	Linearity range of response $\mu g\ mL^{-1}$	LoD $\mu g\ mL^{-1}$	LoQ $\mu g\ mL^{-1}$
Formic	8.54 ± 0.07	y = 0.1194x + 0.5602	0.998	5.00-140	0.32	1.05
Oxalic	8.74 ± 0.08	y = 1.1059x - 0.1895	0.998	2.00-80	0.07	0.23
Succinic	9.75 ± 0.10	y = 0.2427x + 1.1353	0.998	5.00-140	0.42	1.40
Malic	9.98 ± 0.11	y = 0.2129x + 0.8603	0.999	5.00-140	0.37	1.21
Tartaric	10.25 ± 0.11	y = 0.2985x + 1.2136	0.999	5.00-140	0.41	1.35
Acetic	10.85 ± 0.12	y = 0.1868x + 0.3039	0.999	5.00-140	0.34	1.13
Citric	11.10 ± 0.37	y = 0.3134x + 1.1898	0.999	5.00-140	0.37	1.22
Lactic	11.59 ± 0.13	y = 0.3101x + 1.4016	0.998	7.50-210	1.43	4.72
Butiric	11.97 ± 0.14	y = 0.1964x + 0.6268	0.999	5.00-140	0.59	1.94
Benzoic	12.42 ± 0.15	y = 13.859x + 0.0318	0.998	0.25-10	0.003	0.009
PLA	13.55 ± 0.17	y = 7.1518x + 1.7767	0.995	0.25-10	0.007	0.023
H-PLA	14.10 ± 0.18	y = 9.2615x + 1.156	0.998	0.25-10	0.001	0.004

(SpringerScience + Business Media New York, Food Analytical Methods, Capillary Electrophoresis Method Validation for Organic Acids Assessment in Probiotics, 2015, 8, 1335-1340; DOI:10.1007/s12161-014-0018-1, F. Gatea, E. D. Teodor, G. Păun, A. O. Matei, G. L. Radu, original copyright notice) "With kind permission of Springer Science + Business Media."

Table 7. Precision data expressed as RSD% (Standard Deviation/average) for the separation method of 12 organic acids

Organic acid	Intra-assay Precision[a] (%, n = 5)	Inter-assay Precision[b] (%, n = 3)	Inter-assay Precision[c] (%, n − 3)	Inter-assay Precision[d] (%, n = 3)
Formic	0.62	2.16	1.90	1.67
Oxalic	0.45	0.66	0.87	3.58
Succinic	0.61	0.44	1.44	2.80
Malic	4.57	2.93	3.95	0.66
Tartaric	3.00	3.64	4.70	1.86
Acetic	2.75	4.26	4.76	3.77
Citric	4.00	1.23	3.23	2.00
Lactic	3.75	2.55	1.25	1.45
Butyric	3.89	2.24	3.45	2.27
Benzoic	1.18	1.52	1.48	3.36
PLA	1.34	2.24	3.37	4.85
H-PLA	0.96	2.69	1.57	2.47

[a]All acids were at concentration 30 $\mu g\ mL^{-1}$ except : Oxalic acid 80 $\mu g\ mL^{-1}$, Lactic acid 60 $\mu g\ mL^{-1}$, Benzoic acid 6 $\mu g\ mL^{-1}$, PLA 6 $\mu g\ mL^{-1}$, H-PLA 6 $\mu g\ mL^{-1}$.

[b]All acids were at concentration 40 $\mu g\ mL^{-1}$ except : Oxalic acid 30 $\mu g\ mL^{-1}$, Lactic acid 60 $\mu g\ mL^{-1}$, Benzoic acid 4 $\mu g\ mL^{-1}$, PLA 4 $\mu g\ mL^{-1}$, H-PLA 4 $\mu g\ mL^{-1}$.

[c]All acids were at concentration 50 $\mu g\ mL^{-1}$ except : Oxalic acid 35 $\mu g\ mL^{-1}$, Lactic acid 75 $\mu g\ mL^{-1}$, Benzoic acid 5 $\mu g\ mL^{-1}$, PLA 5 $\mu g\ mL^{-1}$, H-PLA 5 $\mu g\ mL^{-1}$.

[d]All acids were at concentration 140 $\mu g\ mL^{-1}$ except : Oxalic acid 80 $\mu g\ mL^{-1}$, Lactic acid 210 $\mu g\ mL^{-1}$, Benzoic acid 8 $\mu g\ mL^{-1}$, PLA 8 $\mu g\ mL^{-1}$, H-PLA acid 8 $\mu g\ mL^{-1}$.

(SpringerScience + Business Media New York, Food Analytical Methods, Capillary Electrophoresis Method Validation for Organic Acids Assessment in Probiotics, 2015, 8, 1335-1340; DOI:10.1007/s12161-014-0018-1, F. Gatea, E. D. Teodor, G. Păun, A. O. Matei, G. L. Radu, original copyright notice) "With kind permission of Springer Science + Business Media."

Table 8. Recovery values obtained for organic acids method. Recovery (%) = ((amount determined – original amount)/amount added) x 100

Organic acid	Added (µg mL^{-1})	Found (µg mL^{-1})	Recovery (%, n = 3)
	20.00	22.43	112.74
Formic	40.00	47.44	108.69
	60.00	83.68	104.65
	20.00	20.54	102.69
Oxalic	40.00	39.56	98.90
	80.00	78.91	98.64
	20.00	22.98	114.88
Succinic	40.00	45.25	113.12
	60.00	92.91	116.73
	20.00	21.79	108.97
Malic	40.00	44.32	110.80
	60.00	87.36	109.20
	20.00	19.51	97.56
Tartaric	40.00	45.37	113.43
	60.00	84.58	105.73
	20.00	18.20	91.01
Acetic	40.00	38.46	96.14
	60.00	85.10	106.37
	20.00	18.72	93.61
Citric	40.00	43.83	109.57
	60.00	87.87	109.84
	30.00	29.44	98.14
Lactic	60.00	62.88	104.80
	120.00	132.91	110.76
	20.00	22.67	113.34
Butyric	40.00	41.60	104.00
	60.00	83.63	104.54
	1.00	1.10	109.91
Benzoic	6.00	5.76	96.04
	10.00	10.19	101.91
	1.00	0.91	90.55
PLA	6.00	6.40	106.59
	8.00	8.28	103.52
	1.00	0.94	94.90
H-PLA	6.00	5.41	90.22
	8.00	9.02	112.87

(SpringerScience + Business Media New York, Food Analytical Methods, Capillary Electrophoresis Method Validation for Organic Acids Assessment in Probiotics, 2015, 8, 1335-1340; DOI:10.1007/s12161-014-0018-1, F. Gatea, E. D. Teodor, G. Păun, A. O. Matei, G. L. Radu, original copyright notice) "With kind permission of Springer Science + Business Media."

The repeatability of the proposed method was studied by repeated injections of the acid organics mixtures (standards) 5 times in the same day (intra-day precision), whereas the reproducibility assimilated to inter-day precision was tested by triplicate injections for 3 different days (Table 7). The results are reported in terms of relative standard deviation (RSD). Intra-day precision values of RSD were in range 0.45 - 4.57%, while the RSD values for inter-day precision were between 0.66 and 4.85%. The results indicate that the proposed method has good precision for both qualitative and quantitative studies of the organic acid and is suitable for the analysis of real samples.

Recovery was estimated by spiking with three different concentrations of standard mixtures of organic acids into one individual culture medium sample diluted x20 (Table 8). The percentage recoveries of organic acids for all three concentrations of standard mixtures were in the range from 90.22 to 116.14%. Because nearly 100% recovery was observed, it can say that this method can be used for the analysis of organic acids in culture media, while avoiding the matrix effect.

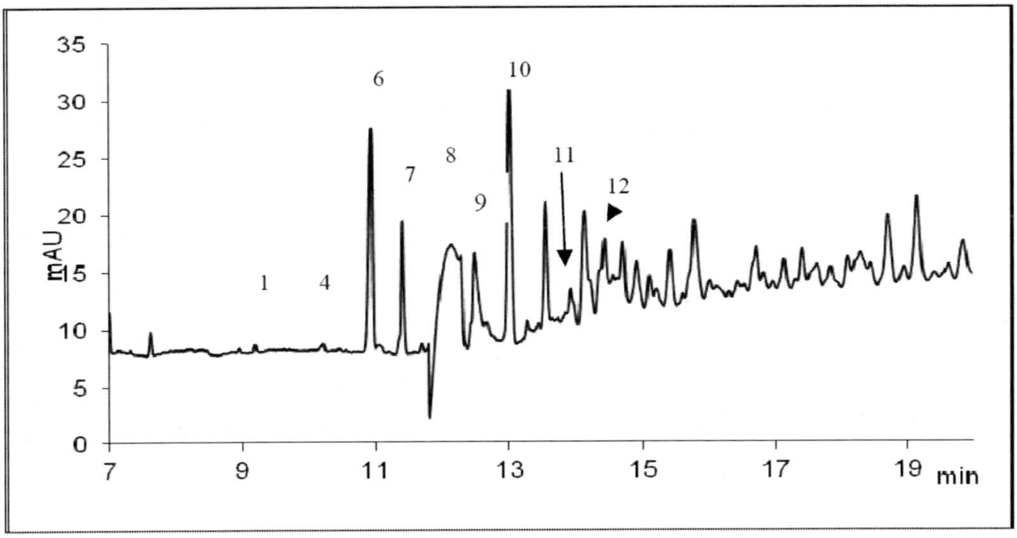

Figure 8. Electropherogram of sample GM3 (Lactobacillus plantarum): 1-formic acid, 4-malic, 6-acetic acid, 7-citric acid, 8-lactic acid, 9-butyric acid, 10-benzoic acid, 11-PLA, and 12-H-PLA.

Table 9. Concentration of organic acids in lactic bacteria culture media

Samples	Organic acid concentration* µg mL^{-1}									
	Formic	Succinic	Malic	Acetic	Citric	Lactic	Butyric	Benzoic	PLA	H-PLA
MRS	118.6	< 0.42	93.16	3055.33	1328.9	373.2	207.8	62.31	17.21	21.40
GM3	66.51	< 0.42	92.66	3364.71	1154.6	5181.2	2066.6	69.47	34.50	26.45
F2a3	< 0.32	79.77	128.9	3145.2	1425.4	3797.9	245.3	80.04	16.31	7.78
E2.4	< 0.32	172.8	63.90	1023.67	299.6	6363.2	160.5	83.14	16.05	17.08
E4.2	< 0.32	< 0.42	70.05	3177.36	1689.9	1975.1	647.1	95.93	15.68	11.28
47.1a	< 0.32	< 0.42	123.16	3670.79	185.5	912.2	497.0	74.93	26.02	13.45
428TS	< 0.32	< 0.42	95.74	3228.45	1851.8	2225.6	365.9	78.08	21.98	11.21
Fta	<0.32	< 0.42	< 0.37	3271.26	1815.4	1016.4	484.2	81.37	23.41	8.47
FFb	< 0.32	< 0.42	< 0.37	3099.86	1862.5	4471.4	227.5	94.27	22.57	9.02

Tartaric acid was absent and oxalic acid in all samples was under detection limit.

* Average of 3 measurements.

Samples Analysis

Different real samples were analyzed for the content of organic acids, respectively lactic bacteria extracts, medicinal plant extracts and wines. The results obtained from the analysis of these various samples are presented in Table 9 and 10, while in Figures 8 and 9 are shown the electropherograms for a supernatant of culture medium produced by a *Lactobacillus plantarum* strain and respectively for an aqueous extract of mint.

Table 10. Concentrations of organic acids in wines and medicinal plants samples

	Concentration μg mL^{-1} ± SD					
Organic acid	White wine	Red wine	Chamomile infusion	Chamomile decoction	Mint infusion	Mint decoction
Oxalic	-	-	-	-	20.18 ± 0.09	55.04 ± 0.09
Succinic	774.23 ± 0.21	613.25 ± 0.15	< LoD	9.98 ± 0.10	5.77 ± 0.06	9.74 ± 0.02
Malic	2183.12 ± 0.16	91.02 ± 0.13	20.45 ± 0.08	43.25 ± 0.06	87.64 ± 0.11	111.53 ± 0.12
Tartaric	1801.10 ± 0.41	1781.13 ± 0.18	-	-	19.02 ± 0.08	24.76 ± 0.10
Acetic	453.04 ± 0.09	545.03 ± 0.12	-	-	-	-
Citric	81.05 ± 0.17	131.02 ± 0.02	17.34 ± 0.22	31.36 ± 0.16	76.43 ± 0.06	104.49 ± 0.16
Lactic	101.21 ± 0.42	7150.12 ± 1.8	-	-	< LoD	-

Speaking about culture media produced by lactic acid bacteria, these samples presented large amounts of lactic and acetic acids (which are recognized for antimicrobial effect [65]) various amounts of phenyllactic acid with the most powerful antimicrobial effect [39], and notable values for benzoic acid with antifungal effect [66]. Citric and butyric acids are also found in significant amounts. Tartaric and oxalic acids were absents or under detection limit, and succinic and formic acids were not detected in all the samples. As could be seen in Table 4, *Lactobacillus plantarum* GM3 presented the largest quantities of analyzed organic acids and could be selected for practical application.

A good selection of LAB (and a criterion should be the content in organic acids) allows the control of mould growth and improves the shelf life of many fermented products and, therefore, reduces health risks due to exposure to mycotoxins.

Generally, chemical studies are focused on developing analytical systems for monitoring or analyzing of organic acids in mammalian cell culture [67], during ripening of cheese [68], or for evaluation as antimicrobial agents against various pathogens [39].

The method was developed for organic acids from bacterial extracts (especially lactic acid bacteria), but can be applied on others extracts or foods. We used the method for plants extracts and wines. The results obtained are presented in Table 10.

As could be observed from Table 10, five short-chain organic acids were identified in wines and four short-chain organic acids were identified in medicinal tea samples, except for lactic acid which is under the detection limit or absent in majority of plants extracts. Tartaric acid is significantly present only in mint extracts and succinic acid is present in mint and chamomile but in low concentrations. Malic and citric acids were always present in significant levels, between 18.3 mg L^{-1} (linden) and 111.5 mg L^{-1} (mint) for malic acid, and between 7.3 mg L^{-1} (linden) and 104.5 mg L^{-1} (mint) for citric acid. It should be noted that all the medicinal tea samples prepared as decoction offered higher concentrations in organic acids compared to samples prepared by infusion. The content in organic acids obtained by CE analysis was difficult to be compared with other data from literature because short-chain organic acids are poorly studied in plant extracts. Anyway, this content is different depending of various factors and is correlated with plant physiology and cultivation conditions.

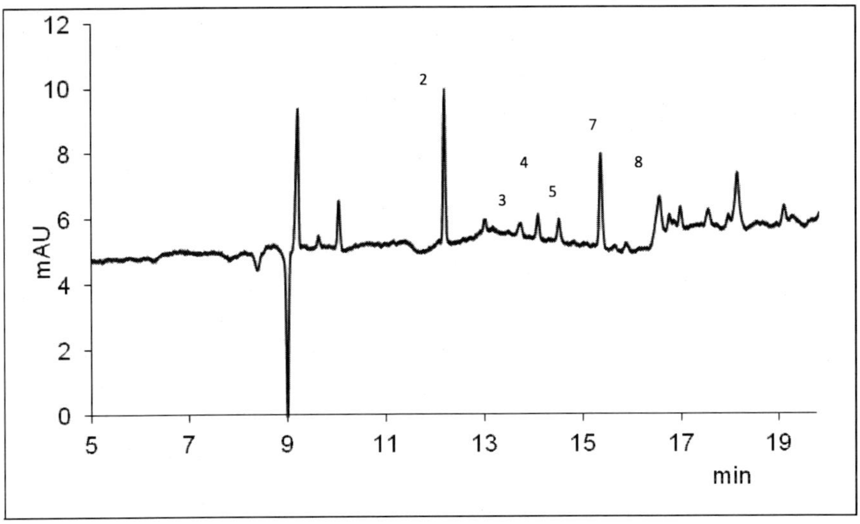

Figure 9. Electropherogram of a mint sample (decoction): 2-oxalic acid, 3-succinic acid, 4-malic acid, 5-tartaric acid, 7-citric acid, and 8-lactic acid.

Tartaric acid was present, as stated in literature [29, 69, 70], in any sort of wine. Malic acid was identified in high concentrations in white wine and lactic acid in red wine. Succinic acid was also present in both sorts of wines in noticeable quantities and citric acid in small quantities.

Quantitative determination of organic acids in different foodstuffs is important for nutritional and flavor reasons and as an indicator of bacterial activity. Our method can simultaneous separate aliphatic small chain organic acids and aromatic derivatives of lactic acid, is simply, rapid, reliable, and with very low consumption of reagents and samples.

CONCLUSION

Two capillary electrophoretic methods for separation and quantification of organic acids and polyphenolic compounds were validated in terms of linearity of response, limit of detection (LoD), limit of quantification (LoQ), precisions (i.e., intra-day, inter-day reproducibility) and recovery. The methods are simply, rapid, reliable and cost effective and can be applied on various real samples.

The first method (for polyphenols quantification) was experimentally improved with a BGE consisting of sodium tetraborate and SDS and is very suitable for separation of polyphenolic compounds in propolis and plants extracts. This simple, reliable and fast CZE method developed and partially validated for simultaneous detection of 20 polyphenolic compounds run in less than 27 min. Regarding all the validation parameters, the method complies with validation requirements and it is suitable for the analysis of selected samples. The results obtained from the analysis of real samples are in correlation with other literature data and bring new information about less studied samples such as Romanian propolis and *Mentha aquatica.*

The second method, an accessible, simple and fast CZE method was partially validated for simultaneous quantification of nine aliphatic and three aromatic organic acids in 15 minutes in fermentation products of lactic bacteria. All the analytes have good linearity in the range 5.00–140 $\mu g\ mL^{-1}$ for formic, succinic, malic, tartaric, acetic, citric and butyric acid, 2.00-80 $\mu g\ mL^{-1}$ for oxalic acid, 7.50-210 $\mu g\ mL^{-1}$for lactic acid and 0.25-10 $\mu g\ mL^{-1}$ for benzoic acid, PLA and H-PLA. The regression coefficients ranged between 0.995 and 0.999, LoD values ranged from 0.001 to 1.43 $\mu g\ mL^{-1}$ and LoQ values from 0.004 to 4.72 $\mu g\ mL^{-1}$, so we conclude that the sensitivity of the method is satisfactory for a capillary electrophoresis method. The percentage recoveries of organic acids for all three mixtures were in the range from 90.22 to 116.14%.

The method was applied successfully on samples obtained from fermentation of several strains of lactic bacteria, wines and plants extracts and could be used generally in foodomics.

REFERENCES

[1] Hurtado-Fernández, E., Gomez-Romero, M., Carrasco-Pancorbo, A., Fernandez-Gutierrez, A. (2010) Application and potential of capillary electroseparation methods to determine antioxidant phenolic compounds from plant food material. *J Pharm Biomed Anal*, 53, 1130–1160.

[2] Zhao, J., Hu, D.-j., Lao, K., Yang, Z.-m., Li, S.-p. (2014) Advance of CE and CEC in phytochemical analysis (2012–2013). *Electrophoresis,* 35, 205–224.

[3] García-Cañas, V., Simo, C., Castro-Puyana, M., Cifuentes, A. (2014) Recent advances in the application of capillary electromigration methods for food analysis and Foodomics. *Electrophoresis,* 35, 147-169.

[4] Fiehn, O., Kopka, J., Dormann, P., Altman, T., Tretheweyand, R. N., Willmitzer, L. (2000) Metabolite profiling for plant functional genomics. *Nat Biotechnol* 18, 1157-1163.

[5] Gout, E., Boisson, A.-M., Aubert, S., Douce, R., Bligny, R. (2001) Origin of the cytoplasmic pH changes during anaerobic stress in higher plant cells. Carbon-13 and phosphorous-31 nuclear magnetic resonance studies. *Plant Physiol,* 125, 912-925.

[6] Naczk, M., Shahidi, F. (2004) Extraction and analysis of phenolics in food. *J Chromatogr A,* 1054, 95-111.

[7] Afanasev, I. B., Dorozhko, A. I, Brodskii, A. V., Kostyuk, V. A., Potapovitch, A. I. (1989) Chelating and free radical scavenging mechanisms of inhibitory action of rutin and quercetin in lipid peroxidation. *Biochem Pharmacol,* 38, 1763–1769.

[8] Blokhina, O., Virolainen, E., Fagerstedt, K. V. (2003) Antioxidants, oxidative damage and oxygen deprivation stress: A review. *Annals of Botany,* 91, 179–194.

[9] Nazari, Q. A., Kume, T., Takada-Takatori, Y., Izumi, Y., Akaike, A. (2013) Protective effect of luteolin on an oxidative-stress model induced by microinjection of sodium nitroprusside in mice. *J Pharmacol Sci,* 122, 109-117.

[10] Taviano, M. F., Marino, A., Trovato, A., Bellinghieri, V., Melchini, A., Dugo, P., Cacciola, P., Donato, P., Mondello, L., Güvenç, A., De Pasquale, R., Miceli, N. (2013) *Juniperus oxycedrus* L. subsp. *oxycedrus* and *Juniperus oxycedrus* L. subsp. *macrocarpa* (Sibth. & Sm.) Ball. “berries” from Turkey: Comparative evaluation of

phenolic profile, antioxidant, cytotoxic and antimicrobial activities. *Food Chem Toxicol,* 58, 22–29.

[11] Giada, M. L. R. Phenolic Acids as Additives in the Food Industry. In: Munné-Bosch S. (Eds.) *Phenolic Acids: Composition, Applications and Health Benefits.* Nova Science Publisher, NY, 2012, pp. 125-145.

[12] Marcucci, M. C. (1995) Propolis: chemical composition, biological properties and therapeutic activity. *Apidologie,* 26, 83–99.

[13] Ghisalberti, E. L. (1979) Propolis: a review. Bee World, 60, 59–84.

[14] Choudhari, M. K., Haghniaz, R., Rajwade, J. M., Paknikar, K. M. (2013) Anticancer activity of Indian stingless bee propolis: an in vitro study. *Evid Based Complement Alternat Med* 2013, ID 842985.

[15] Knekt, P., Jarvinen, R., Seppanen, R., Hellovaara, M., Teppo, L., Pukkala, E., Aromaa, A. (1997) Dietary flavonoids and the risk of lung cancer and other malignant neoplasms. *Am J Epidem,* 146, 223–230.

[16] Soleas, G. J., Grassc, L., Josephy, P. D., Goldberg, D. M., Diamandis, E. P. (2002) A comparison of the anticarcinogenic properties of four red wine polyphenols. *Clin Biochem,* 35, 119–124.

[17] Darvesh, A. S., Bishayee, A. (2013) Chemopreventive and Therapeutic Potential of Tea Polyphenols in Hepatocellular Cancer. *Nutr and Cancer,* 65, 329-344.

[18] Yu, Y., Deng, Y., Lu, B.-m., Liu, Y.-x., Li, J., Bao, J.-k. (2014) Green tea catechins: a fresh flavor to anticancer therapy. Apoptosis, 19, 1-18.

[19] Wang, J., Ho, L., Zhao, Z., Seror, I., Humala, N., Dickstein, D. L., Thiyagarajan, M., Percival, S. S., Talcott, S. T., Pasinetti, G. M. (2006) Moderate consumption of Cabernet Sauvignon attenuates Abeta neuropathology in a mouse model of Alzheimer's disease. *FASEB J,* 20, 2313–2320.

[20] Ho, L., Ferruzzi, M. G., Janle, E. M., Wang, J., Gong, B., Chen, T.-Y., Lobo, J., Cooper, B., Wu, Q. L., Talcott, S. T., Percival, S. S., Simon, J. E., Pasinetti, G. M. (2013) Identification of brain-targeted bioactive dietary quercetin-3-*O*-glucuronide as a novel intervention for Alzheimer's disease. *FASEB J,* 27, 769-782.

[21] Voet, D., Voet, J. G. Biochimie, second edition, De Boeck Universite, Paris, 1998

[22] McCluskey, J., Herdman, L., Skene, K. R. (2004) Iron deficiency induces changes in metabolism of citrate in lateral roots and cluster roots of *Lupinus albus. Physiol Plant,* 121, 586-594.

[23] Jones, D. L., Darrah, R. (1994) Role of root derived organic acids in the mobilization of nutrients from the rhizosphere. *Plant Soil,* 166, 247-257.

[24] Saavedra, L., Barbas, C. (2003) Validated capillary electrophoresis method for small-anions measurement in wines. *Electrophoresis,* 24, 2235-2243.

[25] Arellano, M., Andrianary, J., Dedieu, F., Couderc, F., Puig, P., (1997) Method development and validation for the simultaneous determination of organic and inorganic acids by capillary zone electrophoresis. *J Chromatogr A,* 765, 321-328.

[26] Vaz., F. A. S., Da Silva, P. A., Passos, L. P., Heller, M., Micke, G. A., Costa, A. C. O., De Oliveira, M. A. L. (2012) Optimisation of a Capillary Zone Electrophoresis Methodology for Simultaneous Analysis of Organic Aliphatic Acids in Extracts of *Brachiaria brizantha. Phytochem Anal,* 23, 569-575.

[27] Azevedo, M. S., Pirassol, G., Fett, R., Micke, G. A., Vitali, L., Oliveira Costa, A. C. (2014) Screening and determination of aliphatic organic acids in commercial Brazilian

sugarcane spirits employing a new method involving capillary electrophoresis and a semi-permanent adsorbed polymer coating. *Food Res Intern,* 60, 123-130.

[28] Castineira, A., Pena, R. M., Herrero, C., Garcıa-Martın, S. (2002) Analysis of organic acids in wine by capillary electrophoresis with direct UV detection. *J Food Comp Anal,* 15, 319-331.

[29] Santalad, A., Teerapornchaisit, P., Burakhama, R., Srijaranai, S. (2007) Capillary zone electrophoresis of organic acids in beverages. *LWT-Food Sci Technol,* 40, 1741-1746.

[30] Galli, V., Barbas, C. (2004) Capillary electrophoresis for the analysis of short-chain organic acids in coffee. *J Chromatogr A,* 1032, 299-304.

[31] Fayol-Messaoudi, D., Berger, C. N., Coconnier-Polter, M. H., Lie´vin-Le Moal, V., Servin, A. L. (2005) pH-, lactic acid-, and non-lactic acid-dependent activities of probiotic lactobacilli against Salmonella enterica serovar Typhi- murium. *Appl Environ Microbiol,* 71, 6008-6013.

[32] Lavermicocca. P., Valerio, F., Visconti, A. (2003) Antifungal activity of phenyllactic acid against molds isolated from bakery products. *Appl Environ Microbiol,* 69, 634-640.

[33] Valerio, F., De Bellis, P., Lonigro, S. L., Visconti, A., Lavermicocca, P. (2008) Use of lactobacillus plantarum fermentation products in breadmaking to prevent bacillus subtilis ropy spoilage. *Int J Food Microbiol,* 122, 328–332.

[34] Gerez, C. L., Torino, M. I., Rollán, G., Font de Valdez, G. (2009) Prevention of bread mould spoilage by using lactic acid bacteria with antifungal properties. *Food Control,* 20, 144-148.

[35] Corsetti, A., Gobbetti, M., Rossi, J., Damiani, P. (1998) Antimould activity of sourdough lactic acid bacteria: identification of a mixture of organic acids produced by Lactobacillus sanfrancisco CB1. *Appl Microbiol Biotechnol*, 50, 253-256.

[36] Niku-Paavola, M. L., Laitila, A., Mattila-Sandholm, T., Haikara, A. (1999) New types of antimicrobial compounds produced by *Lactobacillus plantarum. J Appl Microbiol,* 86, 29-35.

[37] Devlieghere, F., Vermeiren, L., Debevere, J. (2004) New preservation technologies: possibilities and limitations. *Int Dairy J,* 14, 273-285.

[38] Dieuleveux, V., Lemarinier, S., Guèguen, M. (1998) Antimicrobial spectrum and target site of D-3-phenyllactic acid. *Int J Food Microbiol,* 40, 177-183.

[39] Rodríguez-Pazo, N., Vázquez-Araújo, L., Pérez-Rodríguez, N., Cortés-Diéguez, S., Domínguez, J. M. (2013) Cell-Free Supernatants Obtained from Fermentation of Cheese Whey Hydrolyzates and Phenylpyruvic Acid by Lactobacillus plantarum as a Source of Antimicrobial Compounds, Bacteriocins, and Natural Aromas. *App Biochem Biotechnol,* 171, 1042-1060.

[40] Srinivas, K., King, J. W., Monrad, J. K., Howard, L. R., Hansen, C. M. (2009) Optimization of Subcritical Fluid Extraction of Bioactive Compounds Using Hansen Solubility Parameters. *J Food Sci,* 74, E342–E354.

[41] Srinivas, K., King, J. W., Howard, L. R., Monrad, J. K. (2010) Solubility of Gallic Acid, Catechin, and Protocatechuic Acid in Subcritical Water from (298.75 to 415.85) K. *J Chem Eng Data,* 55, 3101–3108.

[42] Ferreira, O., Pinho, S. P. (2012) Solubility of Flavonoids in Pure Solvents. *Ind Eng Chem Res,* 51, 6586−6590.

[43] Luo, Z., Murray, B. S., Ross, A. L., Povey, M. J., Morgan, M. R., Day, A. J. (2012) Effects of pH on the ability of flavonoids to act as Pickering emulsion stabilizers. *Colloids Surf B Biointerfaces,* 92, 84-90.

[44] Petr, J., Vitkova, K., Ranc, V., Znaleziona, J., Maier, V., Knob, R., Sevcik, J. (2008) Determination of some phenolic acids in Majorana hortensis by capillary electrophoresis with online electrokinetic preconcentration. *J Agric Food Chem*, 56, 3940–3944.

[45] Fukuji, T. S., Tonin, F. G., Tavares, M. F. M. (2010) Optimization of a method for determination of phenolic acids in exotic fruits by capillary electrophoresis. *J Pharmceut Biomed,* 51, 430–438.

[46] Ballus, C. A., Meinhart, A. D., Grando de Oliveira, R., Godoy, H. T. (2012) Optimization of capillary zone electrophoresis separation and on-line preconcentration of 16 phenolic compounds from wines produced in South America. *Food Res Intern,* 45, 136–144.

[47] Ballus, C. A., Meinhart, A. D., de Souza Campos, F. A., Bruns, R. E., Godoy, H. T. (2014) Doehlert design-desirability function multi-criteria optimal separation of 17 phenolic compounds from extra-virgin olive oil by capillary zone electrophoresis. *Food Chem,* 146, 558–568.

[48] Franquet-Griell, H., Checa, A., Núñez, O., Saurina, J., Hernández-Cassou, S., Puignou, L. (2012) Determination of polyphenols in Spanish wines by capillary zone electrophoresis. Application to wine characterization by using chemometrics. *J Agric Food Chem,* 60, 8340-8349.

[49] Whatley, H. *Basic principles and modes of capillary electrophoresis, clinical and forensic applications of capillary electrophoresis.* In: Pathology and Laboratory Medicine, Petersen JR, Mohammad AA (Eds.), Humana Press, 2001, pp. 21-58.

[50] Tomas-Barberan, F. A., Garcia-Viguera, C., Vit-Oliver, P., Ferreres, F., Tomas-Lorente, F. (1993) Phytochemical evidence for botanical origin of tropical propolis from Venezuela. *Phytochemistry*, 34, 191-196.

[51] Bankova, V. S., De Castro, S. L., Marcucci, M. C. (2000) Propolis: Recent advances in chemistry and plant origin. *Apidologie,* 31, 3-15.

[52] Kujumgiev, A., Tsvetkova, I., Serkedjieva, Y., Bankova, V., Christov, R., Popov, S. (1999) Antibacterial, antifungal and antiviral activity of propolis from different geographic origin. *J Ethnopharmacol,* 64, 235-240.

[53] Uzel, A., Sorkun, K., Oncag, O., Cogulu, D., Gencay, O., Salih, B. (2005) Chemical compositions and antimicrobial activities of four different Anatolian propolis samples. *Microbiol Res,* 160, 189-195.

[54] Kalogeropoulos, N., Konteles, S. J., Troullidou, E., Mourtzinos, I., Karathanos, V. T. (2009) Chemical composition, antioxidant activity and antimicrobial properties of propolis extracts from Greece and Cyprus. *Food Chem,* 116, 452–461.

[55] Mărghitaş, D. A., Dezmirean, D. S., Bobiş, O. (2013) Important developments in Romanian propolis research. *Evid Based Complement Alternat Med* 2013, ID 159392.

[56] Cioanca, O., Mircea, C., Iancu, C., Gille, E., Hăncianu, M. (2013) Intraspecific variability of *Origanum* crop varieties depending on growing conditions: secondary metabolites accumulation. *Sgem2013 Conference Proceedings,* Doi:10.5593/ Sgem2013/Bf6/S25.015.

[57] Danila, A. O., Gatea, F., Radu, G. L. (2011) Polyphenol composition and antioxidant activity of selected medicinal herbs. *Chem Nat Compds,* 47, 22-26.

[58] Radušienė, J., Ivanauskas, L., Janulis, V., Jakštas, V. (2008) Composition and variability of phenolic compounds in *Origanum vulgare* from Lithuania. *Biologja,* 54, 45–49.

[59] Hithamani, G., Ramalakshmi, K. (2013) Microwave assisted extraction of phenolics from *Origanum vulgare. Int J Agric Soil Sci,* 1, 7-12.

[60] Grevsen, K., Fretté, X. C., Christensen, L. P. (2009) Content and Composition of Volatile Terpenes, Flavonoids and Phenolic Acids in Greek Oregano (*Origanum vulgare* L. ssp. hirtum) at different Development Stages during Cultivation in Cool Temperate Climate. *Europ J Hort Sci,* 74, 193-203.

[61] Mišan, A. Č., Mimica-Dukić, N. M., Mandić, A. I., Sakač, M. B., Milovanović, I. Lj., Sedej, I. J. (2011) Development of a Rapid Resolution HPLC method for the separation and determination of 17 phenolic compounds in crude plant extracts. *Cent Eur J Chem,* 9, 133-142.

[62] Teodor, E. D., Badea, G. I., Alecu, A., Calu, L., Radu, G. L. (2014) Interdisciplinary study on pottery experimentally impregnated with wine. *Chem Papers,* 68, 1022-1029.

[63] Truică, G. I., Teodor, E. D., Radu, G. L. (2013) Organic acids assessments in medicinal plants by capillary electrophoresis. *Rev Roum Chim,* 58, 809-814.

[64] Zhu, Q., Xu, X., Huang, Y., Xu, L., Chen, G. (2012) Field enhancement sample stacking for analysis of organic acids in traditional Chinese medicine by capillary electrophoresis. *J Chromatogr A,* 1246, 35-39.

[65] Valerio, F., Favilla, M., De Bellis, P., Sisto, A., de Candia, S., Lavermicocca, P. (2009) Antifungal activity of strains of lactic acid bacteria isolated from a semolina ecosystem against Penicillium roqueforti, Aspergillus niger and Endomyces fibuliger contaminating bakery products. *Syst Appl Microbiol,* 32, 438-448.

[66] Dalié, D. K. D., Deschamps, A. M., Richard-Forget, F. (2012) Lactic acid bacteria – Potential for control of mould growth and mycotoxins: A review. *Food Control,* 21, 370-380.

[67] Alhusban, A. A., Gaudry, A. J., Breadmore, M. C., Gueven, N., Guijt, R. M. (2014) On-line sequential injection-capillary electrophoresis for near-real-time monitoring of extracellular lactate in cell culture flasks. *J Chromatogr A,* 1323, 157-162.

[68] Akalin, A. S., Gönç, S., Akba, Y. (2002) Variation in Organic Acids Content during Ripening of Pickled White Cheese. *J Dairy Sci, 85,* 1670-1676.

[69] Peres, R. G., Moraes, E. P., Micke, G. A., Tonin, F. G., Tavares, M. F. M., Rodriguez-Amaya, D. B. (2009) Rapid method for the determination of organic acids in wine by capillary electrophoresis with indirect UV detection. *Food Control,* 20, 548-552.

[70] Rovio, S., Sirén, K., Sirén, H. (2011) Application of capillary electrophoresis to determine metal cations, anions, organic acids, and carbohydrates in some Pinot Noir red wines. *Food Chem,* 124, 1194-1200.

INDEX

A

B

C

D

E

F

G

H

I

J

K

L

M

N

O

P

Q

R

S